ENERGY 2000–2020: WORLD PROSPECTS AND REGIONAL STRESSES

WORLD ENERGY CONFERENCE

Conservation Commission

ENERGY 2000–2020: WORLD PROSPECTS AND REGIONAL STRESSES

Report co-ordinated and edited by J.-R. Frisch
Chairman of the Regional Energy Balances Study

Translated by P. Ruttley

Graham & Trotman

First published in 1983 by:

Graham & Trotman Ltd
Sterling House
66 Wilton Road
London SW1V 1DE
UK

© World Energy Conference, 1983

Softcover reprint of the hardcover 1st edition 1983

ISBN-13:978-94-010-8981-4   e-ISBN-13:978-94-009-5624-7

DOI: 10.1007/978-94-009-5624-7

Typeset in Great Britain by Herts Typesetting Services Limited, Hertford

CONTENTS

PREFACE

In 1977, at the Xth World Energy Conference in Istanbul, the Conservation Commission presented an initial global study of future trends in energy supply and demand up to 2020. This pioneering work was the origin of large-scale global energy surveys.

At the Munich Conference of 1980, the Commission made more thorough analyses by concentrating particularly on the specific behaviour of Third World Nations.

Today, in New Delhi, in keeping with the tradition established by its previous surveys, the Commission is developing a new global study, the originality of which lies in the procedure adopted in its production.

Rejecting centralised forecasting models which have almost exclusively dominated the minds of researchers, the Commission recommended a decentralised method of approach which was totally different both in its basic principles and in its implementation. Essentially, the inhabitants of the regions were to be given the opportunity of expressing their own views on their energy future, rather than imposing on them an external model and therefore external results.

The World Energy Conference was certainly the organ most suited to promote and bring this approach to a successful conclusion. It was an approach which permitted a significant strengthening of regional long-term analytical thinking to occur. In addition, it offers an overview of energy constraints and resources of the ten great regions of the world which is, if not more correct, at least nearer to the right direction. Beyond the data presented by the study, its main value in fact lies in a better understanding of the real problems.

We must remain more vigilant than ever, during the temporary period of calm which we are living through: although the profitability of short term oil trade is attractive, the thorny realities of the future nevertheless subsist and will surface again as soon as there is a recovery of economic activity.

Oil demand will not decrease, especially under the pressure of the Third World. Coal and nuclear power will experience massive demand in all the industrialised nations. The implementation of vigorous programmes for the rational use of energy also remains a priority.

This global study brings a new perspective on these major challenges. It is all the more striking that it broadly confirms most of the previous analyses.

This study could not have seen the light of day without the contribution and the energy of the fifty experts who agreed to contribute to it. It is they who marked the outlines of the future which we must face without illusion or weakness.

By illuminating an essential aspect of the future, this study gives decision makers and politicians the means of substantially increasing their ability to deal with inevitable long-term uncertainties. If it succeeds in making them aware of these circumstances, which appear more remote but yet are crucial for the stability of the world system, then it will have achieved its purpose.

Marcel Boiteux,
President of the Conservation Commission,
World Enery Conference.

FOREWORD

In every way, the study of regional energy balances was a lively adventure which is not revealed by cold figures.

While most members of the Conservation Commission were convinced from the start that it would be interesting to explore long-term perspectives in a really decentralized manner, there was a certain amount of scepticism at the obvious obstacles which confronted such an enterprise: the absence of finance, the novelty of the approach, doubts at the degree of collaboration of the regions, the lack of time.

The present report is proof that the challenge was met, but not without difficulty or various incidents.

(A)   In the first place, this was a *technological adventure*. Contrary to normal methodological approaches, where the quality of models created "in camera" determines the quality of forecasts without any surprising results, we were entering unknown territory. No results could be determined *a priori*. The very logic of the procedure adopted implied that the global picture would be changing constantly right up to the last regional meeting, and furthermore that this would occur latently, since only the final results would reveal the overall picture.

It was therefore necessary to invent a new, simple and pertinent methodology for information, for consultation and coordination. This could not take the conventional form of an inquiry by questionnaire since this would have doomed our approach based on personal contact and rigorous regional follow-up. Beyond methodological considerations and the difficult collection of basic data, the main challenge was a human one: ultimately, the success of the study depended on people.

Of course, the Central Team bore the heavy responsibility of organising the project, of preparing the work of the decentralised teams, of ensuring that the timetable was complied with, and of exercising its very particular coordinating role. But the success of the study depended mainly on the Regional Working Teams.

(B)   This was a *human adventure*, therefore. What would be the quality of the experts, their availability, their degree of commitment to the project, their spirit of mutual cooperation: such were a few of the initial questions. At the end of the two year study, it must be said that the Regional Teams amply fulfilled their difficult mission. Motivated by the interest of the research itself, sometimes sacrificing holidays, encouraged by their companies or government departments (which had themselves been won over to the idea), the fifty experts directly involved in the task showed a real spirit of cooperation and a profound wish to find agreement beyond their differences of character, culture and even of ideology.

Understanding one another, extracting the maximum from know-how and individual experiences, exploring the future together, assuming the risks of forecasting collectively, pursuing one objective: all this did not occur without profoundly transforming inter-personal relationships. Many of these colleagues became friends. And a journey of two years enabled us to cross the landmarks of life: births, deaths, sufferings and joys, and even military invasions. And all this around a common professional project which became progressively what each made of it.

(C) Finally, it was a *spiritual adventure.* With many others, I am convinced that this entire enterprise which was bound to fail for many reasons could not have been brought to a successful conclusion without the constant help of the Lord. Every day I prayed to Him either by myself or in a group, so that this project should not only be completed but also that it should do so successfully. During the difficult parts when one had to persevere, to keep one's bearings, during all these journeys which contained so many unknowns, during the most laborious phases of study, God was present and effective.

Three visits to the Benedictine sisters of Vanves and to the sisters of the Cénacle de Versailles enabled me to clarify most of the problems involved in the writing of this text in a climate of prayer and peace. Everywhere in the world, during the regional meetings, the local Church accompanied our work and it was associated with the brothers and sisters of the Communauté Emmanuel who persevered in contributing to our work with prayer.

In other words, if this study has any merit this is not due to myself. Everything was made possible by the Holy Spirit who guided me, comforted me, stimulated me during this long journey filled with traps. Thanks be to the Holy Spirit. Thanks also be to the men and women who so generously gave their best to the study.

I would first like to thank the entire World Energy Conference, particularly its Secretary-General Mr Eric Ruttley, the members of the Conservation Commission and the Secretary of the French Committee, M. Paul Auriol, for having allowed me to direct this exciting task right to the end. I should also like to thank the Board of Electricité de France which encouraged this study and assisted it technically and financially. In particular, I should like to thank its President, M. Marcel Boiteux, who initiated this study and who supervised so carefully all its phases. I am grateful to M. Albert Robin and to M. Jacques Lacoste for their enlightened advice, to Dominique Gourmelon for his help during the drawing up of the reference base, and to Isabelle Leullieux for her constant efforts in deciphering the manuscript, in connecting all our correspondents throughout the world, and in presenting as best as possible the results of all this work. I should also like to thank each of the fifty experts, all their assistants and all their respective managements for having agreed to join us in this common task. Finally, I must thank Catherine and my children for their patience during the separation imposed on them by these many journeys, and for their constant prayers that this adventure be not just a matter of figures but more profoundly a matter of hope for all humanity, especially for the poorest.

J.-R. Frisch,
Paris, March 1983.

ABBREVIATIONS

B	balance		PEP	primary energy production
BF	balance factor		POP	population
BU	bunkers		PP	petroleum products
CEC	commercial energy consumption		PR	production
			RWT	regional working team
CEP	commercial energy production		SMF	solid mineral fuels
CO	crude oil		TOE	ton of oil equivalent
E	exports		TW	Third World
EC	energy consumption		TWh	terrawatt hour (T = 10^{12})
EIC	eastern industrialized countries		UR	uranium
			VAW	vegetable and animal wastes
EPP	exports of petroleum products		WIC	western industrialized countries
FW	fuelwood			
G	Giga = billion = 10^9			* * *
GNP	gross national product			
HY	hydropower		R1	North America
I	imports		R2	Western Europe
IC	industrialized countries		R3	Industrialized Countries of the Pacific
inhb	inhabitants			
IRGT	interregional global trade		R4	Eastern Europe
KOE	kilo of oil equivalent		R5	North Africa and Middle East
LDC	developing countries			
M	million = 10^6		R6A	Africa South of the Sahara (excluding South Africa)
NCEC	non-commercial energy consumption			
			R6B	Africa South of the Sahara (including South Africa)
NCEP	non-commercial energy production			
NEU	non-energy uses		R7	South Asia
NG	natural gas		R8	South-East Asia
NS	new energy sources		R9	Centrally Planned Asian Countries
NU	nuclear energy			
O	oil		R10	Latin America
PEC	primary energy consumption			

SUMMARY OF THE STUDY

I. SITUATION OF THE SURVEY

1.1 HISTORICAL BACKGROUND

This survey is in keeping with the continuity of the work of the Conservation Commission. In 1977, an initial worldwide analysis of energy supply and demand for the horizon years of 2000 and 2020 was presented at the 10th World Energy Conference in Istanbul. This initial study was continued in greater depth during the following years. In 1980, in the context of the 11th World Energy Conference in Munich, the Conservation Commission discussed in particular the prospects of the long-term energy development of the LDCs, but it also presented an updated analysis of world-wide production potentials. The question was how to follow up the above studies.

1.2 SPECIFICITY

The project of a survey of the long-term development of regional energy balances had its origin in a reflection on the *comparative advantage* of the World Energy Conference (WEC).

(A) The specificity of the WEC may be summarized in the form of three major features, that is:

(i)   It is a *non-governmental* organization and, consequently, submitted to less constraints than many other international organizations.
(ii) It is *not specialized*, because its activities cover all forms of energy.
(iii) It is organized into 80 national committees, representing energy producers and consumers of North and South, West and East.

(B) It could, therefore, be envisaged to organize a project based on an effective, decentralized and inter-energy regional cooperation. In addition, the analysis could cover not only consumption forecasts but also production and international trade forecasts.

This project was formally approved by the International Executive Council at its Munich meeting in September 1980. The IEC also wanted to stimulate the interest of the national committees in long-term energy forecasting by encouraging practical regional cooperation in this field. Finally, by means of a new methodology, the Survey attenuated the inadequacy of traditional forecasting models in front of a more and more unstable and uncertain future.

1.3 OBJECTIVE

The objective was to present at the 12th Congress of the World Energy Conference, to be held in New Delhi in September 1983, the primary energy balances for 2000 and 2020 of each of the ten regions of the world, comprising forecasts of total consumption, production, as well as interregional trade, by eight sources and according to two divergent economic development scenarios. These regional energy balances were to be established by ten regional groups of experts and would be the first attempt to approach the problem of long-term energy forecasts in a really decentralized way.

1.4 MEANS

(i) *One central team*, essentially composed of the Director of Research, who was active for the duration of the study (3 years).
(ii) *Ten regional groups*, containing a total of *50* experts from all continents (30 from LDCs and 20 from industrial countries) with very diversified backgrounds (11 for energy authorities, 9 for oil, 9 for electricity, 14 for international and regional organizations, 7 for university and research institutions). *Twenty* national committees were directly associated with the work. The regional groups were kept small in the interest of efficiency.

Seventeen international meetings were organized throughout the world during the second half of 1981 and the first half of 1982. In addition, supplementary consultations were made by correspondence. Proposals to revise estimates could be made up to the end of 1982.

Consequently, the final forecasts reflect expert opinion as stated in *1982*.

II. METHODOLOGY

2.1 REGIONALIZATION

The division into 10 regions employed is similar to that of the UN. Above all, geographical situation, climate and predominating economic systems were taken into account.

2.2 REFERENCE PERIOD AND FORECASTING HORIZONS

The following reference period was adopted:
1960–1978
To remain in keeping with the continuity of the Conservation Commission's prior surveys, *2000 and 2020* were adopted as forecasting horizons. *Cross-sections* were made for 1960, 1978, 2000 and 2020. No time series were established.

2.3 AREA OF ANALYSIS

(i)   The analysis is limited to the first link of the energy chains, that is *primary energy.* Energy conversion and secondary vectors (electricity etc.) were not analysed. Also, no analysis of demand by sectors or of useful energy was made. Consequently, the energy balances are not complete, that is, they are limited to primary supplies.
(ii) In the interest of simplification (and also for political reasons) only regional and no national results are presented.

2.4 FORECASTING PROCEDURE

(i)   A compromise ensuring both the coherence and the flexibility of the forecasting procedure had to be found. Initial coherence was ensured by the work of the central team (Phase I), because each RWT was supplied with a *regional file* comprising reference bases for each country, a uniform conversion system, and initial supply and demand forecasts according to two economic and demographic scenarios.
(ii) The forecasting procedure was initially based on an *overall model*, where the future development of commercial energy consumption per capita of each region was related to the future development of its economic growth per capita by means of an income elasticity coefficient. Forecasts of non-commercial consumptions were made on an exogenous basis. Population forecasts were based on UN projections. The GNP forecasts, as well as supply patterns, were based on the

report presented at the Munich Conference for the LDCs, and on recent work of international organizations for the industrial countries. Price-elasticity was not introduced, but the general climate resulting from the two scenarios is implicitly characterized by a gradual upward trend of prices in the long run, that is after the present depression.

(iii) In reality, the regional teams broke to a large extent free from the proposed framework. In particular, a number of regional groups adjusted their statistical reference bases. Also, economic forecasts were often slightly changed, contrary to the initially adopted principle in this respect. In addition, each RWT adopted its own forecasting methodology. For example, R1 adopted a complex simulation model, R6 made estimates based mostly on empirical and judgmental criteria, etc. . . . .

The significant variations between the final (Phase II) and the initially proposed (Phase I) forecasts is one of the criteria for measuring the success of the procedure. On a worldwide scale, the variation is +2% for 1978, −8%/−10% for 2000, and −8%/−12% for 2020.* To a large extent, these variations result from lower forecasts for China. In fact, without R9, these variations are reduced to −2% (I) or −5% (II) for 2000 as well as 2020.

2.5 SCENARIOS

The two scenarios were derived from Scenarios B and C of an OECD 1978 study ("Interfutures"), but these latter scenarios were only used to characterize two diverging types of development.

(i) *Scenario (I)* is called *"normative-cooperation scenario"*; that is it corresponds to the desirable long-term image of a world in which, after the present crisis, economic growth will again reach levels in accordance with the common aspirations of LDCs and industrial countries.

(ii) *Scenario (II)* is called *"increasing international tensions scenario"*; that is it corresponds to the long-term image of a world where, in a more difficult context, rivalries and uncertainties will increase, and, where new lines of division will appear, especially between the more protected industrial world and the more threatened Third World.

To avoid imposing upon the RWT too constraining underlying assumptions, liable to condition their forecasts excessively, the development of extreme scenarios too far away from practical decision-making was excluded. The description of the two scenarios was deliberately limited to their overall climate, that is their adaptation to the conditions of each region was left to the regional groups. Scenarios (I) and (II) should, therefore, be considered as the "rosy" and "grey" versions of a future that will be difficult, one way or another. The plausibility of Scenario (II) in relation to Scenario (I) is, of course, reinforced by the projection of present uncertainties.

2.6 UNITS AND EQUIVALENCES

All results are expressed in *tonne of oil equivalent* (TOE), as defined by its calorific value which was fixed here conventionally at 10500 Mcal. In general, the system of equivalences of UN's World Energy Statistics (1979 Edition) was adopted, except for electricity, for which its equivalence at the production level of thermal plants was defined conventionally as: 1000 kWh = 0.222 TOE = 2330 Mcal.

*In the text "…/…" indicates that the figure on the left of the bar / corresponds to Scenario (I) and that of the right to Scenario (II).

In the absence of specified equivalences the following system was adopted:

1 tonne of crude oil (oil products) = 10000 Mcal = 0.952 TOE
1 tonne of solid mineral fuels = 7000 Mcal = 2/3 TOE
1000 m³ of natural gas = 9000 Mcal = 0.857 TOE
1000 kWh = 0.222 TOE = 2330 Mcal

with

1 TOE = 44 GJ and 1 TJ = 22.8 TOE

III. DEVELOPMENT OF THE WORLD BALANCE

3.1 DEMOGRAPHIC AND ECONOMIC ENVIRONMENT

(i) Total world population could reach 6 billion by 2000, and 7.7 billion by 2020 (versus 4.3 billion in 1978), that is, its average growth rate will decrease from 2% per year for 1960 to 1978, to 1.6% by 2000, and 1.2% by 2020.

(ii) Under Scenario (I), world GNP will continue to grow at a sustained rate of 3.6% by 2000, and 2.9% by 2020, compared to 5% for 1960 to 1978. Under Scenario (II), world GNP growth will slow down more: 2.6% by 2000, and 2% by 2020.

(iii) World GNP per capita will increase from $2000 in 1978 to $3200 (I) or $2600 (II) by 2000, and $4500 (I) or $3000 (II) by 2020.

3.2 ENERGY CONSUMPTION

(i) Total world energy consumption increased from 3.3 GTOE in 1960 to *6.8 GTOE in 1978*. It will reach 10.1–11.7 GTOE by *2000*, and 13.8–18 GTOE by *2020*. These volumes are impressive indeed. In fact, total world consumption could double or more between 1978 and 2020.

(ii) The picture is quite different in terms of average growth rates. In fact, demand growth will slow down considerably: 1.8% to 2.5% by 2000, and 1.6% to 2.1% by 2020, compared to more than 4% between 1960 and 1978. Average annual growth rates will be halved.

(iii) Growth of consumption per capita will be very slow: 1.7–1.9 TOE by 2000, and 1.8–2.3 TOE by 2020, compared to 1.1 TOE in 1960 and 1.6 TOE in 1978. Over more than 40 years per-capita consumption will grow by only 0.2–0.7 TOE by 2020, compared to 0.5 TOE between 1960 and 1978. The average annual growth rate up to 2020 will be less than 1% under Scenario (I) and 0.2–0.3% under Scenario (II), compared to more than 2% for 1960–1978.

(iv) Population growth will play a major part in the growth of world energy needs. In fact, it explains 60% of demand growth over the whole forecasting period under Scenario (I), and even 90% of growth by 2000 (and 75% by 2020) under Scenario (II), versus 50% only for 1960-1978.

3.3 ENERGY SUPPLIES

(i) *Non-commercial* sources covered 17% of world energy demand in 1960, representing 560 MTOE. They covered only 11% in 1978, representing 735 MTOE. The decrease of their part in world demand will continue: 8% by 2000, and 5% by 2020, under Scenario (I), 10% and 8%, respectively, under Scenario (II), though they will represent increasing volumes: 900 MTOE (I) or 1100 MTOE (II) by 2020. At the same time, non-commercial energy consumption per capita will gradually decrease from 0.17 TOE in 1978 to 0.12 TOE (I) or 0.14 TOE (II) by 2020.

(ii) The part of *coal* in world energy demand decreased sharply between 1960 and 1978 (from 36% to 25%), though representing an increase in volume from 1.2 GTOE to 1.7 GTOE. This tendency will be reversed in the future: 28% by 2000 (2.8–3.2 GTOE), and 32% by 2020 (4.4–5.7 GTOE). In fact, as from 2000, coal will become again the world's leading source of energy.

(iii) *Nuclear power* will also massively replace oil in the future. Its part in total world supplies could grow from 2% in 1978 to 8% by 2000 (0.8–1.0 GTOE), and 12–13% by 2020 (1.7–2.3 GTOE).

(iv) *Hydropower* production will increase from 0.4 GTOE in 1978 to 0.6–0.7 GTOE by 2000, and 1.0–1.3 GTOE by 2020; that is it would maintain its part in total world supplies at 6–7%.

(v) The part of *new energies* in world supplies will grow gradually to 3% by 2000 (300 MTOE), and 6% by 2020 (0.8–1.0 GTOE).

(vi) In the long run, *natural gas* will maintain its present part of 17% in world supplies, because world demand will grow considerably: from 1.2 GTOE in 1978 to 1.8–2.2 GTOE by 2000, and 2.4–3.2 GTOE by 2020.

(vii) Finally, *oil's* part in world supplies will decrease from about 40% in 1978 to about 30% by 2000, and about 20% by 2020. Under Scenario (II), world oil demand could reach a ceiling of 2.8 GTOE by 2000, decreasing to 2.4 GTOE by 2020 (1978: 2.7 GTOE). Under Scenario (I), oil substitution will be less successful. Consequently, world demand could increase to 3.4 GTOE by 2000, and 3.6 GTOE by 2020.

3.4 ENERGY PRODUCTION

(i) On a world-wide scale, energy production differs structurally from energy consumption only because of stock changes (assumed as being zero in forecasting), as well as bunkering and non-energy consumptions.* In addition, the adopted forecasting procedure does not permit one to expect the automatic adjustment of supply and demand forecasts, because these forecasts were recorded separately, without iteration.

(ii) The gross differential between total supply and demand remains within a limit of 3–5% of demand, that is 500–640 MTOE over the whole forecasting period (versus 360 MTOE for 1978). The residual differential resulting from the inadequacy of supply and demand projections, is only about 0.5%. In other words, the "natural" adjustment of the forecasts is very satisfactory.

(iii) The greatest differential is between oil production and consumption, because oil is used for the greater part of bunkering and non-energy consumptions. In spite of the anticipated success of all alternative solutions, oil production does not achieve stabilization, except under Scenario (II): from 3 GTOE in 1978 to 3.2–3.9 GTOE by 2000, and 3.0–4.1 GTOE by 2020.

3.5 INTERREGIONAL ENERGY TRADE

(i) All intraregional trade was excluded. The procedure adopted, that is, the separate addition of each region's imports and exports, made a residual differential inevitable.

(ii) The total volume of interregional trade was 1.5–1.6 GTOE in 1978, that is, 22–23% of world energy consumption, supplied as to 92% by crude oil, 5% by

*Variations of refined products offset each other in the world market.

coal, and 3% by gas. Under Scenario (I), this volume will slightly increase: 1.85 GTOE by 2000, and more than 2 GTOE by 2020. Under Scenario (II), it will decrease: 1.5 GTOE by 2000, and 1.35 GTOE by 2020. Nonetheless, the share of interregional trade in world energy consumption will decrease under both scenarios: 15-16% by 2000, and 10-11% by 2020. This decrease will not differ much from one scenario to another. Consequently, in relative terms, interregional energy trade tends to decrease, especially because of the reduction of oil trade.

(iii) The "natural" adjustment of import and export forecasts worked well. The residual differential decreases from 1.2% of world production in 1978 to 0.1-0.3% by 2020 in three cases out of four (10 to 50 MTOE), that is, 2 to 4% of total imports, compared to nearly 6% in 1978. Under Scenario (I), the residual variation for 2020 remains near to its 1978 level (for a volume of 200 MTOE), that is 10% of imports.

(iv) Finally, the development of interregional trade varies very much from one fuel to another. Coal trade will grow very sharply: from 5% of total interregional energy trade in 1978 (70 MTOE) to 17% by 2000, with 320 MTOE (I) or 240 MTOE (II), and 33% by 2020, with 700 MTOE (I) or 440 MTOE (II). The growth of gas trade will be similar, at least up to 2000: from 3% of interregional energy trade in 1978 (50 MTOE) to 13% by 2000, with 270 MTOE (I) or 180 MTOE (II), and 16% by 2020, with 325 MTOE (I) or 230 MTOE (II). Under Scenario (I), interregional oil trade will decrease from 1.5 GTOE in 1978 to 1.3 GTOE by 2000, and 1 GTOE by 2020. Under Scenario (II), the drop will be even sharper: 1 GTOE by 2000, and 0.7 GTOE by 2020. In fact, oil trade will represent only about 50% of interregional energy trade by 2020, compared to 92% in 1978, and still about 70% in 2000.

IV. NORTH–SOUTH DYNAMICS

4.1 DEMOGRAPHIC AND ECONOMIC ENVIRONMENT

(i) The Third World's (LDCs) share in total world population will grow regularly: 67% in 1962, 72% in 1978, 77% in 2000, and 80% in 2020. This means that the Third World's population will increase by 3 billion by 2020, versus 300 million for the industrialized countries (ICs). In other words, it will reach 4.7 billion by 2000, and 6.2 billion by 2020.

(ii) The Third World's share in world GNP slightly increased between 1960 and 1978: from 14.5% to 17.5%. It will increase rather more in the future: 22–25% by 2000, and 28–33% by 2020. Under Scenario (I), the average growth rates of the Third World's GNP will be 5.4% per year up to 2000, and 4.4% up to 2020. Under Scenario (II), its average annual growth rates will be 3.9% and 3.2%, respectively (1960–1978: 5.9%). The average annual growth rates of the industrialized countries will decrease sharply in the future: 3.2% by 2000, and 2.3% by 2020, under Scenario (I), and 2.4% and 1.6%, respectively, under Scenario (II) (1960–1978: 4.8%).

(iii) The ratio between the per-capita GNPs of the Third World and the industrial countries will remain practically unchanged: 8% in 1960 as well as 1978, 10/8% by 2000 and 12/10% by 2020. This is indeed a far way from "catching up".

4.2 ENERGY CONSUMPTION

(i) The development of consumption volumes will be impressive. The demand of the industrialized countries will increase from 5 GTOE in 1978 to 7–8 GTOE by 2000, and 9–11 GTOE by 2020. The Third World's demand will increase from 1.7 GTOE in 1978 to 3–4 GTOE by 2000 and 5–7 GTOE by 2020. This means that the Third World's share in world energy demand will increase from 24% in 1978 (23% in 1960) to 34% (I) or 31% (II) by 2000, and 40% (I) or 35% (II) by 2020.

(ii) None the less, in per capita terms, the ratio between the energy consumption of South and North will hardly improve: 15% in 1960, 13% in 1978, 15% (I) or 13% (II) in 2000, and 17% (I) or 13% (II) in 2020. In fact, the Third World may at best expect to double its energy consumption per capita by 2020 under Scenario (I) (1.2 TOE versus 0.55 TOE in 1978, or +0.6 TOE). However, under Scenario (II), its per capita consumption will be hardly 0.8 TOE by 2020, that is higher by 0.2 TOE only than in 1978. During the same time, the per capita consumption of the industrialized countries will increase by +2.7 TOE (I) or +1.5 TOE (II), and this in spite of much lower growth rates. The difference between the average per capita consumptions of North and South was 2.2 TOE in 1960 and 3.8 TOE in 1978, but it will be almost 5–6 TOE by 2020. The relative situations of North and South are less unbalanced under the higher-growth scenario.

4.3 ENERGY SUPPLIES

(i)   *Non-commercial energy* consumptions will remain the Third World's domain. The Third World's share in world consumption increased to about 90% in 1978, but it could find again its 1960 level of 80% in 2000 as well as 2020. On the other hand, the share of non-commercial energy in total Third World demand decreased regularly from 56% in 1960 to 37% in 1978. This will continue in the future: 19% (I) or 28% (II) by 2000, and 10% (I) or 18% (II) by 2020. Consumption volumes will slightly increase: from 0.6 GTOE in 1978 to 0.7 GTOE (I) or 0.9 GTOE (II) in 2000 as well as 2020. Fire-wood consumption will increase from 0.4 GTOE in 1978 to 0.5 GTOE (I) or 0.6 GTOE (II) by 2000, but it will fall again to 0.4 GTOE (I) or 0.5 GTOE (II) by 2020. Consequently, the environmental impact of fire-wood consumption remains very great. The fire-wood consumption of the industialized countries will slightly increase: about 0.2 GTOE in 2000 as well as 2020, compared to 0.1 GTOE in 1978.

(ii) Development in the field of commercial sources will also be marked by sharp differences between North and South. *Coal* will be in great demand as one of the preferred alternative energy sources for oil substitution in the North. Its share of total supplies decreased from 40% in 1960 to 25% in 1978, but it will increase significantly in the future: 31% by 2000 (2.2–2.4 GTOE versus 1.3 GTOE in 1978), and 38% by 2020 (3.4–4.1 GTOE). Coal will not play the same part in the South. First of all, because of the expansion of coal in China, its share in total supplies did not decrease significantly between 1960 and 1978. Coal thus always covered 23% of the Third World demand during this period. Its share will be 20–22% only during the forecasting period but with fast growing volumes: 0.6–0.8 GTOE by 2000 (1978: 0.4 GTOE), and 1–1.6 GTOE by 2020.

(iii) The future development of *nuclear power* will also be very different in North and South. It will be the North's second alternative energy source for oil substitution. Its share in total supplies will increase from 3% in 1978 to possibly 11% by 2000 (0.8–0.9 GTOE), and 17–18% by 2020 (1.5–1.9 GTOE). Nuclear power's share will be necessarily very limited in the South. Its share in total supplies will be only 2% by 2000, and 3–5% by 2020 (0.2–0.4 GTOE).

(iv) The development of *new energies* will be rather similar in North and South. It will even be slightly more pronounced in the Third World. New energies will represent 2–3% of total suplies by 2000, and 5–7% by 2020 in North and South, with volumes of 0.5–0.6 GTOE for the North, and 0.3–0.4 GTOE for the South, by 2020.

(v)   *Hydraulic power's* share in the North's total supplies will decrease from 6% in 1978 (0.3 GTOE) to 5% by 2020 (about 0.5 GTOE). However, the expansion of hydraulic power is one of the South's main potentials. This expansion will be particularly significant after 2000. Hydraulic power covered 6% of the Third World's total energy demand in 1978, with a production volume of 0.1 GTOE. Its share in total supplies will increase to 7% by 2000 (0.2–0.3 GTOE), and 11% by 2020 (0.5–0.8 GTOE). In fact, the Third World will represent 50–60% of the world's hydropower production by 2020.

(vi) *Natural gas* will mainly develop in the South: from 6% in 1978 to 10–12% by 2000 (0.3–0.5 GTOE), and 14-16% by 2020 (0.7–1.1 GTOE). Natural gas represented already 20% of the North's total supplies in 1978 (1.1 GTOE). It will barely maintain its share at this level up to 2000 (with 1.5–1.7 GTOE). As from 2000, the share of natural gas will gradually decrease to 19% in 2020 (1.7–2 GTOE).

(vii)   As far as *oil* is concerned, the combined success of all potential alternative
solutions will decrease its share in total supplies from 43% in 1978 (1960: 33%),
with a volume of 2.2 GTOE, to 26% by 2000 (1.9–2 GTOE), and 13% by 2020
(1.2–1.4 GTOE) in the North. This means that oil will become the North's
second energy source by 2000, and the fourth by 2020.

The evolution of oil's share in the South's total supplies cannot be the same.
It will inevitably increase: from 28% in 1978 (0.5 GTOE) to 36% (I) or 30% (II)
by 2000, with volumes of 1.4 GTOE and 0.9 GTOE, respectively. Even if oil's
share will be reduced to 30% (I) or 26% (II) by 2020, that is, more or less its
present level, the volumes involved will be much greater: 2.2 GTOE (I) or 1.2
GTOE (II). Consequently, the Third World will represent more than one half of
the world's oil consumption by 2020. Actually, the Third World's share in world
oil demand will grow dramatically in the future: from 18% in 1978 (1960: 13%)
to 41% (I) or 33% (II) by 2000, and 60% (I) or 50% (II) by 2020. It is obvious that
this development will have significant strategic consequences for the long-term
policies of oil producers.

Oil constraints in the Third World (excluding R9, the Asian countries with
centrally planned economies because of the importance of coal in R9) will be
even more serious. In fact, oil will still cover 33% (I) or 27% (II) of total needs of
the Third World (without R9) by 2020 (versus 37% in 1978), with consumption
levels that will be 3 to 5 times higher than now.

4.4 ENERGY PRODUCTION

(i)   As it was shown above, the Third World's share in world energy demand will
increase considerably, but its share in world supply will remain quite stable:
42% in 1978 (3 GTOE), between 40% and 45% in 2000 and 2020, with much
higher volumes of course: 5.8–8.2 GTOE by 2020.

(ii)   The North's share in world coal supply will remain overwhelming over the
whole forecasting period: 73–78% (1978: 77%). It is the same with nuclear
power, where the North's share will represent 90% (I) or 93% (II) of world
production in 2000, and 83% (I) or 90% (II) in 2020 (1978: 98%).

(iii)   The North's share in world hydropower production will decrease sharply: from
76% in 1978 to 40–50% only in 2020. The North's share in world natural gas
supplies will also fall sharply, especially up to 2000: from 87% in 1978 to
68–75% in 2000, and 56–64% in 2020. The future development of new energies
in North and South will be rather balanced. The North's share will be about 55%
of world supplies in 2020. Non-commercial sources will, of course, remain the
domain of the South, with a share of 77–80% in 2020.

(iv)   The South will reinforce its dominating position in oil. Its share in world
production will increase from 58% in 1978 to 60–65% by 2000, and 67–72% by
2020. The volume of the South's oil supply will increase from 1.8 GTOE in 1978
to 1.9–2.6 GTOE by 2000, and 2–3 GTOE by 2020. This means that its oil
production will increase by +0.2/+1.2 GTOE between 1978 and 2020, while its
oil consumption will grow by +0.7/+1.7 GTOE. This strong growth of the
South's oil consumption will, of course, create new tensions. The oil production
of the North will level off at 1.3–1.4 GTOE by 2000 (1.3 GTOE in 1978), but will
gradually decrease to 1–1.2 GTOE by 2020.

(v)   A comparison of North and South production and consumption patterns by
fuels shows great divergences. Coal supply and demand were balanced in the
North in 1978. There will be a supply surplus of about 100 MTOE by 2000, but a
deficit of 20 MTOE by 2020. The North's gas deficit will grow gradually from 30
MTOE in 1978 to 150 MTOE by 2000, and 200 MTOE by 2020, when it will
represent 6–7% of world gas production. Finally, the North's very great oil

deficit of about 1 GTOE in 1978 will gradually decrease to 0.6 GTOE by 2000, and 0.2-0.3 GTOE by 2020, mainly because of the South's rising demand and the success of the North's oil substitution efforts.

V. REGIONAL PROSPECTS

R1 NORTH AMERICA

Population and GNP

North America represented 6% of the world's population in 1978, but 26% of world GNP. This discrepancy will be slightly diminished in the future owing to a significant slow-down of economic growth: 17% (I) or 21% (II) of world GNP by 2020. Nonetheless, North America's GNP per capita will remain very high: $14000–17500 by 2020, versus less than $10000 in 1978.

Energy consumption

(i)   North America has always been a great energy consumer. It represented 30% of world consumption in 1978 (1960: 35%), for a volume of over 2 GTOE.

      Nevertheless, it is expected that North America's energy demand will grow at an average rate of 1% up to 2000, instead of 3% for 1960–1978. This means that it will represent only 2.5 GTOE by 2000. By 2020, it could be 3–3.5 GTOE, representing 20% of world demand.

(ii)   North America's energy consumption per capita is by far the highest in the world: 8.3 TOE in 1978. But the very sharp slow-down of the growth of this consumption (+0.3% instead of +0.2% for 1960–1978) should limit its expansion to 9–10 TOE only by 2020.

Energy supplies

(i)   At present North America is the world's first consumer of oil, natural gas, as well as hydraulic and nuclear power. It is also the world's second coal consumer. Hydrocarbons represent more than 70% (nearly 1.5 GTOE) of its total energy consumption.

(ii)   North America's energy supply pattern has not changed significantly since 1960, but it will undergo deep changes over the whole forecasting period. Coal consumption could increase to 800–900 MTOE by 2000, and 1500–1800 MTOE by 2020, compared to 360 MTOE in 1978, that is, coal could cover 50% of total regional needs by 2020, with a volume almost equivalent to today's world coal production. The expansion of nuclear power will also be considerable: 400–500 MTOE (or 1800–2200 TWh), that is, 13% of total energy demand. The same applies to new energies: 100 MTOE by 2000, and 300 MTOE by 2020.

      Under these conditions, the volume of hydrocarbon consumption should decrease. Natural gas consumption will not decrease massively: from 530 MTOE in 1978 to 400–460 MTOE by 2020. But oil demand will fall considerably: from 900 MTOE in 1978 to about 500 MTOE by 2000, and 200 MTOE by 2020, when oil will represent only 7% of total energy demand, compared to 21% in 2000 and 45% in 1978.

(iii)   North America's objectives on the field of oil substitution are extremely ambitious. In fact, R1 would become the world's first consumer of coal and new energies as from 2000. It would even represent one third of the world's coal demand by 2020. On the other hand, it would only be the world's fifth (I) or seventh (II) oil consumer.

Energy production

(i)   North America was the world's first energy producer in 1978, when it represented 24% of the total world supply. It will keep this position also in the future, mainly because of the development of its coal potential. In fact, coal supply will increase from less than 400 MTOE in 1978, to 1 GTOE by 2000, and 1.7–2 GTOE by 2020, when it will represent one half of total regional production (1978: 22%). Coal will also represent 70% of the total growth of regional production. Under these conditions, North America's energy production will increase from 1.7 GTOE in 1978 to 2.6–2.8 GTOE by 2000, and 3.5–4 GTOE by 2020. None the less, except for coal and new energies, R1 will have lost by 2020 its predominant position as far as all other commercial sources are concerned.

(ii)   In fact, North America will barely maintain the present level of its hydrocarbon production up to 2000, and it will ineluctably decline to only 0.8–0.9 GTOE in 2020, versus 1.1. GTOE in 1978.

Interregional energy trade

(i)   The needs of North America's domestic energy market are such that the quantities available for export are minimal for the time being. In fact, R1's exports represent now only 2% of interregional exports, while its imports represent 22% of total interregional imports (350 MTOE, essentially crude oil). The region's energy trade deficit increased continuously in the 70s, to reach −320 MTOE in 1978.

(ii)   This deficit will be practically eliminated by 2000. The region's energy trade balance will even be positive by 2020 (+135 MTOE). This extremely favourable development is due to two factors. On the one hand, regional coal exports will increase sharply: from 30 MTOE in 1978 to 160–180 MTOE in 2000 as well as 2020. On the other hand, crude oil imports will drop sharply: from 350 MTOE in 1978 to 120–160 MTOE by 2000, and 40–50 MTOE by 2020.

Under these conditions, tensions in the world oil market would be reduced considerably, because the region's share in world crude-oil imports would decrease from 23% in 1978 to 5% in 2020. In addition, the region's share of 40% in world coal exports would be maintained over the whole forecasting period.

Questions

• Is such a development of coal production (and consumption) conceivable?
• Are the forecasts for nuclear power and new energies realistic?
• Consequently, will such a reduction of the share of oil really be feasible?

R2 WESTERN EUROPE

Population and GNP

The average age of Western Europe's population is rising at an accelerated rate. Total population will increase to less than 500 million by 2020 (versus 420 million in 1978).

The region represented 10% of world population in 1978, but 31% of world GNP, that is more than any other region. Western Europe's share in world GNP will also remain the highest in the future, though it will decrease to 22% by 2020. GNP per capita will increase from $6000 in 1978 to $11000-15000 by 2020.

Energy consumption

(i)   Western Europe's total energy consumption was 1.2 GTOE in 1978, representing 18% of world demand. But the growth of the region's consumption will considerably slow down in the future: from +3.6% between 1960-1978 to slightly over +1% as from 2000. Under these conditions, regional energy demand will be 1.7-1.9 GTOE by 2000, and 2.1-2.4 GTOE by 2020.

(ii)   The region's energy consumption per capita is rather "sober" compared to that of the other industrial regions: less than 3 TOE in 1978. Because of the slower growth of demand, per capita consumption will be 4-5 TOE by 2020, that is, it would then still be far lower than that of North America.

Energy supplies

(i)   Western Europe's energy supply pattern underwent deep changes between 1960 and 1978, mainly because of the increasing penetration of oil, whose share in total demand increased from 28% to 51%, as well as gas (from 2% to 14%) while coal's share fell from 55% to 21%. Hydrocarbons represented two thirds of total supplies in 1978 (versus 30% in 1960). Western Europe's share in the world market of hydrocarbons was 20% in 1978.

(ii)   The region's supply pattern will also undergo deep changes in the future. This development will be characterised by a decrease of oil's penetration to one third of total needs by 2000 (when oil consumption will level off at its 1978 level of 630 MTOE), and to one fifth of total needs by 2020 (400-450 MTOE). This will be accompanied by a more gradual decrease of the share of natural gas. These deep changes will be made possible by the simultaneous success of the expansion of coal and, above all, nuclear power. In fact, coal's share in total supplies will rise again to 30% by 2020 (600-700 MTOE versus 260 MTOE in 1978). Nuclear power's share will grow from 4% in 1978 to 13% by 2000, and 25% by 2020 (500-600 MTOE or 2300-2800 TWh).

Energy production

(i)   Western Europe's energy production capacities are limited: 9% of world supply in 1978 (630 MTOE). 50% of its needs were covered by imports in 1978. In spite of the expansion of nuclear power, the region's share in world energy production will remain at its 1978 level of 10%, representing 1 GTOE in 2000 and 1.5 GTOE in 2020.

(ii)   Nuclear power will represent 35-40% of regional energy production by 2020. In fact, Western Europe will be the world's first producer of nuclear power by that date. The share of all other sources will decrease, except for the new energies (100 MTOE by 2020): from 35% in 1978 to less than 25% by 2020 for coal, and from 24% to 7% for natural gas. Oil production will reach its ceiling with 160 MTOE before the end of this century, representing 15% of total supply in 2000, as in 1978. It will decrease to 110-130 MTOE by 2020 (90 MTOE in 1978). Nuclear power and new energies together will represent nearly 50% of regional supply in 2020.

Interregional energy trade

(i)   Western Europe has always been a net energy importer. It was the world's first energy importer in 1978, representing 45% of world imports, and 3% only of world exports. Its energy trade deficit was of the order of 660 MTOE in 1978, crude oil representing 90% of this deficit.

(ii) This situation will hardly improve in the future. In fact, Western Europe's energy trade deficit would reach 1000 MTOE (I) or 700 MTOE (II) by 2020, representing still 50% of world imports. None the less, the region's energy import pattern will undergo deep changes by 2020, mainly because of the increase of the shares of coal and gas to 35% and 20%, respectively, while coal and gas together represented only 10% of 1978 imports. This improved diversification will lead to the reduction of crude imports from 660 MTOE in 1978 to 350–400 MTOE by 2020. However, Western Europe's crude imports will then still represent 40–50% of world imports, versus 44% in 1978. This development implies inevitably increasing volumes for regional coal and natural gas imports by 2020: 260–370 MTOE and 130–200 MTOE, respectively. In other words, Western Europe will represent 50–60% of world coal and natural gas imports by 2020.

Questions

• Will this massive substitution of oil by coal be feasible in the proportions as expected?

• Will the expansion of nuclear power, which is the indispensable key component of the region's energy policy, materialize?

• Is the persistence of a very great energy trade deficit not a matter of concern, especially in view of the considerable tensions resulting from this deficit in the world markets of all fossil fuels?

• However, within the narrow limits imposed by its insufficient energy supply, has the region much other choice?

R3 INDUSTRIALIZED COUNTRIES OF THE PACIFIC

Population and GNP

In spite of the near-stagnation of its population growth (3% of world population) R3 will maintain its 1978 share in world GNP over the whole forecasting period (12% versus 6% in 1960), because of an economic growth stimulated by that of Japan. Under these conditions, R3's GNP per capita will be the highest in the world before 2000, to reach a level of $17000–25000 by 2020.

Energy consumption

(i) Owing to the lack of natural resources, Japan has always managed its energy consumptions in a very efficient way. Nevertheless, R3's energy demand increased at the very high average rate of +7% per year between 1960 and 1978, because its economic growth rate was the world's highest during this period (+7.5% per year). But the slow-down is already beginning and will be characterized by considerably lower rates in the future. Energy consumption will increase from 400 MTOE in 1978 to 600–700 MTOE by 2000, and 750–920 MTOE by 2020, when it will represent 5–6% of world demand, as in 1978. The income-elasticity of R3's energy demand will be comparable to that of Western Europe, as in the past.

(ii) R3's energy consumption per capita "caught up" with that of Western Europe between 1960 and 1978 (3 TOE in 1978). It will be higher than that of Western Europe in the future, in spite of slowed down growth rates: 5–6 TOE by 2020, versus 4–5 TOE for Western Europe.

Energy supplies

Colonized by oil as from 1960, R3's oil dependence is still the highest in the world: 62% (250 MTOE) in 1978, versus 32% in 1960. The share of coal in total supplies dropped accordingly: from 51% in 1960 to 21% in 1978.

The energy policies adopted in R3 should reduce the share of oil to one third of total needs by 2000, and 20% by 2020, that is to proportions very similar to those prevailing in Western Europe. Under these conditions, coal will represent one third of total supplies by 2020, while the share of natural gas will increase from 6% in 1978 to 18-19% by 2020, and that of nuclear power from 4% to 17-22%. The success of this strategy depends mainly on the expansion of coal, whose consumption will increase from 80 MTOE in 1978 to 250-300 MTOE by 2020. Oil demand is already at its ceiling and will decrease to 210-225 MTOE by 2000, and 160-185 MTOE by 2020.

Energy production

R3's energy dependence is now the highest in the world (62% in 1978). Its energy production is mostly coal, which represented nearly 50% (70 MTOE) of its 1978 supply. The development of R3's (mainly Austalia's) coal resources will lead to a production volume of 240-290 MTOE by 2020, representing nearly 50% of regional energy supply, as in 1978. At the same time, the production of nuclear power in Japan will increase from 14 MTOE in 1978 to 130-200 MTOE in 2020. Regional production of natural gas will also grow from 10 MTOE to 50-60 MTOE during the same period, while R3's oil production will practically double (from 25 MTOE to 45-50 MTOE). The combined effect of these developments would be the increase of total regional energy production to 530-680 MTOE by 2020, versus 150 MTOE in 1978. Under these conditions, R3's energy supply would grow faster than its demand and represent 4% of world production.

Interregional energy trade

(i) Because of the extent of its demand and its insufficient supply, R3 exports practically no energy at present. It even had to import 285 MTOE in 1978 (250 MTOE in the form of crude oil), representing 18% of world imports.

(ii) In spite of the growth of R3's energy production, its energy trade deficit will remain at its 1978 level of 280 MTOE over the whole forecasting period. Therefore, its oil imports alone will still be of the order of 150 MTOE in 2020 (15-20% of world imports). On the other hand, its coal exports will increase from 7 MTOE in 1978 to about 70 MTOE by 2000, and 100-120 MTOE by 2020. R3's share in world coal exports will grow from 10% in 1978 to 20-25% by 2020. However, its coal imports will also increase from 20 MTOE in 1978 to 100-120 MTOE by 2020, when they represent 15-25% of world imports, making R3 the world's second coal importer.

R3 will also remain the world's second gas importer: 25-40% of world imports between 2000 and 2020, versus 30% in 1978.

(iii) R3's importance as an energy exporter will increase considerably, because its share in world exports will grow from 1% in 1978 to 4% by 2000, and 7-8% by 2020. Nevertheless, in spite of the great diversification of its fuel demand, R3's total import dependence will be aggravated, because its share in world imports will be 19-25% by 2020.

This shows to which degree the region will depend on a very open world energy market, even if this dependence will gradually decrease by 2020, when regional supplies will cover 70-75% of demand.

Questions

- Will the extent and the rate of the development of Australia's coal (and gas), as well as Japan's nuclear power, be sufficient to reduce regional oil demand and energy dependence as expected?
- And will this eliminate the risk of a higher energy trade deficit?

R4 EASTERN EUROPE*

Population and GNP

Eastern Europe's population growth will slow down in a similar way to that of the western countries: 460 million by 2020. But the region's economic growth will be relatively sustained, unlike that of these countries: 2–4% in the future.

Owing to these more favorable prospects, R4 will maintain its share in world GNP at a level of 16–17%, while its share in world population will decrease from 9% in 1978 to 6% in 2020. Eastern Europe's GNP per capita will also grow faster than that of the western countries. In fact, it will represent 80% of that of these countries by 2020, versus 60% in 1978, and hardly 50% in 1960.

Energy consumption

(i)   Eastern Europe was the world's second energy consumer in 1978 (over 1.4 GTOE), with a share of 21%. It will practically maintain this share over the whole forecasting period, with volumes of 2.2–2.4 GTOE by 2000, and 2.8–3.4 GTOE by 2020.

(ii)   Eastern Europe's GNP per capita was lower than that of Western Europe in 1978, but its energy consumption per capita was higher: 3.8 TOE versus 2.9 TOE. Owing to Eastern Europe's higher growth rates, this difference will become even greater in the future. In fact, R4's energy consumption per capita will increase to 5.1–5.8 TOE by 2000, and 6–7.4 TOE by 2020, that is, it will be multiplied by 1.5–2 between 1978 and 2020, while it doubled between 1960 and 1978. In other words, in the most favorable case, it will double again between 1978 and 2020, but over a period twice as long because of an average growth rate cut by half.

Energy supplies

Eastern Europe's energy supply pattern has undergone deep changes since 1960, though they were less pronounced than in the western countries. Oil's penetration at the expense of coal has grown much less. At the same time, the massive expansion of natural gas has also contained oil's penetration.

Because of its enormous reserves, R4 is the world's first coal consumer, with nearly 600 MTOE in 1978, representing 40% of regional energy needs. Eastern Europe's coal demand will also increase in the future (but under more difficult conditions) to 780–930 MTOE by 2020. However, R4's share in world coal demand will decrease from 34% in 1978 to 15–18% by 2020.

Natural gas is another great asset of the region. Demand will increase from 330 MTOE in 1978 to 700–800 MTOE by 2000, and 950–1100 MTOE by 2020. In fact, R4 will be the world's first natural gas consumer even before 2000, with a share of 35–40% of world demand. The expansion of nuclear power will also be considerable: from 12 MTOE in 1978 to 180–240 MTOE by 2000, and 450–600 MTOE by 2020, when R4 will be the world's number two in this field.

*All forecasts for R4 were established under the responsibility of the central team and noted by the RWT.

Combined with the development of hydropower and new energies, this great expansion of alternative sources will limit the growth of Eastern Europe's oil demand to 550-600 MTOE in 2000, and 420-530 MTOE in 2020, versus 400 MTOE in 1978. Oil's share in regional supplies will decrease from 28% in 1978 to 15-16% by 2020, but Eastern Europe will remain the world's second oil consumer. It is probable that this reduction of oil's penetration will pose problems, in spite of the pressure resulting from the region's tight oil resource base.

Energy production

Eastern Europe is the world's second energy producer: 1.6 GTOE in 1978. In addition, it is number one for coal (35% of world supply), as well as number two for oil (19%) and gas (29%). Energy production will grow at a slightly lower rate than needs. However, the volumes involved are impressive: 2.8-3.6 GTOE by 2020. Also, supply and demand patterns will be more and more identical. This is in accordance with the region's objective of strict self reliance in the field of energy. As from 2000, R4 will be the world's first gas producer (nearly 40% of world supply). It will be number two for nuclear power by 2020 (27% of world supply), but it will be only number two for coal (35-38%), as well as number three (I) or four (II) for oil, before 2000.

Interregional energy trade

(i)   In view of the principle of self-sufficiency, as well as the importance of intraregional trade, only limited quantities of energy are available for Eastern Europe's interregional energy trade: 100 MTOE for exports (7% of world exports), and 35 MTOE for imports (2% of world imports). Consequently, the region has a positive energy trade balance of 65 MTOE. In addition, exports represent a well-balanced mix of the three fossil fuels.

(ii)   Scenario (I), based on international cooperation, favours the simultaneous growth of Eastern Europe's exports and imports, but with higher growth rates for exports, which would represent 8% of world exports by 2000, and 10% by 2020, with volumes of 150 MTOE and 190 MTOE, respectively. In addition, gas will play an increasing part in R4's exports as from 2000, representing 120 MTOE by 2020, or 33% of world gas exports (number one). Coal exports will also grow, but to a lesser extent. On the other hand, there will be no oil exports as from 2000. Eastern Europe's energy imports will level off at about 80 MTOE (4-5% of world imports), oil representing 75-80%. Finally, the surplus of R4's interregional trade balance will increase from 65 MTOE in 1978 to 70 MTOE by 2000, and 110 MTOE by 2020.

(iii)   Under Scenario (II), Eastern Europe's energy exports will be limited to gas and coal (85-90 MTOE). Imports will be 75 MTOE by 2000, and 60 MTOE by 2020, almost all in the form of crude oil. Eastern Europe's energy trade balance will remain only slightly positive: 10 MTOE by 2000, and 30 MTOE by 2020, because the region's oil deficit will increase to −70 MTOE by 2000, and −50 to −60 MTOE by 2020. Eastern Europe expects to compensate its oil imports by its gas and coal exports.

Questions

• May it be safely assumed that the growth of energy demand will be maintained at rates much higher than those of the western countries?
• Will the simultaneous — and considerable — expansion of gas, coal and nuclear production materialize as expected?
• Is the risk of an outside oil dependence not a matter of concern in the context of the region's system?

R5 NORTH AFRICA AND MIDDLE EAST

Population and GNP

R5's share in world population will increase from 4% in 1978 to 6% by 2020, when its total population (465 million) will be equal to that of Eastern Europe. But R5 has already the highest GNP per capita of all Third World regions: $1600 in 1978. This will increase at very fast rates to $3400 (II) or $6000 (I) by 2020. Consequently, its growth will be very different according to the scenarios. R5's share in world GNP is still far lower than that of Latin America, but it will increase from 3.5% in 1978 to 6.5-8% by 2020.

Energy consumption

R5's share in world energy consumption is still very low: 2% in 1978, for a volume of 130 MTOE. Nevertheless, non-commercial sources have already been largely abandoned. The growth rates of the region's commercial energy consumption will be very high in the future, especially under Scenario (I): 7.5% by 2000, and 4.5% by 2020. The volumes involved will be considerable: 380-550 MTOE by 2000, and 700-1300 MTOE by 2020, when R5 is the world's fifth or sixth energy consumer, while it was only number 10 in 1978. Its share in world energy consumption will increase accordingly: from 2% in 1978 to 5-7% in 2020.

At the same time, the growth of R5's energy consumption per capita will also be very fast: from 0.7 TOE in 1978 to 1.5-2.8 TOE in 2020. In fact, the growth rate of the region's per capita consumption will be the highest in the world over the whole forecasting period.

Energy supplies

The share of hydrocarbons in R5's total energy consumption increased from 60% in 1960 to more than 80% in 1978. This is not surprising. In fact, the region has a vast hydrocarbon resource base, while it has no significant hydraulic or coal reserves. The share of new energies in R5's total supply will become significant only after 2000: 4%-5% by 2020. It is the same with nuclear power: 5-7% of total supply to 2020. But the development of new energies and nuclear power will be essentially in preparation of the post-oil era. In fact, hydrocarbons will represent 85-90% of regional demand by 2000, and still 75-80% by 2020. The growth of natural gas supply will be particularly remarkable. Its share in regional supply will increase from 22% in 1978 (30 MTOE) to 35% by 2020 (250-400 MTOE), when R5 becomes the world's third consumer, with 10-14% of the world market. The volume of regional oil demand will increase very considerably in the future (from 80 MTOE in 1978 to 300-600 MTOE in 2020), but oil's penetration will be contained by the expansion of natural gas. Consequently, oil's share in the region's energy balance will decrease from 60% in 1978 to 40-45% in 2020. Nevertheless, R5 will represent 12-17% of world demand of crude oil by 2020, versus 3% in 1978.

Energy production

(i)   R5 is the world's first oil producer (40% of world production in 1978) with a volume of nearly 1200 MTOE, and its fourth natural gas producer (60 MTOE in 1978).

R5's crude production forecasts for 2020 were fixed at 1000 MTOE (II) or 1300 MTOE (I), and this without any major problems. Consequently, it may be assumed that the region will conserve more or less one third of the world market of crude oil.

(ii)   Because of the expansion of R5's production of natural gas from 60 MTOE in 1978 to 300 MTOE (II) or 500 MTOE (I) by 2020, this source will represent 20%

of the regional production in the long run, while oil's share will decrease from 93% in 1978 to 65%.

(iii)   None the less, R5's share in world energy production will continuously diminish in the future: from 18% in 1978 to 10% in 2020, because, outside hydrocarbons, the region has no significant alternatives.

Interregional energy trade

The growth of R5's energy demand will gradually reduce the region's export surplus. This surplus was very considerable in 1978: 1 GTOE, representing two thirds of world energy exports. The region's energy exports will gradually decrease to 630–700 MTOE by 2020, representing 40–50% of world exports. At the same time, the volume of R5's coal imports will assume significant proportions: 35–60 MTOE by 2020, that is, 8% of world coal imports. The region will still represent 70–80% of world oil exports by 2020, as in 1978. However, under 2020 (I), R5's share in world gas exports will be only 10%, versus nearly 40% in 2000. Under these conditions, the positive trade balance of R5 will remain considerable: 800–1000 MTOE by 2000, and 600–650 MTOE by 2020.

Questions

• Are the economic and energy growth forecasts not rather optimistic?
• Will the present depression in the world oil market not accentuate the trend to give priority to domestic needs? And, above all, will it not have a negative impact on the future economic development of a region which will remain very dependent on the fluctuations of the international oil market?
• Will natural gas be able to take over from oil in such a massive and easy way as expected?
• Will the supply tensions on the international coal market, as well as the possible financial problems (nuclear power and new energies), not discourage efforts to prepare the post-oil era?

R6A AFRICA SOUTH OF THE SAHARA (excluding South Africa)

Population and GNP

The average growth rate of Black Africa's population has been, and will be, high: about 3% per year. Consequently, the region's population will increase from 320 million in 1978 to more than *one billion by 2020*. In fact, R6A will be the world's third region in terms of population even before 2000, while it was only number seven in 1978.

However, the region is also one of the poorest in the world. This will not change much in the future, because the average annual growth rate of R6A's GNP per capita will be of the order of 1% up to 2000, and 0.5% after that date. Under these conditions, its GNP per capita would be only $500-600 by 2020, versus a little less than $400 in 1978.

Energy consumption

(i)   *Black Africa's energy consumption pattern is marked by the very high share of traditional and the exceptionally low share of commercial sources.* Non-commercial energy demand was nearly 100 MTOE in 1978, representing 80% of total needs, and 13% of world NCEC,* Black Africa's NCEC will continuously grow in the future, contrary to developments in the rest of the Third World. It could reach 200-240 MTOE by 2020 (22% of world NCEC), representing still one half of total regional needs at that time.

(ii) Black Africa's CEC* was less than 30 MTOE in 1978, representing 0.4% only of world CEC, or the equivalent of Norway's energy demand in that year (with a population of 4 million only). However, R6A's CEC will grow rapidly in the future and could be 170-240 MTOE by 2020, or 1.5% of world CEC at that time.

(iii) Black Africa's energy consumption per capita was very low in 1978: less than 0.4 TOE, of which only 0.1 TOE (85 KOE) represented CEC. It will grow at a very low rate: around 0.4 TOE by 2020 (0.39-0.43 TOE). None the less, in spite of the low growth of total energy consumption, the substitution of non-commercial by commercial sources will be very efficient: CEC per capita will grow to 0.16-0.25 TOE by 2020, while NCEC will decrease from 0.3 TOE to 0.18-0.22 TOE during the same time.

Energy supplies

(i) Black Africa's energy supplies are now dominated by *firewood*: 85 MTOE in 1978, representing 70% of total needs. This will not change significantly in the future, especially under Scenario (II): 160-200 MTOE by 2020, representing 34-48% of total regional needs, but nearly 30% of world firewood supply. Consequently, the region will be by far the world's first firewood consumer. Vegetable waste will maintain its share of 8-10% in total supplies (35-40 MTOE) over the whole forecasting period.

(ii) Under Scenario (I), R6A's oil supply could cover 25% of total needs by 2020 (about 115 MTOE), and 15% under Scenario (II), versus 14% in 1978. Coal, natural gas, nuclear power, and new energies could diversify the region's supply pattern towards the end of the forecasting period, with shares of 4% for coal (15-20 MTOE), 4-6% for gas (15-30 MTOE), and 5-7% for new energies (25-30 MTOE), as well as a nuclear power production representing 5-10 MTOE by 2020. Apart from oil, Black Africa's greatest energy asset is its *hydropower* potential. Hydropower's share in total supplies increased from 1% in 1960 to 5% in 1978 (6 MTOE), and could reach 11-15% by 2020 (45-70 MTOE), that is 200-320 TWh (30 TWh only in 1978). Therefore, the region's share in world CEC could be 1% by 2000, and 1.5% by 2020.

Energy production

Oil is Black Africa's greatest energy asset: 113 MTOE in 1978 (4% of world production), representing more than one half of total regional energy supply, more than firewood (40%). Regional oil could be 150-160 MTOE by 2000, and 180-250 MTOE by 2020, that is 5% and 6%, respectively, of world production. The only other commercial sources with significant shares in world production in 2020 will be hydropower (3%) and new energies (3-4%). The region's share in total world energy production will remain limited to 3-4%.

Interregional energy trade

Black Africa's present contribution to interregional energy trade is limited to its *oil* exports: 100 MTOE in 1978, representing 7% of world exports. Prospects for significant coal and natural gas exports seem hardly favourable.* In addition, growing regional oil demand will reduce quantities available for exports:

---

* NCEC = Non-Commercial Energy Consumption
  CEC = Commercial Energy Consumption

* The recent projects of Nigeria, Cameroon and Congo in the field of LNG exports were postponed *sine die*.

120-130 MTOE by 2020, representing 14-17% of world exports at that time. Regional demand growth will also lead to increased energy imports. The combined effect of these developments will be that Black Africa's net energy exports will level off at their 1978 volume of about 100 MTOE. The maintenance of this export surplus is essential for the region.

Questions

- Can the region be satisfied by these very low per-capita consumption levels in the long run?
- Will it be possible to multiply firewood consumption by 2 or 3 up to 2020?
- Will it be possible to develop hydropower at the expected rate?
- Will it be possible to maintain net crude exports at the expected high level in the long run?

R6B AFRICA SOUTH OF THE SAHARA (including South Africa)

Population and GNP

The addition of South Africa to R6A has a relatively slight impact in terms of population: 350 million in 1978, and 1150 million by 2020, when R6B represents 15% of world population, versus 8% in 1978. R6B's GNP per capita could increase from $480 in 1978 to $620-750 by 2020, but R6B's share in world GNP will hardly be higher than 2.5-3% by 2020 (2% in 1978).

Energy consumption

(i) The growth of R6B's energy consumption will be very high: from less than 180 MTOE in 1978 to 350-400 MTOE by 2000, and 700-900 MTOE by 2020, representing 3% and 4-5%, respectively, of world demand. Non-commercial sources still play a very great part: 100 MTOE in 1978, covering still more than one half of total needs. But the growth rates of commercial energy consumption will be higher in the future: 4-5% per year. Under these conditions, commercial sources would cover 65-80% of total needs by 2020.

(ii) The growth of energy consumption per capita will be very slow: from 0.5 TOE in 1978 to 0.6-0.75 TOE by 2020. But R6B's per capita consumption pattern will undergo deep changes, because CEC will grow from 0.28 TOE in 1978 to 0.60-0.38 TOE by 2020 at the expense of NCEC, which will decrease from 0.22 to 0.17-0.21 TOE.

Energy supplies

(i) Firewood is R6B's dominating source of energy: 50% of total supplies in 1978. But its share will be reduced regularly to 18-30% by 2020, though the volumes involved will increase considerably: 160-200 MTOE (85 MTOE in 1978).

(ii) Coal is already R6B's second supply source (40 MTOE), representing one quarter of total supplies in 1978. Regional coal demand will reach 260-380 MTOE by 2020, repreenting 6-7% of world demand, versus 3% in 1978. The expansion of the other commercial sources, including nuclear power, will also be significant. The combined effect of the expansion of non-oil commercial sources, especially coal, as well as the resistance of the traditional sources, will contribute to contain the growth of oil's penetration: 60 MTOE (II) or 130 MTOE (I) by 2020 (versus less than 30 MTOE in 1978), representing 15% (as in 1978) under Scenario (I), and 10% under Scenario (II).

Energy production

Two thirds of R6B's energy production concerns commercial sources. Coal, oil and hydropower are the region's great energy assets. R6B's energy production will increase from a little over 50 MTOE in 1978 (3% of world production) to 150–200 MTOE by 2000 (5–6% of world production), and 300–500 MTOE by 2020 (7–8% of world production), that is, volumes comparable to those of the present production of China or the USA. A great part of this output will be used for the production of synfuels. R6B's oil production will be 180–250 MTOE by 2020, representing 6% of world production at that time, versus 4% in 1978. In addition, the region's shares in the world production of other sources will be significant by 2020: 5% (45–70 MTOE) for hydropower (versus 2% in 1978), 4–5% for new energies (40 MTOE).

In fact, R6B would represent 6% and 5%, respectively, of world energy supply and demand by 2020.

Interregional energy trade

R6B's net energy exports amounted to more than 80 MTOE in 1978, because of crude and coal exports of 100 MTOE and 10 MTOE, respectively. The region's net coal exports will increase regularly to 50–80 MTOE by 2020, while net oil exports will be maintained at a level of 120–130 MTOE between 2000 and 2020.

R6B's total energy balance will improve even more on the long run, with net exports of 140–150 MTOE in 2000, and 150–165 MTOE in 2020.

Under these conditions, R6B's share in world energy exports will increase from 7% in 1978 to 11–13% in 2020, that is it will remain the *world's second energy exporter* (170–210 MTOE), its third coal exporter (before Eastern Europe and China), as well as its second crude exporter, over the whole forecasting period.

Questions

- Will commercial energy consumptions grow at the expected rates?
- Will it be possible to multiply fire-wood consumption by 2 or 3 up to 2020?
- Will coal production targets be achieved?
- Will the part played by R6B in the international energy market be as important as expected?

R7 SOUTH ASIA

Population and GNP

South Asia is one of the two most populated regions in the world. Its population will double between 1978 to 2020, that is, it will increase from 850 millon (20% of world population) to 1650 million (21%). But this region will represent only 2% of world GNP over the whole forecasting period. Its GNP per capita will also be the lowest in the world, at least up to 2020 ($310–440), versus less than $180 in 1978.

Energy consumption

(i) R7's energy consumption is still very much based on traditional sources, which represented 55% of total needs in 1978 (16% of world NCEC). The region's NCEC pattern is rather well-balanced: 70 MTOE for fire-wood in 1978 (the world's third consumer), and 50 MTOE for vegetable and, in particular, animal waste (second consumer). NCEC will increase from 120 MTOE in 1978 to 210–230 MTOE by 2020, when R7 will represent 21–24% of world NCEC. The region's share of world CEC will increase from 2% in 1978 (hardly 100 MTOE) to 3% by 2020 (340–370 MTOE). However, commercial sources will cover 60–70% of total regional needs by 2020.

South Asia's share in world demand will increase very gradually from 3% in 1978 to 4% by 2000.

(ii) The region's situation and prospects in the field of energy consumption per capita are rather poor. In fact, its per capita consumption is, and will remain, the lowest in the world. It increased from 0.21 TOE in 1960 to 0.25 TOE only in 1978. Future commercial sources will play a greater part: 0.20-0.35 TOE by 2020, while NCEC per capita will remain practically unchanged: less than 0.15 TOE.

Energy supplies

South Asia's abundant but low-grade coal resources are its greatest energy asset. Coal covered 23% of total needs in 1960 as well as 1978 (nearly 50 MTOE). Regional coal supply could increase to 150-250 MTOE by 2020, representing 3-4% of world supply, but 26-31% of regional energy needs. The expansion of hydrocarbons should be moderate: 19-24% of total supplies in 2000 as well as 2020, versus 14% in 1978. This expansion will be mainly due to the development of regional gas supply, whose share in total demand will increase from 2% in 1978 to 6-8% by 2020, while oil demand will be limited to 75-130 MTOE in 2020, versus 30 MTOE in 1978. Hydropower supplies will represent 40-55 MTOE by 2020, that is, volumes comparable to those of new energies (and gas) by that time. This expansion of commercial sources will not be sufficient to contain the growth of animal and vegetable waste demand from 50 to 110 MTOE by 2020, when R7 is the world's number one in this field. The simultaneous growth of firewood supply, from 70 MTOE in 1978 to 110 MTOE by 2020, is a matter of concern. In fact, South Asia will be the world's second firewood consumer.

Energy production

R7's coal production, which is only used to cover regional needs, will increase from 50 MTOE in 1978 to 150-250 MTOE in 2020, representing 3-5% of world coal supply at that time. Regional gas production will also increase considerably: from less than 10 MTOE in 1978 to 55-80 MTOE by 2020, when it represents 10% of regional energy supply and 2% of the world market. It is also hoped that regional oil production will reach 50-85 MTOE by 2020.

Under these conditions, R7 should continue to balance supply and demand in the long run, especially as non-commercial sources will still represent 28-40% of regional production in 2020.

Interregional energy trade

South Asia is, and will be, a marginal actor only in the world energy market. Its imports, limited to crude oil, represented 1% of world energy imports in 1978. This share will increase to 3% by 2020, when its share in world exports is only 1%. The region's gas exports (10 MTOE in 2000, and 20 MTOE in 2020) will not be sufficient to compensate its crude imports, the share of which in regional oil needs will decrease from 75% in 1978 to 50% in 2020. Consequently, South Asia will have a small energy trade deficit: 30-50 MTOE in 2020, versus 20 MTOE in 1978. Nevertheless, the region will continue to be practically self-sufficient, because its total interregional energy trade will represent only 5% of its demand, versus nearly 10% in 1978.

Questions

- Will the very low levels of energy consumption be acceptable in the long run?
- Will it be possible to double non-commercial energy production by 2020?
- Will it be possible to increase coal and hydrocarbon production to such a large extent?
- Will the energy trade deficit remain marginal?

R8 SOUTH-EAST ASIA

Population and GNP

R8 is a Third-World region which is apparently on the whole very privileged. Its share in world population of 8% will not change significantly between 1978 and 2020, when it has a population of 630 million. Its share in world GNP will increase regularly from 2.5% in 1978 to 4.5–5.5% by 2020, that is, its GNP growth rate will be the second highest in the world, after that of R5 (North Africa and Middle East). The region's GNP per capita will also grow considerably: from $650 in 1978 to 1700–3000 by 2020.

Energy consumption

R8's energy consumption could increase from 220 MTOE in 1978 (3% of world demand) to 650–1000 MTOE by 2020 (5–6% of world demand), with rather high growth rates over the whole forecasting period, though they will be more moderate after 2000. Energy consumption per capita increased from 0.4 TOE in 1960 to 0.65 TOE in 1978 and may reach 1–1.6 TOE by 2020.

Energy supplies

(i)   Non-commercial sources play a great part in R8's energy supplies: 100 MTOE in 1978 (nearly 50% of total needs), representing 14% of world NCEC. In addition, R8's NCEC per capita was the highest in the world in 1978: 0.3 TOE. R8's NCEC will not grow much by 2020: 110–140 MTOE (13% of world NCEC), 0.2 TOE per capita, and 10–20% of total regional supplies. Firewood consumption will almost level off at a ceiling of 60–70 MTOE. The growth of vegetable waste will be very slow: from 40 MTOE in 1978 to 60–70 MTOE by 2020.

(ii) The growth of R8's CEC will be much higher: from 2% of world CEC in 1978 (as well as 1960) to 4% by 2000, and 5% by 2020. The share of oil in R8's total consumption is the highest in the Third World. In fact, it increased from less than 20% in 1960 to more than 40% in 1978 (nearly 100 MTOE). The future growth of R8's oil demand varies very much from one scenario to the other: 250 MTOE by 2000, and 310 MTOE by 2020 (I), but 130 MTOE by 2000, and 80 MTOE by 2020 (II). Consequently, under Scenario (I), R8's share in world oil demand will increase considerably: from 4% in 1978 to 8% by 2020.

(iii) It is expected that the expansion of coal (and natural gas) will reduce R8's dependence on oil, because regional coal demand will increase from 14 MTOE in 1978 (7% of total needs) to 170–260 MTOE by 2020 (one quarter of total needs). Regional gas demand will grow from less than 10 MTOE in 1978 to 100–150 MTOE by 2020, representing then 15% of total needs. Hydropower supply will increase considerably: from 5 MTOE in 1978 (1% of world supply) to 60–80 MTOE by 2020 (6% of world supply). Nuclear power will also represent 20–35 MTOE by 2020 (2–4% of world supply). Finally, new energies will supply 40–50 MTOE by 2020. Under these conditions, R8 would arrive at a very satisfactory diversification of its energy supplies.

Energy production

South-East Asia's prospects in the field of energy production give rise to concern, because its oil production, which is now the region's best energy asset with an output of 100 MTOE in 1978, will reach its ceiling by 2000 (85–150 MTOE), and decrease to 50–120 MTOE by 2020, particularly slowed down under Scenario (II). This means that regional oil supply will cover only 8–16% of total needs by 2020 (43% in 1978).

On the other hand, coal production will be ten times higher in 2020 (130–150 MTOE) than in 1978, when coal represents 20% of regional energy supply, versus 5%

in 1978. However, it is the growth of natural gas supply that will be decisive: from less than 20 MTOE in 1978 to 65–100 MTOE by 2000, and 130–200 MTOE by 2020, that is, it may double between 2000 and 2020, when it would represent 20–25% of regional supply and 5–6% of the world market. Nevertheless, the region's share in world energy production will level off as from 2000 at 4% (3% in 1978).

Interregional energy trade

(i)   The growing imbalance between regional supply and demand will, of course, have an increasingly negative impact on the region's energy trade balance, especially under Scenario (I), where its deficit will become chronic and grow continuously: about −120 MTOE by 2000, and −300 MTOE by 2020, while it was near to zero in 1978. Under Scenario (II), this deficit will be limited to −50 MTOE by 2000 and −35 MTOE by 2020.

(ii) South-East Asia was already a significant actor in interregional energy trade in 1978: 5% of world exports (mostly crude oil), and 6% of world imports (with 85 MTOE for crude oil alone). The region's future energy trade will be influenced by the combined effect of two new problems. On the one hand, coal imports will rise sharply under Scenario (I): 40 MTOE by 2000, and 130 MTOE by 2020, representing 18% of world coal imports, making South-East Asia the world's second coal importer. Under Scenario (II), coal imports will be limited to 20 MTOE. On the other hand, oil exports will practically disappear (70 MTOE in 1978), while oil imports will grow to 130 MTOE by 2000, and 225 MTOE by 2020, under Scenario (I), representing 22% of world oil imports, making South-East Asia the world's second oil importer. Under Scenario (II), the growth of oil imports will be limited to 70 MTOE by 2000, and 45 MTOE by 2020, representing 6–7% of world oil imports.

Only natural gas will be available for exports: 30–50 MTOE by 2020 (14% of world gas exports), making South-East Asia the world's third gas exporter.

Consequently, South-East Asia will not be able to cover its energy consumption by regional production.

Questions

• Will the encouraging economic development prospects really result in such high consumption levels?
• Will alternative commercial sources be able to replace oil in such a well-balanced and massive way?
• Is there no risk of disappointment as far as the simultaneous development of coal and gas production is concerned? And will it be possible to find sufficient quantities for coal imports in the world market?
• Will the trade deficit be acceptable at the high levels expected under Scenario (I)?

R9 ASIAN COUNTRIES WITH CENTRALLY PLANNED ECONOMIES

Population and GNP

(i)   R9's population is more than one billion, that is, one out of every four human beings already lives in this region. However, it may be hoped that lower growth rates will limit the region's population to 1.6 billion by 2020 (less than South Asia's).

(ii) R9 is a poor region. GNP per capita was only $240 in 1978. It will remain at this level up to 2000 under Scenario II, while it will slightly increase under Scenario (I). After 2000, it will grow significantly: $335 by 2020 under Scenario (II), or double, that is $770, under Scenario (I). Regional GNP will increase from G$250 in 1978 to G$530–1200 by 2020, representing still about only 3% of world GNP.

Energy consumption

R9's total energy consumption was 640 MTOE in 1978, representing 9% of world demand (the world's number four), versus less than 300 MTOE in 1960. Because of its economic problems up to 2000, as well as lower income-elasticities than in the other Third-World regions, regional demand will grow at a slower rate in the future: 1–1.4 GTOE by 2020, representing 7–8% of world consumption. R9's energy consumption per capita will increase from 0.6 TOE in 1978 to 0.9 TOE by 2020 under Scenario (I). It will remain practically at its 1978 level under Scenario (II), whose more pessimistic results are very accentuated for R9.

Energy supplies

(i) Coal plays an exceptionally large part in R9's supplies: it covers almost one half of its total needs. In fact, R9 is the world's third coal consumer, with a share of 18% of world demand (300 MTOE). Coal will maintain its share in regional supply (50%) also in the long run, with volumes representing 500 MTOE by 2000, and 700 MTOE by 2020, under Scenario (I). This 2020 volume corresponds to the present supply level of the western countries and would represent 12% of the world coal market. Under Scenario (II) regional supply would be 500 MTOE by 2020.

(ii) The non-commercial sector is another great supply source of R9. In fact, R9 is number one in this field for firewood (26% of world supply), as well as animal and vegetable waste (38%). R9's NCEC represented 30% of world NCEC in 1978, and one third of regional supplies.

   The non-commercial sector's 1978 volume will be maintained up to 2000, but it will fall rapidly to 80–160 MTOE by 2020 (6–16% of total supplies), because of environmental and production constraints.

(iii) Oil's penetration increased at a fast rate between 1960 and 1978: from 10 MTOE to 90 MTOE (14% versus 3% of total needs). Oil demand could represent 200 to 300 MTOE by 2020 (20% of total needs) after a particularly strong expansion up to 2000: 130–210 MTOE.

R9 will represent 8% of the world oil market by 2020, versus 3% only in 1978. Natural gas could cover up to 10% of total needs by 2020, for a volume of 90–160 MTOE (5% of the world market). Hydropower production could also increase to 60–150 MTOE by 2020 (5–10% of world production), versus 14 MTOE in 1978. However, nuclear power and new energies will represent only 1% (II) or 2% (I) of the region's energy balance (10–30 MTOE each).

Energy production

The overriding strategic principle of R9's policy is to preserve its energy independence on the supply side. In fact, in R9, supply largely conditions demand. Coal production will be increased to 500–800 MTOE by 2020, in part for exports. Oil production will also be stimulated and may represent 200–400 MTOE by 2020. Gas production will be increased especially after 2000. This growth of fuel production will result in net energy exports which will strengthen in the course of time, especially under Scenario (I).

Interregional energy trade

R9's top priority is self-sufficiency, but energy exports will be developed, because they are indispensable for the balance of payments. This policy will make R9, a significant energy exporter: 1% of world energy exports in 1978, 8% by 2000, and 15% by 2020, under Scenario (I). In fact, R9 will be the world's second energy exporter by 2020, while it was only the eighth in 1978. R9's exports will benefit to the maximum from the opening of world markets which characterizes Scenario (I): 120 MTOE by 2000, and 280 MTOE by 2020. At the same time, it will practically have no energy imports. Under Scenario (II), R9's energy exports will be limited to 30–35 MTOE by 2020 (2% of world exports).

The high growth of R9's energy exports under Scenario (I) will be very well balanced between the different fossil fuels by 2020: 15% of world coal exports (the world's number three), 11% for oil (number three), and 28% for gas (number two).

Questions

- Will R9's economic development not produce some positive surprises?
- Will R9 be able to multiply its coal production by 2.5 by 2020? Will it be able to do even better?
- Should and could R9's non-commercial energy consumption be further extended?
- Under Scenario (I), will R9 really be able to produce such a great export surplus for each of the three fossil fuels?

R10 LATIN AMERICA

Population and GNP

Latin America's population will increase from 350 million in 1978 to 800 million by 2020 (8% and 10%, respectively, of world population). The region's economic growth prospects are promising. Its GNP per capita could increase from $1400 in 1978 to $3000–5000 by 2020. Latin America has also the highest GNP in the Third World: G$500 in 1978. Its GNP will be G$2400–4000 by 2020, representing 10–12% of world GNP, versus less than 6% in 1978.

Energy consumption

(i)   Latin America's total demand growth between 1960 and 1978 (from 90 MTOE to 290 MTOE) was based exclusively on the growth of its commercial energy consumption. The growth of the region's CEC will also be responsible for the future growth of total energy consumption to 0.8–1 GTOE by 2000, and 1.5–2.2 GTOE by 2020, representing 11–12% of world demand (350 MTOE in 1978), the world's number four (versus 5% only in 1978, as well as 1960).

(ii) Latin America was the only Third World region whose per-capita energy consumption represented an average of 1 TOE in 1970. This consumption could increase to 1.4 TOE (II) or 1.8 TOE (I) by 2000, and even 2 TOE (II) or 2.7 TOE (I) by 2020, that is, volumes not far from Western Europe's 1960 and 1978 per-capita consumption levels, according to the scenario.

Energy supplies

(i)   Oil has been Latin America's first supply source for a long time, covering almost one half of total needs, that is, more than in any other Third World region. The region's oil demand was 170 MTOE in 1978, and 50 MTOE for gas and hydropower (14% of total needs), but the growth of these latter supply sources was particularly high between 1960 and 1978. Latin America is already the world's third producer of hydropower (13% of world production). Owing to the high growth of Brazil's fuel alcohol supply, the region is already a significant producer of new energies (2% of total needs). It represents also 1% of the world's nuclear power production.

(ii) Latin America's energy supply pattern will not change much up to 2000, except for the continuous reduction of the share of non-commercial sources: from 42% in 1960 to about 20% in 1978, and 7-10% by 2000, as well as the growing importance of new energies (8% of total needs by 2000). However, demand volumes will increase considerably. Oil demand will be 360-470 MTOE by 2000, with 13-14% of the world market (number four). Under Scenario (I), R10 will be the world's second hydropower producer (155 MTOE), with 22% of world production, versus 21% under Scenario (II). Latin America will also represent one quarter of the world market of new energies by 2000 (60-85 MTOE).

(iii) By 2020, Latin America will be the world's first oil consumer (500-700 MTOE with 35% of regional supplies, but 20% of the world market). Hydropower's growth will be impressive: 300-430 MTOE, representing 30% of the world market (number one), and 20% of regional supplies. Its share in the world gas market will be 8-9% (200-280 MTOE). Supply of new energies will represent 160-230 MTOE (20-23% of world supply). In fact, R10 will be the world's number two in this field. Finally, Latin America's nuclear power supply will increase to 60-145 MTOE (5% of world production).

Coal will play a large part in Latin America's energy supply as from 2000. It will cover 11-14% of regional needs by 2020, with a demand of 170-300 MTOE (5% of the world coal market).

In brief, the growth of Latin America's energy supply will be high in all sectors, but it seems that this will not result in higher energy dependence.

Energy production

Latin America's energy balance will remain positive, without any major problems, over the whole forecasting period. In addition, its supply and demand patterns will adjust almost perfectly. As the regions with centrally planned economies, R10 is, and will be, practically self-sufficient. Its share in world energy production will increase from 6% in 1978 to 11-12% by 2020. It will be the world's third energy producer after the two giants by 2020, while it was only number six in 1978.

Interregional energy trade

Consequently, Latin America does not play a large part in interregional energy trade. Its export surplus will increase from 4 MTOE in 1978 to 30 MTOE by 2020, but its oil exports will decrease gradually from about 70 MTOE in 1978 to 20 MTOE by 2020. It plays a marginal part in the world coal market (10-20 MTOE). However, Latin America expects to develop its gas exports: 30-40 MTOE by 2020 (11-14% of world gas exports), that is, it would be the world's fourth gas exporter. The region will reduce its import dependence continuously: from nearly 20% in 1978 to 7% by 2000, and 3-4% by 2020.

Questions

- Will Latin America sustain such a fast economic and energy development?
- Will Latin America be able to develop simultaneously oil, hydropower, coal, and new energies at the expected high rates?
- Will Latin America so easily preserve its overall energy independence?

VI. GENERAL OBSERVATIONS

6.1 ENERGY SITUATION AND LONG-TERM OUTLOOK

The depressed general climate characterizing the world economy and the energy sector since the second oil shock is not favourable to long-term outlook, especially as the instability of oil prices may invalidate the results of any forecasting effort. Nevertheless, two fundamental aspects emerge clearly:

(i)   It is not easy to deny the correlation between energy growth and economic growth. This is shown by the negative impact of the present economic situation on energy demand, with its grave consequences for the Third World.

(ii) It is to be feared that the temporary lower pressure on demand and, consequently, on supply may also encourage a short-sighted and less constraining implementation of energy policies.

(iii) However, it is a fact that present long-term energy consumption forecasts are much lower than those presented at the Munich Conference three years ago: by 1–2 GTOE in 2000, and by 2–4 GTOE by 2020. Scenario (I) is based on the prior Scenario (C), which was considered as pessimistic in 1980. Scenario (II) is much more pessimistic than Scenario (I). The variation of population forecasts explains about 10% of the reduction for 2000, and about one third of the reduction for 2020. In addition, worse estimates for economic prospects explain about 15% of the reduction for 2000 as well as 2020. The combined effect of these two factors explains about 25% of the variation between the forecasts for 2000 and almost one half of the variation between the forecasts for 2020.

6.2 LONG-TERM READJUSTMENT OF THE ENERGY SYSTEM

(i)   *Income elasticities* of total energy demand will decrease continuously, especially in the Third World. Nevertheless, there could be a certain increase of the North's income elasticities after 2000, that is, after their pronounced decrease up to that year.

(ii) The development of *energy intensity* (KOE/$ GNP) will be similar. The ratios will decrease regularly in the North and the South, but especially up to 2000. Therefore, it seems that the structural readjustment of the world energy system will essentially take place by 2000, under the predominant pressure of the North.

(iii) A trend towards *diversification* also characterizes the development of the world energy system. In fact, in 1960 five of the ten regions depended on one energy source for more than one half of their supplies, while this was the case for only three regions in 1978. One region only will be in this situation by 2000, and none by 2020. At the same time, the world's ten principal consumption centres* represented 71% of world energy demand in 1960, but only 65% in 1978. In addition, their share in world demand will decrease to 52% by 2000, and 46% by 2020. The development of the world's ten principal energy production centres was, and will be, very similar.

(iv) The regions of the North will diminish their *energy dependence*, especially North America, but the energy dependence of most regions of the South will increase, especially as far as South-East Asia is concerned, even if the South's total energy balance will remain positive.

(v) The development of each region's share in world demand will follow that of each region's share in world GNP, but *production* patterns will be much more *stable*, and also much more independent of the variations of the economic situation.

(vi) The trend of interregional energy trade is to decrease in the long run. In combination with the growing energy indepedence of the regions, as well as the increasing diversification of their supplies, this trend shows that the regions will be more and more self-sufficient. This will, of course, attenuate tensions in interregional energy trade.

6.6 ESSENTIAL REGIONAL QUESTIONS

The following general observations may be made on the basis of a summary of the separate regional analyses (cf. chapter V):

(i) One half of the 23 essential questions is focused on supply, while one quarter concerns demand, and another quarter energy trade.

(ii) Uncertainties about future *coal supply* are a matter of particular concern. In fact, these concerns are expressed in one quarter of the questions. The problems related to *energy trade deficits* come next (three questions), followed by concern about *nuclear power* (in the North), and the persistent dependence on *non-commercial sources* (in the South). The potential insufficiency of *oil exports* is a matter of concern for two regions. Finally, on the demand side, persistent *low demand* (two regions) and the *possibilities of increasing coal demand* (two regions) are primary matters of concern.

6.4 FUNDAMENTAL REGIONAL CHANGES

These are the total result of the developments concerning the great consumption and production centres.

(i) On the demand side, there is, above all, the continuous decline of North America's influence, and the appearance of new consumption centres: R10 (Latin America) and R5 (North Africa and Middle East).

(ii) On the supply side, there is the remarkable decrease of R5's importance, but also the increasing importance of Latin America.

(iii) In the field of interregional energy trade, R5's export capacities will decrease very significantly, while North America seems to be able to rebalance its energy trade as from 2000. In addition, South-East Asia's energy trade will become unbalanced.

6.5 TOWARDS A TRIPOLAR WORLD

The ten regions may be grouped in three super-regions: SR1 comprises the industrialized countries (R1 + R2 + R3 + R4 + South Africa). SR2 represents the more

*A consumption centre being defined by the consumption of a given source in one region (for example, gas in R2).

advanced Third World (R5 + R8 + R10), and SR3 the less advanced Third World facing persistent problems (R6 + R7 + R9). The grouping of forecasts by these three super-regions shows that the relative decline of SR1 (in terms of its shares in world GNP and world energy consumption) benefits almost exclusively SR2, at the expense of SR3, though the population growth of these two super-regions will be about equal.

In terms of growth in volume (GNP, energy consumption, and energy production), the strong position of SR1, that is the industrialized countries, is confirmed. In fact, SR1 represents 65–70% of the growth of world GNP between 1960 and 2020, 56–62% for consumption, and 54–61% for production (since 1978), while SR1's population growth represents only 11% of world population growth, between 1960 and 2020. SR2's share in each of these factors is between 24% and 29% (including for population growth). However, SR3's shares are 60% for population growth, 7% for GNP, and 14–18% for energy supply and demand.

The future of the two South's zones appears to be very contrasted.

6.6 ENERGY PROSPECTS

(i) Energy prospects are characterized by a trend towards more precarious situations. In fact, significant energy scenarios are more and more "grey" scenarios. In appearance, slower growth prospects facilitate the attenuation of supply and demand tensions. However, they also increase relative variations and make it more difficult to reduce existing development handicaps, especially for the poorest. In addition, though present energy consumption forecasts are lower than prior estimates, they still represent considerable volumes. A total energy demand of 400 to 500 GTOE has to be covered up to 2020, and that in an economic context which is, on the whole, more uncertain.

The trend of these slower growth prospects is to crystallize existing positions which are based on already acquired advantages. On the whole, the industrialized countries are well off. It seems that only Latin America will join them sooner or later. All the other Third-World regions seems to be condemned, to a varying extent, to remain more or less disadvantaged. Even the Middle East will have to face increasing problems, though its initial position is comfortable.

(ii) As far as energy sources are concerned, the unanimous faith in coal may actually have a negative impact, especially in view of the growing uncertainties about the expansion of nuclear power. Coal is universally sought-after, but can it, in fact, fulfil the pressing demands made on it: a multiplication by 2.5–3 by 2020? Nevertheless, the expansion of nuclear power and coal, combined with the other alternative solutions, will be indispensable if the growth of oil needs is to be really contained. Under Scenario (II), the mobilization of impressive volumes of all commercial and non-commercial sources will hardly make it possible to start off the decrease oil demand in quantitative terms. Under Scenario (I), it will not even reach its ceiling during the forecasting period. Under these conditions, lower oil prices can only be a temporary matter. Even with moderate rates of growth for demand, the proved reserves of crude should be exhausted before 2020 and the additional reserves some years later. The situation is no more relaxed as far as natural gas and uranium reserves are concerned (with the exception, for the latter, of a dramatic breakthrough of the breeder option).

Consequently, the present climate of ease must not delay the implementation of the policies in the fields of the rational use of resources and oil substitution. Otherwise, it will be even more difficult to manage successfully the world's energy future.

VII. LIST OF COUNTRIES

R1 NORTH AMERICA

CANADA
PUERTO RICO
U.S.A.

R2 WESTERN EUROPE

AUSTRIA
BELGIUM
CYPRUS
DENMARK
FINLAND
FRANCE
GERMANY (Fed. Rep.)
GREECE
ICELAND
IRELAND
ISLANDS: Channel Islands
Isle of Man
Greenland
(Gibraltar)
ISRAEL
ITALY
LUXEMBOURG
MALTA
NETHERLANDS
NORWAY
PORTUGAL
SPAIN
SWEDEN
SWITZERLAND
TURKEY
UNITED KINGDOM
YUGOSLAVIA

R3 INDUSTRIALIZED COUNTRIES OF THE PACIFIC

AUSTRALIA
JAPAN
NEW ZEALAND

R4 EASTERN EUROPE

ALBANIA
BULGARIA
CZECHOSLOVAKIA
GERMANY (Dem. Rep.)
HUNGARY
POLAND
ROMANIA
USSR

R5 NORTH AFRICA/MIDDLE EAST

ALGERIA
BAHRAIN
EGYPT
IRAQ
IRAN
JORDAN
KUWAIT
LEBANON
LIBYA
MOROCCO
OMAN
QATAR
SAUDI ARABIA
SUDAN
SYRIAN ARAB REP.

TUNISIA
UNITED ARAB EMIRATES
YEMEN ARAB REP.
YEMEN DEM. REP.

R6 SUB-SAHARAN AFRICA

<u>SOUTH AFRICA</u>
ANGOLA
BENIN
BOTSWANA
BURUNDI
CAMEROON
CAPE VERDE
CENTRAL AFRICAN REP.
CHAD
COMOROS
CONGO
DJIBOUTI
EQUATORIAL GUINEA
ETHIOPIA
GABON
GAMBIA
GHANA
GUINEA
GUINEA-BISSAU
IVORY COAST
KENYA
LESOTHO
LIBERIA
MADAGASCAR
MALAWI
MALI
MAURITANIA
MAURITIUS
MOZAMBIQUE
NAMIBIA
NIGER
NIGERIA
REUNION
RWANDA
SAO TOME & PRINCIPE
SENEGAL
SEYCHELLES
SIERRA LEONE
SOMALIA
SWAZILAND
TANZANIA
TOGO
UGANDA
UPPER VOLTA
ZAIRE
ZAMBIA
ZIMBABWE

R7 SOUTH ASIA

AFGHANISTAN
BANGLADESH
BHUTAN
INDIA
MALDIVES
NEPAL
PAKISTAN
SRI LANKA

R8 SOUTH-EAST ASIA

BRUNEI
BURMA
FIJI
HONG KONG
INDONESIA
ISLANDS: Solomon
      Guam
      French Polynesia
      Tonga
      Western Samoa
      New Hebrides
      Vanatu
      New Caledonia
      Gilbert-Kiribati
      Pacific US Trust
      American Samoa
MACAO
MALAYSIA
PAPUA NEW GUINEA
PHILIPPINES
SINGAPORE
SOUTH KOREA
TAIWAN
THAILAND

R9 CENTRALLY PLANNED ASIA

CHINA
KAMPUCHEA
LAO
MONGOLIA
NORTH KOREA
VIETNAM

R10 LATIN AMERICA

ARGENTINA
BAHAMAS
BELIZE
BERMUDA
BOLIVIA
BRAZIL

CHILE	Dominica
COLOMBIA	Barbados
COSTA RICA	Antigua
CUBA	Santa Lucia
DOMINICAN REP.	St. Kitts
ECUADOR	Grenada
EL SALVADOR	
FRENCH GUYANA	JAMAICA
GUATEMALA	MEXICO
GUYANA	NICARAGUA
HAITI	PANAMA
HONDURAS	PARAGUAY
ISLANDS: Guadeloupe	PERU
St. Vincent	SURINAME
Martinique	TRINIDAD/TOBAGO
Virgin Islands	URUGUAY
Netherlands Antilles	VENEZUELA

DEMOGRAPHY FORECASTS

	LEVELS (Minhb)				RATES OF GROWTH (% /year)		
	1960	1978	2000	2020	1960-1978	1978-2000	2000-2020
R1	200,9	245,0	296	349	1,11	0,9	0,8
R2	357,1	417,2	463	489	0,87	0,5	0,3
R3	107,5	132,3	149	152	1,16	0,6	0,1
R4	313,9	372,4	420	462	0,95	0,6	0,5
R5	111,4	183,8	320	465	2,82	2,5	1,9
R6	219,5	347,8	681	1 152	2,59	3,1	2,7
R7	560,4	848,6	1 284	1 644	2,33	1,9	1,2
R8	220,6	337,6	503	631	3,03	1,8	1,1
R9	714,5	1 032,0	1 359	1 576	2,06	1,3	0,7
R10	209,6	344,5	564	802	2,80	2,2	1,8
WIC	681,8	822,2	959	1 071	1,05	0,7	0,6
IC	995,7	1 194,6	1 379	1 533	1,02	0,65	0,5
TW	2 019,7	3 066,6	4 660	6 189	2,35	1,9	1,4
TW-R9	1 305,2	2 034,6	3 301	4 613	2,50	2,2	1,7
WORLD	3 015,4	4 261,6	6 039	7 722	1,94	1,6	1,2

TOTAL ECONOMIC GROWTH

- LEVELS AND RATES OF GROWTH -

	LEVELS (G $ 78)						RATES OF GROWTH (% /year)				
	1960	1978	2000		2020		1960-1978	1978-2000		2000-2020	
			I	II	I	II		I	II	I	II
R1	1 162	2 109	4 048	3 633	6 073	4 941	4,0	3,0	2,5	2,0	1,5
R2	1 294	2 596	5 031	4 059	7 528	5 505	4,0	3,0	2,0	2,0	1,5
R3	224	1 017	2 359	1 949	3 865	2 625	8,8	3,9	3	2,5	1,5
R4(*)	471	1 366	(3 237)	(2 617)	(5 846)	(3 888)	6,1	(4,0)	(3)	(3,0)	(2,0)
R5	61	292	1 048	779	2 824	1 577	9,1	6,0	4,6	5,1	3,6
R6	78	166	437	394	870	712	4,3	4,5	4,0	3,5	3,0
R7	77	150	365	294	722	506	3,8	4,1	3,1	3,5	2,5
R8	63	220	824	580	1 848	1 048	7,2	6,2	4,5	4,2	3,0
R9	97	247	558	332	1 217	528	5,3	3,8	1,35	4,0	2,35
R10	180	483	1 685	1 222	4 082	2 435	5,6	5,8	4,3	4,5	3,5
WIC	2 698	5 766	11 553	9 745	17 695	13 259	4,3	3,2	2,4	2,15	1,55
IC	3 169	7 132	14 790	12 362	23 541	17 147	4,6	3,4	2,5	2,35	1,65
TW	538	1 314	4 802	3 497	11 334	6 618	5,1	6,1	4,5	4,4	3,2
TW-R9	441	1 067	4 244	3 165	10 117	6 090	5,0	6,5	5,1	4,4	3,3
WORLD	3 707	8 446	19 592	15 859	34 875	23 765	4,7	3,9	2,9	2,9	2,0

(*) R4 : economic assumptions ex-post related to the energy forecasts by the central team

ECONOMIC GROWTH PER CAPITA

- LEVELS AND RATES OF GROWTH -

	LEVELS ($ 78/p.c.)						RATES OF GROWTH (% /year)				
			2000		2020			1978-2000		2000-2020	
	1960	1978	I	II	I	II	1960-1978	I	II	I	II
R1	5 786	8 608	13 661	12 274	17 391	14 158	2,8	2,1	1,6	1,2	0,7
R2	3 623	6 223	10 866	8 767	15 395	11 258	3,1	2,6	1,6	1,8	1,2
R3	2 086	7 680	15 780	13 035	25 480	17 300	7,5	3,3	2,4	2,4	1,4
R4 (*)	1 500	3 668	(7 707)	(6 230)	(12 654)	(8 415)	5,1	(3,4)	(2,4)	(2,5)	(1,5)
R5	544	1 588	3 275	2 434	6 073	3 391	6,1	3,3	2,0	3,1	1,7
R6	356	478	642	578	755	618	1,7	1,3	0,9	0,8	0,3
R7	137	177	284	229	439	308	1,4	2,2	1,2	2,2	1,2
R8	287	652	1 639	1 153	2 960	1 661	4,7	4,3	2,6	3,0	1,8
R9	136	239	411	244	772	335	3,2	2,5	0,1	3,2	1,6
R10	857	1 402	2 988	2 167	5 091	3 036	2,8	3,5	2,0	2,7	1,7
WIC	3 957	7 013	12 047	10 162	16 522	12 380	3,2	2,5	1,7	1,6	1,0
IC	3 183	5 970	10 725	8 964	15 356	11 185	3,6	2,7	1,9	1,8	1,1
TW	266	428	1 030	750	1 831	1 069	2,7	4,1	2,6	2,9	1,8
TW-R9	338	524	1 286	959	2 193	1 320	2,5	4,2	2,8	2,7	1,6
WORLD	1 229	1 982	3 244	2 626	4 516	3 078	2,7	2,3	1,3	1,7	0,8

(*) R4 :   economic assumptions ex-post related to the energy forecasts by the central team

ENERGY CONSUMPTIONS (Mtoe)

- 1960 -

(Mtoe)	SMF	O	NG	HY	NU	NS	FW	VAW	CEC	NCEC	PEC
R1	249	491	335	57	-	-	35	5	1132	40	1172
R2	359	183	10	61	-	-	31	7	613	38	651
R3	60	38	1	15	-	-	2	1	114	3	117
R4	341	125	53	13	-	-	37	9	532	46	578
R5	1	18	2	1	-	-	4	7	22	11	33
R6	26	9	-	1	-	-	56	8	36	64	100
R7	27	9	1	2	-	-	50	30	39	80	119
R8	6	15	3	2	-	-	43	20	26	63	89
R9	128	9	1	4	-	-	87	61	142	148	290
R10	6	66	12	8	-	-	60	7	92	67	159
WIC	690	715	346	133	-	-	70	13	1884	83	1967
IC	1031	840	399	146	-	-	107	22	2416	129	2545
TW	172	123	19	18	-	-	298	133	332	431	763
TW-R9	44	114	18	14	-	-	211	72	190	283	473
WORLD	1203	963	418	164	-	-	405	155	2748	560	3308

ENERGY CONSUMPTIONS (Mtoe)

Table : 5

- 1978 -

(Mtoe)	SMF	O	NG	HY	NU	NS	FW	VAW	CEC	NCEC	PEC
R1	358	914	528	129	77	2	30	5	2008	35	2043
R2	260	627	173	97	40	-	16	14	1197	30	1227
R3	83	247	24	27	14	-	1	1	395	2	397
R4	575	404	334	42	12	-	30	16	1367	46	1413
R5	2	79	29	6	-	-	8	10	116	18	134
R6	43	27	1	7	-	-	86	13	78	99	177
R7	48	29	5	13	1	-	71	47	96	118	214
R8	14	92	7	5	-	-	63	38	118	101	219
R9	306	89	10	14	-	-	126	93	419	219	638
R10	12	168	50	51	2	6	55	12	289	67	356
WIC	740	1799	725	254	131	2	49	21	3651	70	3721
IC	1315	2203	1059	296	143	2	79	37	5018	116	5134
TW	386	473	102	95	3	6	407	212	1065	619	1684
TW-R9	80	384	92	81	3	6	281	119	646	400	1046
WORLD	1701	2676	1161	391	146	8	486	249	6083	735	6818

Table : 6

ENERGY CONSUMPTION FORECASTS (Mtoe)

- 2000 (I) -

(Mtoe)	SMF	O	NG	HY	NU	NS	FW	VAW	CEC	NCEC	PEC
R1	898	552	526	179	288	111	84	10	2554	94	2648
R2	477	629	281	134	243	55	23	29	1819	52	1871
R3	200	225	96	35	109	15	1	2	680	3	683
R4	715	585	770	70	240	25	25	10	2405	35	2440
R5	10	335	155	17	9	7	9	13	533	22	555
R6	130	64	10	21	6	13	135	24	244	159	403
R7	129	85	30	27	8	9	95	78	288	173	461
R8	110	247	50	20	35	10	65	47	472	112	584
R9	510	209	77	50	7	6	122	82	859	204	1063
R10	62	472	160	155	32	85	49	21	966	70	1036
WIC	1695	1416	903	349	645	186	110	42	5194	152	5346
IC	2410	2001	1673	419	885	211	135	52	7599	187	7786
TW	831	1402	482	289	92	125	473	264	3221	737	3958
TW-R9	321	1193	405	239	85	119	351	182	2362	533	2895
WORLD	3241	3403	2155	708	977	336	608	316	10820	924	11744

ENERGY CONSUMPTION FORECASTS (Mtoe)

- 2000 (II) -

(Mtoe)	SMF	O	NG	HY	NU	NS	FW	VAW	CEC	NCEC	PEC
R1	810	514	487	174	273	92	81	8	2350	89	2439
R2	421	588	230	130	211	45	24	30	1625	54	1679
R3	190	208	90	30	82	11	1	3	611	4	615
R4	635	545	670	65	180	20	25	15	2115	40	2155
R5	7	218	104	13	4	6	10	15	352	25	377
R6	104	37	5	16	6	13	152	27	181	179	360
R7	102	60	17	20	6	6	98	80	211	178	389
R8	100	132	30	17	20	12	70	55	311	125	436
R9	375	128	28	29	3	3	190	82	566	272	838
R10	44	358	120	128	19	62	60	25	731	85	816
WIC	1517	1317	807	335	571	151	108	43	4698	151	4849
IC	2152	1862	1477	400	751	171	133	58	6813	191	7004
TW	636	926	304	222	53	99	578	282	2240	860	3100
TW-R9	261	798	276	193	50	96	388	200	1674	588	2262
WORLD	2788	2788	1781	622	804	270	711	340	9053	1051	10104

ENERGY CONSUMPTION FORECASTS (Mtoe)

- 2020 (I) -

(Mtoe)	SMF	O	NG	HY	NU	NS	FW	VAW	CEC	NCEC	PEC
R1	1774	234	460	199	484	318	91	15	3469	106	3575
R2	730	446	303	171	615	115	24	42	2380	66	2446
R3	290	185	165	40	200	36	1	3	916	4	920
R4	930	530	1100	130	600	80	20	10	3370	30	3400
R5	66	595	449	26	95	55	9	14	1286	23	1309
R6	383	126	30	72	25	39	158	37	675	195	870
R7	246	128	60	54	24	57	107	107	569	214	783
R8	260	310	150	80	80	40	55	57	920	112	1032
R9	700	290	160	150	30	30	50	30	1360	80	1440
R10	300	732	280	437	145	228	30	30	2122	60	2182
WIC	3157	875	928	412	1314	484	117	61	7170	178	7348
IC	4087	1405	2028	542	1914	564	137	71	10540	208	10748
TW	1592	2171	1129	817	384	434	408	274	6527	682	7209
TW-R9	892	1881	969	667	354	404	358	244	5167	602	5769
WORLD	5679	3576	3157	1359	2298	998	545	345	17067	890	17957

Table : 9

ENERGY CONSUMPTION FORECASTS (Mtoe)

- 2020 (II) -

(Mtoe)	SMF	O	NG	HY	NU	NS	FW	VAW	CEC	NCEC	PEC
R1	1486	209	404	189	398	291	91	12	2977	103	3080
R2	605	418	232	165	505	95	25	44	2020	69	2089
R3	250	160	140	35	130	29	2	4	744	6	750
R4	780	420	940	100	450	50	25	10	2740	35	2775
R5	41	287	246	20	33	39	11	19	666	30	696
R6	260	63	15	46	15	40	202	41	439	243	682
R7	147	75	34	40	12	34	114	114	342	228	570
R8	170	80	100	60	40	50	65	73	500	138	638
R9	480	196	90	60	10	10	110	50	846	160	1006
R10	170	520	200	288	60	162	40	40	1400	80	1480
WIC	2585	787	776	390	1043	425	120	61	6006	181	6187
IC	3365	1207	1716	490	1493	475	145	71	8746	216	8962
TW	1024	1221	685	513	160	325	540	336	3928	876	4804
TW-R9	544	1025	595	453	150	315	430	286	3082	716	3798
WORLD	4389	2428	2401	1003	1653	800	685	407	12674	1092	13766

Table :10

ENERGY CONSUMPTION PER CAPITA

COMMERCIAL AND NON-COMMERCIAL

(toe/p.c.)	COMMERCIAL CONSUMPTIONS						NON-COMMERCIAL CONSUMPTIONS					
	1960	1978	2000		2020		1960	1978	2000		2020	
			I	II	I	II			I	II	I	II
R1	5,63	8,20	8,63	7,94	9,94	8,53	0,20	0,14	0,32	0,30	0,30	0,30
R2	1,71	2,87	3,93	3,51	4,87	4,13	0,11	0,07	0,11	0,12	0,13	0,14
R3	1,06	2,98	4,56	4,10	6,02	4,89	0,03	0,02	0,02	0,03	0,03	0,04
R4	1,69	3,67	5,73	5,03	7,29	5,93	0,15	0,12	0,08	0,10	0,07	0,08
R5	0,20	0,63	1,66	1,10	2,77	1,43	0,10	0,10	0,07	0,08	0,05	0,07
R6	0,17	0,22	0,36	0,27	0,59	0,38	0,29	0,29	0,23	0,26	0,17	0,21
R7	0,07	0,11	0,22	0,16	0,35	0,21	0,14	0,14	0,14	0,14	0,13	0,14
R8	0,12	0,35	0,94	0,62	1,46	0,79	0,28	0,30	0,22	0,25	0,18	0,22
R9	0,20	0,41	0,63	0,42	0,86	0,54	0,21	0,21	0,15	0,20	0,05	0,10
R10	0,44	0,84	1,71	1,30	2,65	1,75	0,32	0,19	0,13	0,15	0,07	0,10
WIC	2,77	4,44	5,41	4,90	6,69	5,61	0,12	0,09	0,16	0,16	0,17	0,17
IC	2,43	4,20	5,51	4,94	6,88	5,71	0,13	0,10	0,14	0,14	0,13	0,14
TW	0,17	0,35	0,69	0,48	1,05	0,64	0,21	0,20	0,16	0,19	0,11	0,14
TW - R9	0,15	0,31	0,72	0,51	1,12	0,67	0,21	0,20	0,16	0,18	0,13	0,15
WORLD	0,91	1,43	1,79	1,50	2,21	1,64	0,19	0,17	0,15	0,17	0,12	0,14

ENERGY CONSUMPTION PER CAPITA

- TOTAL AND RATES OF GROWTH -

	TOTAL CONSUMPTION (toe/p.c.)						RATES OF GROWTH (% /year)				
			2000		2020			1978-2000		2000-2020	
	1960	1978	I	II	I	II	1960-1978	I	II	I	II
R1	5,83	8,34	8,95	8,24	10,24	8,83	2,0	0,3	-0,1	0,7	0,3
R2	1,82	2,94	4,04	3,63	5,00	4,27	2,7	1,5	1,0	1,1	0,8
R3	1,09	3,00	4,58	4,13	6,05	4,93	5,8	1,9	1,5	1,4	0,9
R4	1,84	3,79	5,81	5,13	7,36	6,01	4,1	2,0	1,4	1,2	0,8
R5	0,30	0,73	1,73	1,18	2,82	1,50	5,1	4,0	2,2	2,5	1,2
R6	0,46	0,51	0,59	0,53	0,76	0,59	0,6	0,7	0,2	1,3	0,5
R7	0,21	0,25	0,36	0,30	0,48	0,35	1,0	1,7	0,8	1,4	0,8
R8	0,40	0,65	1,16	0,87	1,64	1,01	2,7	2,7	1,3	1,7	0,7
R9	0,41	0,62	0,78	0,62	0,91	0,64	2,3	1,0	0	0,8	0,2
R10	0,76	1,03	1,84	1,45	2,72	1,85	1,7	2,7	1,6	2,0	1,2
WIC	2,89	4,53	5,57	5,06	6,86	5,78	2,5	0,9	0,5	1,0	0,7
IC	2,56	4,30	5,65	5,08	7,01	5,85	2,9	1,2	0,8	1,1	0,7
TW	0,38	0,55	0,85	0,67	1,16	0,78	2,1	2,0	0,9	1,6	0,8
TW-R9	0,36	0,51	0,88	0,69	1,25	0,82	2,0	2,5	1,4	1,8	0,9
WORLD	1,10	1,60	1,94	1,67	2,33	1,78	2,1	0,9	0,2	0,9	0,3

ENERGY PRODUCTION (Mtoe)

- 1978 -

(Mtoe)	SMF	O	NG	HY	NU	NS	FW	VAW	CEP	NCEP	PEP
R1	380	574	527	129	77	2	30	5	1689	35	1724
R2	220	92	151	97	40	-	16	14	600	30	630
R3	71	25	10	27	14	-	1	1	147	2	149
R4	599	575	344	42	12	-	30	16	1572	46	1618
R5	1	1182	59	6	-	-	8	10	1248	18	1266
R6	53	113	1	7	-	-	86	13	174	99	273
R7	49	12	8	13	1	-	71	47	83	118	201
R8	12	101	18	5	-	-	63	38	136	101	237
R9	307	104	12	14	-	-	126	93	437	219	656
R10	8	240	52	51	2	6	55	13	359	68	427
WIC	721	691	688	254	131	2	49	21	2487	70	2557
IC	1320	1266	1032	296	143	2	79	37	4059	116	4175
TW	380	1752	150	95	3	6	407	213	2386	620	3006
TW-R9	73	1648	138	81	3	6	281	120	1949	401	2350
WORLD	1700	3018	1182	391	146	8	486	250	6445	736	7181

Table : 13

ENERGY PRODUCTION FORECASTS (Mtoe)

- 2000 (I) -

(Mteo)	SMF	O	NG	HY	NU	NS	FW	VAW	CEP	NCEP	PEP
R1	1067	570	502	179	288	111	84	10	2717	94	2811
R2	305	158	121	134	243	55	23	29	1016	52	1068
R3	180	35	30	35	109	15	1	2	404	3	407
R4	790	595	835	70	240	25	25	10	2555	35	2590
R5	5	1340	261	17	9	7	9	13	1639	22	1661
R6	188	180	10	21	6	13	135	24	418	159	577
R7	132	55	40	27	8	9	95	78	271	173	444
R8	70	150	100	20	35	10	65	47	385	112	497
R9	530	315	85	50	7	6	122	82	993	204	1197
R10	55	541	189	155	32	90	49	21	1062	70	1132
WIC	1732	763	653	349	645	186	110	42	4328	152	4480
IC	2522	1358	1488	419	885	211	135	52	6883	187	7070
TW	800	2581	685	289	92	130	473	264	4577	737	5314
TW-R9	270	2266	600	239	85	124	351	182	3584	533	4117
WORLD	3322	3939	2173	708	977	341	608	316	11460	924	12384

Table : 14

ENERGY PRODUCTION FORECASTS (Mtoe)

- 2000 (II) -

(Mtoe)	SMF	O	NG	HY	NU	NS	FW	VAW	CEP	NCEP	PEP
R1	966	573	475	174	273	92	81	8	2553	89	2642
R2	300	166	130	130	211	45	24	30	982	54	1036
R3	165	32	25	30	82	11	1	3	345	4	349
R4	665	520	720	65	180	20	25	15	2170	40	2210
R5	3	1070	190	13	4	6	10	15	1286	25	1311
R6	146	150	5	16	6	13	152	27	336	179	515
R7	104	40	27	20	6	6	98	80	203	178	381
R8	80	85	65	17	20	12	70	55	279	125	404
R9	385	158	32	29	3	3	190	82	610	272	882
R10	40	400	144	128	19	65	60	25	796	85	881
WIC	1571	771	630	335	571	151	108	43	4029	151	4180
IC	2236	1291	1350	400	751	171	133	58	6199	191	6390
TW	618	1903	463	222	53	102	578	282	3361	860	4221
TW-R9	233	1745	431	193	50	99	388	200	2751	588	3339
WORLD	2854	3194	1813	622	804	273	711	340	9560	1051	10611

ENERGY PRODUCTION FORECASTS (Mtoe)

- 2020 (I) -

(Mtoe)	SMF	O	NG	HY	NU	NS	FW	VAW	CEP	NCEP	PEP
R1	1958	484	460	199	484	318	91	15	3903	106	4009
R2	360	115	103	171	615	115	24	42	1479	66	1545
R3	290	50	60	40	200	36	1	3	676	4	680
R4	1000	520	1200	130	600	80	20	10	3530	30	3560
R5	8	1310	515	26	95	55	9	14	2009	23	2032
R6	468	250	30	72	25	39	158	37	884	195	1079
R7	254	85	80	54	24	57	107	107	554	214	768
R8	130	120	200	80	80	40	55	57	650	112	762
R9	780	400	300	150	30	30	50	30	1690	80	1770
R10	290	822	335	437	145	240	30	30	2269	60	2329
WIC	3061	649	623	412	1314	484	117	61	6543	178	6721
IC	4061	1169	1823	542	1914	564	137	71	10073	208	10281
TW	1477	2987	1460	817	384	446	408	274	7571	682	8253
TW-R9	697	2587	1160	667	354	416	358	244	5881	602	6483
WORLD	5538	4156	3283	1359	2298	1010	545	345	17644	890	18534

ENERGY PRODUCTION FORECASTS (Mtoe)

- 2020 (II) -

(Mtoe)	SMF	O	NG	HY	NU	NS	FW	VAW	CEP	NCEP	PEP
R1	1654	443	404	189	398	291	91	12	3379	103	3482
R2	345	132	102	165	505	95	25	44	1344	69	1413
R3	240	45	50	35	130	29	2	4	529	6	535
R4	800	400	1000	100	450	50	25	10	2800	35	2835
R5	6	990	350	20	33	39	11	19	1438	30	1468
R6	314	180	15	46	15	40	202	41	610	243	853
R7	151	50	54	40	12	34	114	114	341	228	569
R8	150	50	130	60	40	50	65	73	480	138	618
R9	510	200	100	60	10	10	110	50	890	160	1050
R10	170	562	240	288	60	170	40	-40	1490	80	1570
WIC	2541	620	556	390	1043	425	120	61	5575	181	5756
IC	3341	1020	1556	490	1493	475	145	71	8375	216	8591
TW	999	2032	889	513	160	333	540	336	4926	876	5802
TW-R9	489	1832	789	453	150	323	430	286	4036	716	4752
WORLD	4340	3052	2445	1003	1653	808	685	407	13301	1092	14393

INTER REGIONAL ENERGY EXCHANGES (Mtoe)

- 1978 -

(Mtoe)	SMF			CO			NG			TOTAL		
	E	I	B	E	I	B	E	I	B	E	I	B
R1	27	5	22	2	343	-341	1	2	-1	30	350	-320
R2	1	41	-40	54	656	-602	-	22	-22	55	719	-664
R3	7	19	-12	-	252	-252	-	14	-14	7	285	-278
R4	26	1	25	58	26	32	17	8	9	101	35	66
R5	-	1	-1	1002	-	1002	27	-	27	1029	1	1028
R6	9	-	9	97	22	75	-	-	-	106	22	84
R7	-	-	-	-	22	-22	3	-	3	3	22	-19
R8	-	2	-2	73	85	-12	11	-	11	84	87	-3
R9	1	-	1	13	1	12	-	-	-	14	1	13
R10	-	5	-5	75	66	9	-	-	-	75	71	4
WORLD	71	74	-3	1374	1473	-99	59	46	13	1504	1593	-89

INTER REGIONAL ENERGY EXCHANGE FORECASTS (Mtoe)

- 2000 (I) -

(Mtoe)	SMF			CO			NG			TOTAL		
	E	I	B	E	I	B	E	I	B	E	I	B
R1	169	-	169	2	162	-160	-	24	-24	171	186	-15
R2	-	172	-172	40	565	-525	-	160	-160	40	897	-857
R3	75	95	-20	-	218	-218	-	66	-66	75	379	-304
R4	75	-	75	-	70	-70	80	15	65	155	85	70
R5	-	5	-5	930	-	930	95	-	95	1025	5	1020
R6	55	2	53	126	30	96	-	-	-	181	32	149
R7	1	-	1	-	42	-42	5	-	5	6	42	-36
R8	-	40	-40	-	127	-127	50	-	50	50	167	-117
R9	20	-	20	100	3	97	-	-	-	120	3	117
R10	3	10	-7	40	40	-	25	2	23	68	52	16
WORLD	398	324	74	1238	1257	-19	255	267	-12	1891	1848	43

INTER REGIONAL ENERGY EXCHANGE FORECASTS (Mtoe)

- 2000 (II) -

(Mtoe)	SMF			CO			NG			TOTAL		
	E	I	B	E	I	B	E	I	B	E	I	B
R1	156	-	156	2	120	-118	-	12	-12	158	132	26
R2	-	121	-121	40	511	-471	-	100	-100	40	732	-692
R3	65	90	-25	-	201	-201	-	65	-65	65	356	-291
R4	30	-	30	-	70	-70	55	5	50	85	75	10
R5	-	4	-4	757	-	757	75	-	75	832	4	828
R6	40	2	38	120	20	100	-	-	-	160	22	138
R7	-	-	-	-	30	-30	5	-	5	5	30	-25
R8	-	20	-20	-	67	-67	35	-	35	35	87	-52
R9	10	-	10	25	2	23	-	-	-	35	2	33
R10	3	7	-4	30	30	0	20	-	20	53	37	16
WORLD	304	244	60	974	1051	-77	190	182	8	1468	1477	-9

INTER REGIONAL ENERGY EXCHANGE FORECASTS (Mtoe)

- 2020 (I) -

(Mtoe)	SMF			CO			NG			TOTAL		
	E	I	B	E	I	B	E	I	B	E	I	B
R1	184	-	184	2	51	-49	-	-	-	186	51	135
R2	-	370	-370	20	399	-379	-	200	-200	20	969	-949
R3	120	120	0	-	162	-162	-	105	-105	120	387	-267
R4	70	-	70	-	60	-60	120	20	100	190	80	110
R5	-	58	-58	665	-	665	41	-	41	706	58	648
R6	80	5	75	129	40	89	-	-	-	209	45	164
R7	4	-	4	-	61	-61	10	-	10	14	61	-47
R8	-	130	-130	-	225	-225	50	-	50	50	355	-305
R9	80	-	80	100	5	95	100	-	100	280	5	275
R10	10	20	-10	20	20	0	40	-	40	70	40	30
WORLD	548	703	-155	936	1023	-87	361	325	36	1845	2051	-206

Table   : 21

INTER REGIONAL ENERGY EXCHANGE FORECASTS (Mtoe)

- 2020 (II) -

(Mtoe)	SMF			CO			NG			TOTAL		
	E	I	B	E	I	B	E	I	B	E	I	B
R1	168	-	168	2	36	-34	-	-	-	170	36	134
R2	-	260	-260	20	346	-326	-	130	-130	20	736	-716
R3	100	110	-10	-	136	-136	-	90	-90	100	336	-236
R4	20	-	20	-	50	-50	70	10	60	90	60	30
R5	-	35	-35	548	-	548	83	-	83	631	35	596
R6	50	4	46	122	20	102	-	-	-	172	24	148
R7	-	-	-	-	37	-37	10	-	10	10	37	-27
R8	-	20	-20	-	45	-45	30	-	30	30	65	-35
R9	30	-	30	-	5	-5	-	-	-	30	5	25
R10	10	10	0	15	15	0	30	-	30	55	25	30
WORLD	378	439	-61	707	690	17	223	230	-7	1308	1359	-51

Part I

PRESENTATION AND ANALYSIS OF RESULTS

Chapter I

SITUATION OF THE STUDY

1.1 HISTORICAL INTRODUCTION

Apart from the economic upheavals and consequential soul-searching which it caused, the first oil shock of 1973 was not without influence on the development of new forecasting methodology in the field of energy.

Indeed, before the nineteen seventies "prospective" was still a new word and serious attempts to explore long-term world energy trends were very rare. Helped by necessity, a decisive turning point was reached in 1974. Wide-ranging energy studies began to appear.* The World Energy Conference itself did not remain on the sidelines of this development and greatly contributed to it by its activities.

As a result of the IXth Conference at Detroit (September 1974), the Conservation Committee was established, setting itself the principal goal of presenting a global analysis of energy demand and supply up to 2020 to the *Xth Conference.* This pioneering task was entrusted to a British Team from the University of Cambridge under the direction of Dr Richard Eden.

Its report, which represented an important step in the understanding of long-term energy developments, was thus discussed at a round-table meeting at Istanbul in September 1977.[3] Within the context of several economic scenarios and of a global model, it provided:

(i) projections of future demand for eleven regions of the world;
(ii) projections of the technologically accessible global production potential (as determined by a series of working groups, each specialising in a different energy source);
(iii) elements of preliminary reflexion on energy saving policies;
(iv) extrapolated supply strategies on the basis of the studies made.

The analytical work of the Commission was to continue beyond that initial stage and to define its terms of reference more precisely for the *XIth Conference* at Munich. As a result of the British study, it was decided to examine the prospects for developing countries in greater depth by devoting a special study to them, and to prepare a policy paper on the use of demand models, as well as to place production prospects in their present-day context and to better situate energy saving concepts and policies. In September 1980,

*The author is well acquainted with ref 36, for example.

3

therefore, studies No. 11, 44, 45 and 46 were presented which concerned these various questions.

In particular, the study "Third World Energy Horizons 2000-2020", enabled one to estimate the energy consumption of that period and its corresponding supply patterns within the context of three differentiated scenarios and thanks to the introduction of a new methodology for the forecasting of total income elasticities.

The merit of this study was to fill an obvious gap in the global analyses which were then proliferating and which, in almost every case, concentrated on the problems of industrialised countries to the detriment of an analysis of the specificities of the Third World (except, for example, ref. 37).

After this new stage, the next step was to consider the nature of the follow-up to be given to these world-wide efforts. What could the World Energy Conference contribute over and above all this, knowing that at the same time, a strong research drive on this subject had developed in the large international organisations (OECD, the World Bank and the UN), as well as in the petroleum companies (Exxon, BP), and the specialised institutes (IIASA, MIT, SRI, etc...), each of which enjoyed financial and human resources far in excess of that of the Conference? Out of these thoughts was born the research project on the Long-Term Evolution of Regional Energy Balances (the "LTEREB" Project).

1.2 SPECIFICITY

The projected study had, first of all, to go beyond similar research being done elsewhere, to respect those constraints peculiar to the Conference (especially budgetary constraints) and, above all, to try to take advantage of its *comparative advantage* against other international organisations. These advantages are threefold and may be summarised as follows:

(A) As a *non-specialised organisation,* the World Energy Conference covers a very large energy field since, by bringing together professionals from all areas, it can deal with all forms of energy without exception.
(B) Structured as it is in a network of 80 National Committees it can group together the majority of the large energy producing and consuming countries whether they be situated in the North, the South, the West or the East: the Conference has no dominant geopolitical affiliation.
(C) Finally, as a *non-governmental organisation,* the Conference is less subject than others to those official pressures and constraints which very often act as a brake on the launching of transnational studies.

Starting from this brief analysis, it was possible to envisage the definition of an original and specific research project adapted to this privileged position.

(i) The exhaustive nature of the energy field covered encouraged the pursuit of *"inter-energy"* research.
(ii) The existence of the network of National Committees authorised the setting up of a *decentralised forecasting process.*
(iii) Finally, the organisation's neutrality, allowing it to transcend political bias, also enabled it to begin a process of *regional cooperation* based on the collaboration of experts expressing purely personal views.

One could also hope to *gain in analytical depth* compared to the Munich studies since these tended essentially to concentrate on consumption forecasts, combining them with a survey of corresponding supply patterns and a survey of potential supply sources in the Third World survey without for all that defining precisely the future levels of production and of international trade.

As a result, Mr Boiteux, the President of the Conservation Commission, was able to present the study project to the Programme Committee at its meeting in Munich in September 1980:

"As far as the new activities of the Commission are concerned, I do not think that it would be very useful to repeat the studies made up to 1977. A large number of organisations are engaged in studies in this area. On the other hand, it would, in my view, be very useful to plan a study of energy balances of the world's great regions. There are few studies in this field. Furthermore, the World Energy Conference is particularly well-placed to perform such a task, since it can call upon the resources of the National Committees.

I do not have much faith in the value of a questionnaire which few National Committees would answer. This is why I propose to perform this study in two phases:

(a) the drawing up by a group of international or regional experts of uniform regional energy balances based on identical overall hypotheses;
(b) the draft regional energy balances thus formulated would then be submitted to *ad hoc* regional committees, composed of representatives of the National Committees of the relevant regions which would examine them critically and which would be responsible for the establishment of definite balances.

I should like to stress that the main advantages of such a study is that it would take full account of developing countries, since most of the studies in this area up to now have been made by industrialised nations."

The Programme Committee decided to recommend to the International Executive Council that it should approve this proposed study. This the Council did at its meeting of the 6th September 1980, and the conduct of the project was entrusted to the author.

By this study it was hoped not only to arrive at quantitative forecasts which would be nearer to local perceptions and factual situations, but also, by the very fact of its existence, to reach the following three aims (although these qualitative objections had not necessarily been made explicit):

(i) *to stimulate the interest of the various regions, in energy forecasting work* (even in those regions most deficient in research means), and by that to engender long-term analyses allowing one to see national energy planning efforts in their regional context;
(ii) *to promote a policy of forecasting cooperation* between National Committees (and regional organisations) which could only be of benefit for the different participants and to the Conference itself.
(iii) *to counteract the bankruptcy of most forecasting models by a new approach.* Faced with a very uncertain future, traditional methodologies cannot manage to encompass efficiently the new complexities and risks, mainly based as they are on an analysis of the past and being all the more

reliable in so far as current trends are regular in pattern. From that time onwards, it was both *pertinent* and *opportune* to seek to inaugurate an approach adapted to the troubled overall situation which the global energy system will face for a long time to come.

1.3 THE OBJECTIVE OF THE STUDY

1.3.1. Formulation of the objective

To present to the XIIth conference (New Delhi, September 1983) the primary energy balances for 2000–2020 of the 10 large regions covering the entire globe, which would contain a forecast of total energy consumption according to eight commercial and non-commercial supply sources. This forecast would be linked to a forecast of the regional production of these eight sources and of the corresponding interregional trade, in the context of two contrasted economic development scenarios. These balances would be the fruit of the work of ten groups, composed of experts from the regions and thus being the first global attempt to produce a really decentralised approach to the problem of long term energy forecasts.

1.3.2. Forecasting procedure

On the basis of the initial proposition made at Munich and of the objective defined above, a forecasting method adapted to the needs of the task in hand was used. It was organised according to the diagram on the next page. The following points will be noted:

(i) The decentralised stage, which is the heart of the study, and which gives it all its specificity, was mainly conducted from mid-1981 to mid-1982: 17 meetings took place in all the 5 continents to carry out this task (cf. A3).

(ii) To prepare and place the work of the regional working groups in context, and to take precautions against a possible failure of one of them, a long preparatory effort had to be undertaken:

(a) the establishment of terms of reference for 160 countries based on the same principles as those used in Munich for the terms of reference established there;[2]

(b) the formulation of 4 projected balances for each region (2 scenarios, 2000–2020), to provide a basis for the RWT's analysis (cf. A8).

This important centralised effort was conceived solely with the aim of ensuring the proper functioning of the RWT and the overall coherence of the hypotheses used.

(iii) Finally it was still necessary to coordinate and synthesise the results obtained, to organise them and to make a report of the conclusions. This was the goal of the third stage, which occurred from the summer of 1982 until Spring 1983.

As regards the timetable initially worked out, 3 months were gained in relation to the first stage and 6 months on the second. This allowed the RWTs to correct the commentaries made by the Central Team on regional results (Part III of the Report).

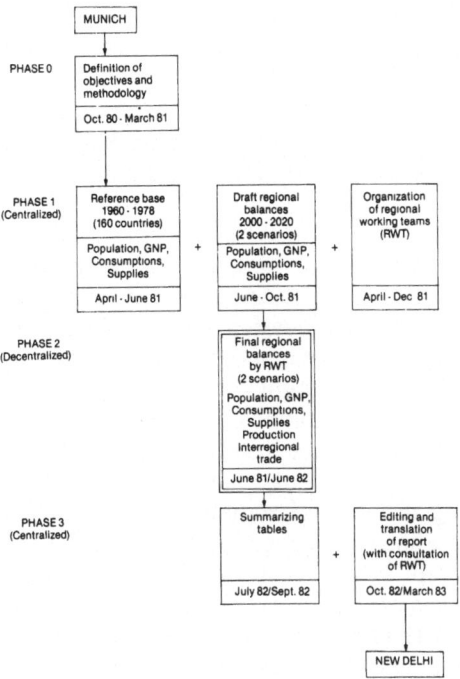

1.3.3. Means

A number of working parties were set up:

(a) A very small *Central Team* which was responsible for overall coordination under the supervision of the Conservation Commission. It was composed of the Project Director (for the duration of the project) and of a Research Assistant (for six months).
(b) Ten regional *working groups,* bringing together 50 experts with the effective support of 20 national committees (cf. A2).
(c) *Various contacts* made in relation to the work of the Central Team; these included, for instance, the members of the Conservation Commission, the members of the Committee of the Developing Countries, experts from the UN or from the World Bank, etc.

We should like to note, for the sake of information, that the organisation and the performance of the decentralised second phase necessitated the sending of more than 260 letters and telexes, and the making of numerous telephone calls to all parts of the world.

1.4 PRELIMINARY REMARKS

In order to gain the maximum benefit from this study, it is important to be aware of its scope and to be conscious of a certain number of preliminary points.
(a) *The abundance of statistical data* provided in the tables of results (more than 10,000 items of information) must not overwhelm the reader. Although

this data did not emerge from the "computer's black box", but were all considered and calculated one by one, it is necessary to see beyond their apparent precision. All these data are, of course, useful for a delineation of future trends which is not purely qualitative, but one cannot pretend to predict each element of each statistical table exactly. On a timescale of 20 or 40 years, such precision would be entirely illusory: to express a certain and unique truth in an uncertain and abundant world, would be unthinkable. A forecaster does not have a crystal ball.

By this mass of figures, our aim was rather to *highlight the main constraints* which the world and its regions will be facing in the future, and which transcend the differences of theories and of points of view (bearing in mind that each forecasting horizon engenders a different category of problems and therefore of decisions to be taken).

(b) Every effort was made to maintain an *equilibrium between the coherence and the realism* of forecasts. At first, the results of the first stage gave us a set of apparently very coherent balances. But experience clearly demonstrated that taking account of regional conceptions meant giving a clear priority to decentralised concepts, even if, during the course of this, the overall consistency of hypotheses was somewhat weakened. This slide towards *flexibility* was the price to be paid to really obtain the interest of regional experts for the proposed task, and, moreover, to give substance to the essential preliminary objectives of *specificity* and *realism.* In any case, it was repeatedly noted that for all that initial coherence appeared an attractive goal, it also in fact hid a number of approximations and lacunae in the information collected by international bodies.

(c) Although we should like to stress that this *priority* placed on *decentralised thinking* gives the present study its originality, it may nevertheless have introduced a certain bias in the results formulated. Perhaps this bias was not as all-pervasive as one might have feared, knowing from experience that regional estimates have a tendency to sin through excess in forecasting compared with views expressed from the centre (which take greater account of compensation mechanisms). However, a certain deformation in the above sense cannot be entirely excluded.

(d) Conversely, *the gloomy climate of 1981–1982* marked by the economic recession and the general pessimistic atmosphere cannot fail to have influenced conceptions of the long-term economic future. The forecasts provided in this study therefore reflect views expressed between mid 1981 and mid 1982, but subject to revision until the end of 1982.

(e) It was decided at the outset that the future would be structured around *two decisional scenarios:* these two scenarios exclude extreme extrapolations and constitute *desirable and probable variants* of the same situation rather than being two clearly different evolutions. This procedure restricts the field correspondingly and omits situations which, while being less plausible, would have radically different consequences. Thus what is explored here is a future that has been deliberately simplified and schematised. Moreover, since it is characterised by reference to the *situation in two years only,* (2000 and 2020), it does not give any indications as to the intervening paths of development.

(f) At the same time, it is difficult to measure the effective *influence* of the *Central Team* in the formulation of regional forecasts. Although the central team was theoretically neutral in Stage 2, the need to participate closely with regional activities to explain and retain control on the overall progress of the study, added to the need to bring the work of the teams to a conclusion of substance and to preserve a certain coherence, no doubt resulted in an effect on the final results which was not insignificant.

(g) The composition of the RWT themselves was a determining factor. For reasons of efficiency, the decentralised teams were kept small and it was therefore difficult to make their membership reflect the entire range of regional opinions. In the choice of experts, priority was therefore given to their level of competence, their motivation and their availability. This cannot of course be sufficient to ensure a perfect representation of the regional opinions, all the more so as the underlying policy was that each expert would express his opinion within the group on a purely personal basis without committing his organisation, a policy which was thought to be the only one capable of producing data.

(h) Finally, the study's *aim* was *deliberately limited* at the outset to make its realisation possible, knowing that one would already have difficulties in producing forecasts from the most general of aggregates.

The field of investigation had to be circumscribed to the essential regional data, which accepted the abandonment of other most interesting elements (for example, sectoral patterns of consumption), if one wanted to have a reasonable hope of bringing the task in hand to a conclusion. The experience obtained in performing this task amply confirmed the correctness of this intuition, which enabled one to keep to the original objectives.

(i) This study does not pretend to answer all the questions which can be asked about energy policies — far from it. Firstly, we have only provided *regional estimates* which partly elude national problems. Secondly, we have placed ourselves at the first link of the energy chain, that of *primary supplies*, without explicit reference to transformation vectors (such as electricity) or to final uses sectors.

It is thus a very particular category of problems which is elucidated by this study, a category which is essential, moreover, since it can usefully serve to place the policies of the region's countries in context. However, one should beware of transposing the methodology and the results on the level of a national analysis. Indeed, the explanation of long term national perspectives requires other tools and much finer detail about the energy system.

1.5 INSTRUCTIONS FOR USE

(Advice from the author to the reader)

1. The reader should not hesitate to ignore the order of the chapters and to extract the final results in advance, since the chapters were designed to be used autonomously. Professional pride aside, we would suggest that this study is more one to be *consulted* than one to be digested at one reading.

2. Short-hand expressions are always a little abusive in principle; however, a number have been used:

(a)   *The Third World*
      This expression here covers a precise number of countries: Latin America, and Africa (except for South Africa), and Asia (except for Japan and Israel) and Oceania (except for Australia and New Zealand).

(b)   *The Industrialised Countries*
      This expression designates all the other countries of the world: North
      America and Western Europe and Eastern Europe and the industria-
      lised countries of the Pacific and South Africa.

(c)   *Commercial Energies*
      This expression is used to designate all solid mineral fuels and oil
      (conventional or non-conventional) and natural gas and hydraulic power
      and nuclear power and new energies.

(d)   *Non-commercial energies*
      This is used to describe fuel wood and vegetable and animals waste (in
      other words, traditional uses of the biomass). This use of the term is
      made even if it is well known that in many countries fire-wood is already
      an individual market commodity.

(e)   *New Energies*
      This expression covers all the renewable new energies without distinc-
      tion: solar, wind, tidal, wave, sea thermal, new uses of biomass (e.g.
      production of alcohol and biogas), etc.

(f)   In the text, the two terms *"consumption"* and *"demand"* are amalga-
      mated. By contrast, the terms *"supply"* and *"production"* are differen-
      tiated: by supply is meant the quantities of energy which will cover
      supply in the primary balance; the term "production" characterises the
      local level of energy production.

(g)   The term "the past" is always used to mean the reference period of the
      study: 1960–1978.

(h)   The word "actually" or "today" designates the year 1978 and not the
      present year (1983).

(i)   Where two figures appear in the text separated by a "/" without other
      specifications, this first figure refers to Scenario I and the second to
      Scenario II.

(j)   Equally "I" and "II" refer to Scenario I and to Scenario II.

(k)   The terms "in the year 2000" (or "before" or "after" 2000) do not
      necessarily mean that the event in question or that trend focuses
      precisely in the year 2000. The numerical value 2000 constitutes a
      reference point of passage even if that event referred to occurs a little
      before or a little after 2000.

3. It will also be remembered that all the values expressed in "tons of oil
equivalent" ("TOE") correspond to a theoretical TOE, equivalent to *10500 M
calories;* however, a great number of international studies use a term of
reference of 10000 M calories. The average differential introduced by this
convention is thus -5% for the present study: this should be borne in mind
when one wishes to make rapid comparisons between the results of this study
and those of other global studies.

4. We also wish to remind the reader that the term *"energy consumption"* is
here used in a rather *restrictive* sense, since it excludes bunkers, stock
variations, and non-energy uses.

5. With regard to abbreviations, the reader is referred to the table at the front
of the book.

Chapter II

GLOBAL EQUILIBRIUM AND NORTH–SOUTH DYNAMICS

2.1 THE FUNDAMENTAL FACTORS: DEMOGRAPHY AND ECONOMIC GROWTH

The growth of energy consumption is linked to demography directly and massively and indirectly to economic growth, to the extent that energy demand will more or less closely match the development of the productive sector and the growth of needs.

These two factors are far from being the only two factors: however, they are known to play a determining role in the shaping of energy demand. In this study, they have been introduced as external variables and will therefore be briefly commented upon.

2.1.1. Demography

Demographic hypotheses will be found in Annex 6. The following essential points will be remembered:

(A) The world population, which grew from 3 to 4.3 billion inhabitants between 1960 and 1978, could reach *6 billion in 2000* and *7.7 billion in 2020.* This remarkable growth corresponds, however, to a certain reduction in demographic stresses since the global growth rate, which was about 2% p.a. during the past years, will fall to 1.6% by the year 2000 and 1.2% beyond.

(B) *The Third World,* which has been the driving force behind demographic growth for a long time, will increase its share of the world total very regularly: 67% in 1960, 72% in 1978, 77% in 2000 and 80% in 2020. 4 out of 5 of the planet's inhabitants from 2000 will live in what are developing countries today (as against 2 out of 3 in 1960). This is because the difference in growth rates between Industrialised Countries and Third World countries, which was already substantial in the past (1% against 2.35%), will continue in the future (0.65% against 1.9% by 2000 and 0.52% against 1.4% beyond). The population of industrialised countries will only increase by a little more than 300 million from 1978 to 2020, while that of the Third World will increase by more than 3 billion.

It may be helpful to note the significative evolution of the classification of regions in terms of population.

Table 1 CLASSIFICATION OF REGIONS BY POPULATION

		1978	2020
Industrial Countries	R1	8	9
	R2	3	6
	R3	10	10
	R4	4	8
Third World	R5	9	7
	R6	5	3
	R7	2	1
	R8	7	5
	R9	1	2
	R10	6	4

Among the industrialised countries, Western Europe and Eastern Europe held a good rank in 1978, being 3rd and 4th respectively. In 2020, they occupy places 6 and 8, while all the regions of the Third World go up to the intervening places (except for R9 which is removed from the first place by R7).

2.1.2. Economic growth

Economic forecasts which are linked to energy forecasts are set out in detail in Annex 7.

(A) The two scenarios clearly differentiate economic growth: in Scenario (I) the volume of world GNP will continue to grow at a sustained rate: 3.65% from 1978 to 2000 and 2.9% up to 2020, as against, it must be admitted, 5% between 1960 and 1978.

By contrast the slowdown is more substantial in Scenario II, since, there, the pattern of growth is a moderate pace of 2.65% between 1978 and 2000 and of 2% beyond (therefore almost 1% behind the figures in Scenario I).

From that fact, the global GNP total starting from $8900 G in 1978 will reach $35000 G in 2020 according to (I) and only $24000 G according to (II).

(B) While between 1960 and 1978 the share of the Third World in world production had laboriously increased from 14.5% to 17.5%, the two scenarios notably inflect this trend. The scenarios forecast development prospects which are far more intense for the Third World than for Industrialised Countries within the overall context of a world economic downturn.

Thus, the Third World would grow, according to (I), to producing 25% of world GNP in 2000 and 33% in 2020. According to (II), the shift is slower but none the less substantial: 22% in 2000 and 28% in 2020.

In addition, it should be noted that certain areas (Latin America, North Africa, Middle East: R10 and R5), which only made up 20% of the Third World's population, will contribute by themselves more than 60% of the Third World's growth between 1978 and 2020.

Even within the Third World, therefore, future trends differ widely from region to region.

(C) An *analysis per capita* is also instructive: indeed, the movements of masses tend to hide individual realities. The average inhabitant of the world had $1230 in 1960, and more than $2000 in 1978.

According to the scenarios, average world per capita "wealth" will increase to \$3200 in 2000 and to \$4500 in 2020 according to (I); however, according to (II), his situation will improve much more moderately, since an individual's wealth will only just exceed \$2600 in 2000 and \$3000 in 2020.

The gap between the inhabitants of the industrialised countries and that of the Third World is significant. In the beginning, in 1960, the average of GNP of the Third World was 8% of that of the industrialised countries. In 1978, this ratio remained stable at 8% for all intents and purposes. No substantial improvement is therefore revealed, at least from the statistics alone. Of course, forecasts predict an increase but in proportions that remain very modest: 2000, the ratio will be 8% or 10% according to Scenarios (II) and (I), and 10–12% in 2020. In so far as this type of analysis is of any relevance, this would indicate that one is far from a "catching up" of the industrialised countries by those of the Third World.

As for the growth rates which explain these evolutions, it will be noted that they reach a ceiling of 4.3% and 3.2% in Scenario (I) for the most favoured region (as against 7.5% between 1960 and 1978). In Scenario (II), the ceiling is 2.6% and 1.8%. The slowdown is therefore very marked and enables one to understand that the (measuredly) optimistic impression given by Scenario (I) is relative. Despite a rate of economic development which will operate an average rate not far from that of the past (3.4% between 1978 and 2000, and 2.4% beyond, as against 3.5% from 1960 to 1978), the inhabitant of the Third World will hardly reach the economic level of a West European before the Second World War in 2020, thus 80 years late!

2.2 WORLD ENERGY CONSUMPTION

2.2.1. Consumption totals and consumption per capita

(A) *World energy consumption* as defined and measured by the study's criteria (cf. p178) reached 3.3 billion TOE in 1960, and 6.8 GTOE in 1978.

According to the scenarios, it will reach levels of 10.1 to 11.7 GTOE in 2000, and of 13.8 to 18 GTOE in 2020. These quantities are double and nearly treble those obtaining today.

(B) The quantities involved are therefore very important: +3.3 GTOE and +4.9 GTOE from 1978 to 200, and between +3.7 and +6.3 GTOE between 2000 and 2020. According to Scenario (II), total increases will be in the range of past growth rates (+3.5 GTOE between 1960 and 1978), while, according to Scenario (I), after 2000 they will grow to up to twice the levels of the past.

(C) The translation of the quantities in *terms of growth rates,* however, gives us a totally different picture of the future. Indeed, the corresponding rates of growth reflect a very substantial slowdown in the growth of the world's consumption, even within the framework of Scenario (I): *1.8 to 2.5% p.a. between 1978 and 2000, and 1.6 to 2.1%* between *2000* and *2020,* as against *4.1%* between 1960 and 1978. The rates observed would therefore only be of the order of half the *rates* of the past.

(D) Moreover, for all that the demographic effect is filtered and analyses made on a per capita basis, the overall picture appears very different in terms both of levels and rates:

(i)   The average inhabitant of the world consumed 1.1 TOE in 1960 and 1.6 TOE in 1978.

(ii) According to Scenario (II), he will only dispose of barely 1.7 TOE in *2000* and 1.8 TOE in *2020*. According to Scenario (I) (although it is more favourable to growth), the figures are: barely 2 TOE in *2000* and 2.3 TOE in *2020*. In other words, over a period of more than forty years, there will be an increase of 0.2 to 0.7 TOE as against +0.5 TOE in the past 18 years.

(iii) As for the corresponding levels, the rates are very reduced: 0.9% growth per capita and per annum between 1978 and 2000 and 2020 according to (I); 0.2 to 0.3% during the same period accroding to Scenario (II) as against 2.1% from 1960 to 1978.

In all cases the slowdown is impressive.

(E) Demography is of prime importance in the shaping of world demand. Table 2 reveals this clearly.

Table 2   AVERAGE ANNUAL GROWTH RATES (%)

Entire World	1960/1978	1978/2000	2000/2020
Population	1.9	1.6	1.2
Per Capita Energy Consumption Scenario I Scenario II	2.1	0.9 0.2	0.9 0.3
Total Energy Consumption Scenario I Scenario II	4.1	2.5 1.8	2.1 1.6

The growth of the world population accounted for almost half the growth of energy consumption in the past; it will be responsible for about 60% of continued increase, according to Scenario (I), and for 90% from 1978 to 2000 and for 75% from 2000 to 2020, according to Scenario (II). One need hardly add that demography was and will be even more of a determining factor in world energy demand. This is illustrated by the following calculation:

Assuming a population stabilised at its 1978 level (4.26 billion), and according to the hypotheses elaborated in this study (namely keeping the *per capita* levels for 2000 and 2020 resulting from the forecast), world consumption will only reach 8.3 GTOE in 2000 and 10 GTOE in 2020 according to Scenario (I) (starting, it will be remembered from 6.8 GTOE in 1978). In Scenario (II), it will barely reach 7.1 and 7.6 GTOE in 2000 and 2020 respectively, thus presenting an almost nil overall increase.

2.2.2. The dynamics of North–South consumption

But one cannot confine oneself to an analysis expressed solely in total world figures. In particular, the study of the relations between North and South, Industrialised World and Third World is extremely instructive.

(A) The overall quantities involved are becoming considerable. In relation to the industrialised countries, despite the slowdown of growth, the amounts are: 5.1 GTOE in 1978 to 7.1 to 7.8 GTOE in 2000, and 9 to 10.7 GTOE in 2020. For the developing countries, the figures are: 1.7 GTOE in 1978, 3.1 to 4 GTOE in 2000 and 4.8 to 7.2 GTOE in 2020. The Third World's share of world demand which has almost stagnated from 1960 to 1978 (23 to 24%) will see a substantial increase by 2000 (34 to 31%), increasing further up to 2020 (40 to 35%).

(B) But if the situation of average world per capita consumption is indeed what has been described, this implies a radical disparity between the situations in the North and South from the very start.

In 1960, the world average consumption of 1.1 TOE was made up as to 2.6 TOE per capita in the North and 0.4 TOE in the South (even if one is careful to include non-commercial energies), the relative North to South consumption ratio being 15%. In 1978, the gap widened further: 4.3 TOE for the North as against 0.55 in the South, the ratio being 13%.

(C) What is the result of the above, in terms of forecasting? The growth rates of the North slow down considerably: 1.2% and 1.1% before and after 2000 according to Scenario (I); 0.8% and 0.7% according to Scenario (II), as against 2.9% in the past. In the South, the erosion of growth rates is weaker: 2% and 1.6% according to Scenario (I), before and after 2000; but 0.9% and 0.8% according to Scenario (II), thus enjoying growth rates which are similar to those of the North. Consequently, the ratio will become gradually more favourable according to Scenario (I): 15% in 2000, and 17% in 2020, and will maintain itself at 13% according to (II). No spectacular evolution will occur, therefore, except for the slow improvement or the actual termination of any change in the North to South ratio, which had worsened in the past.

Moreover, far from decreasing, the gap stated in absolute TOE terms, will deepen profoundly. It was +2.2 TOE in favour of the inhabitants of the North in 1960. It rose to +3.75 TOE in 1978, and it will reach +4.8 TOE in 2000, according to Scenario (I) and +4.4 TOE according to Scenario (II); in 2020, it will be near to +6 TOE according to (I) and +5 TOE according to (II).

(D) Despite all the efforts made to develop and despite growth rates which may sometimes appear ambitious, the South will only make up a little or none of the North's "lead" in consumption. In any case, it cannot expect to match it. The South's situation is less unacceptable in Scenario (I) where its inhabitant enjoys the prospect of doubling his average level of consumption in forty years (1.16 TOE in 2020 as against 0.55 in 1978). However, it remains precarious in Scenario (II), where the increase will hardly exceed 40% (0.78 TOE in 2020).

During the same period, the inhabitant of the North will witness his average consumption increase to +2.7 TOE by 2020, according to (I) and to +1.5 TOE, according to (II), despite all the brakes created by circumstances and the policies implemented. This disproportion is flagrant, even shocking. In particular this is because the effect of initial levels is at its strongest here: despite lower growth, the eight times higher starting level produces a much greater increase in terms of quantities. For instance, a rate of 2% over twenty years increases the original in "stock" by about 50%; but 4.3 TOE increased by 50% gives a figure of nearly 6.5 TOE (+2.2 TOE), while 0.55 TOE increased by 50% only gives a result of 0.8 TOE (+0.25 TOE). Consequently, even when account is taken of climatic differences in heating demand, any theory of the South "catching up" the North appears very illusory and instead seems to be a will-o'-the-wisp.

It must be realised that, even in the more favourable case of Scenario (I), the average inhabitant of the Third World in 2020 will hardly reach the level of consumption of West European in 1950 (seventy years later!), and that is a statement which does not even mention the gap compared with North America.

Table 3 illustrates the relative discrepancies by a comparison of the average per capita consumption in the South and the North, and then with North America (the world's leading consumer). Parts (3) and (4) of Table 3 compare per capita consumption rates of South Asia (the lowest per capita consumer) with North America and Western Europe.

Table 3 RELATIVE NORTH/SOUTH CONSUMPTION

Comparison average per capita consumptions	1960	1970	2000		2020	
			I	II	I	II
(1) North South	6.7	7.8	6.6	7.6	6.0	7.5
(2) N. Am. South	15.3	15.2	10.5	12.3	8.8	11.3
(3) N. Am. South Asia	27.8	33.4	24.9	27.5	21.3	25.2
(4) W. Eur. S. Asia	8.7	11.8	11.2	12.1	10.4	12.2

All the ratios deteriorated between 1960 and 1978, only the second remaining constant. According to our forecasts, ratio (1) will hardly improve in Scenario (II) but will do so more noticeably in Scenario (I), where the 1960 level is once again reached in 2000. Compared with North America, the improvement is more substantial in ratios (2) and (3), but almost non-existent compared to Western Europe in (4).

In all cases, development Scenario (I) (which is faster and allows for more important differential efforts) leads to less unbalanced situations.

2.2.3. Commercial and non-commercial consumptions

A significant regrouping can be made by using the figure for total consumption of energy in two sub-totals: commercial energies on the one hand, and non-commercial energies on the other (in so far as these will experience very differentiated evolutions).

A. NON-COMMERCIAL (OR TRADITIONAL) ENERGIES

It should be remembered that these consist of fuel wood and wood waste, vegetable waste from cereals and other vegetables as well as animals. In 1960, they represented in total 17% of world energy demand, corresponding to 560 MTOE estimated. In 1978, their share fell to 11% (to 735 MTOE estimated).

This relative regression can only keep on in the future it seems,* and will do so all the more rapidly as the general economic situation is favourable to the development of modern substitute energies. Therefore, in Scenario (I), the non-commercial share is only 8% in 2000 and 5% in 2020. By contrast, in Scenario (II) (with its more difficult context), these traditional energies resist that trend better: 10% in 2000 and 8% in 2020.

---

*Modern uses of the biogas were here placed under the heading "new energies", that is to say, commercial energies.

Expressed in terms of quantities, non-commercial energies will develop further despite the efforts made to limit their growth; they will stabilise around 900 MTOE in scenario (I) and even reach 1.1 GTOE in 2020 (II).

Growth rates decrease markedly, especially beyond 2000. The entire NCEC (Non-Commercial Energy Consumption) grew by an average of 1.5% between 1960 and 1978. It will continue at the same rate until 2000 according to Scenario (II), to collapse beyond that date to 0.2%, while in Scenario (I), the decrease is already in motion and more accentuated: 1% up to 2000 and −0.2% beyond,

Expressed per capita, the level of consumption will experience a regular decline: about 0.19 TOE per average world inhabitant in 1960; 0.17 TOE in 1978; 0.15 TOE in 2000 (I) and 0.12 in 2020. According to Scenario (II), the level which remains steady at 0.17 TOE in 2000 decreases to 0.14 TOE in 2020.

Against the trends of the past, it would seem that these rates of consumption will be revitalised in *industrialised nations*, at least in the Western nations.

The forecasts translate this relative change of direction (caused by the need to use these energies better, especially in the domestic context) through the medium of more elaborate technologies. While these traditional consumptions decreased in absolute terms from 129 to 116 MTOE between 1960 and 1978 in the group of industrialised countries, they will reach 190 MTOE in 2000 and 210 MTOE in 2020.

The situation is different in the *Third World* where demographic pressure and the high cost of modern substitute sources will primarily cause this type of consumption to develop up to 2000 and to perpetuate itself thereafter, no doubt beyond what is ecologically desirable in many regions.

The aggregation of these two movements lead, on a global scale, to an inflexion of that past tendency which devolved a continually growing proportion of this market to the Third World (79% in 1960, 89% in 1978). The Third World will see its share limited to 80 to 82% in 2000 and 77 to 80% in 2020.

Finally, the inhabitant of the Third World will rely less and less on these sources to satisfy his energy consumption: in 1960, they covered 56% of its total need, in 1978, only 37%. According to Scenario (I), they decrease to 19% in 2000 and 10% in 2020; according to Scenario (II), where they remain better represented, they nevertheless decrease to 28% and then 10%.

B. COMMERCIAL ENERGIES

Contrary to non-commercial sources, commercial energies will experience continuous growth in proportions which are complementary to those cited above: while they supplied 83% of the world's demand in 1960, they will cover 89% in 1978 and will reach 92% in 2000 and 95% in 2020 (I), and 90% in 2000 and 92% in 2020 (II).

As for the quantities involved, they will be very large: 6 GTOE in 1978, and, according to Scenario (I), almost 11 GTOE in 2000 and 17 GTOE in 2020; according to Scenario (II), 9 GTOE in 2000 and almost 13 GTOE in 2020.

However, these quantities correspond to markedly reduced growth rates: 4.5% between 1960 and 1978 and 2.7% and 2.3% before and after 2000 according to Scenario (I). Over the same periods, they will be only 1.8% and

1.7% according to Scenario (II). The decrease is also corroborated when the CEC (Commercial Energy Consumption) is examined on a per capita basis. From 2.5% in the past, growth rates indeed decrease to 1% in (I) between 1978 and 2020, and in (II) down to 0.2% up to 2000 and 0.4% thereafter.

Here again, North and South are very clearly differentiated. In the first place they are differentiated by their rates of growth. Third World commercial consumption had already, in the past, developed at a proportionately faster rate: 4.3% between 1960 and 1978 as against 3.1%. The general downturn does not affect the two areas with the same intensity. If, for the North, the rate falls to 1.25% and then 1.1% in (I), in Scenario (II), the rate stagnates at 0.75% for the length of the period under consideration. By contrast, the South will change its growth rate better: 3.2% and then 2.1% in (I), 1.5% and then 1.4% in (II). The South's rates become twice those of the North. Despite this, and for the same reason as that of the effect of starting levels, the absolute gap measured in TOE widens, even if it decreases a little in relative terms.

In terms of TOE, the North's CEC increased from 2.4 TOE in 1960 to 4.2 TOE in 1978 (or +1.8 TOE). In 2000, it will increase further by 1.3 TOE (I), and 0.7 TOE (II), while between 2000 and 2020, it will undergo a new increase of the same magnitude; +1.3 TOE (I) and + 0.8 TOE (II). At the same time, the South's per capita CEC only increased by +0.2 TOE in the past; it will grow by only a mere +0.3 TOE up to 2000 (I), and by a little over 0.1 TOE (II). Beyond that date, growth will be hardly any greater: +0.35 TOE (I) and +0.15 TOE (II).

From that fact, the absolute North-South gap which was +2.3 TOE in 1960 and +3.85 TOE in 1978 will go up to +4.8 TOE in 2000 (I) and +4.5 TOE in 2000 (II); and +5.8 TOE in 2020 (I) and +5.1 TOE in 2020 (II). From that fact, will the consumer of the South really have a say with regard to commercial sources?

However, the South will be called upon to play an increasingly large role on the world market for commercial energies by virtue of its demographic importance. Indeed, in 1960, its demand constituted only 12% of the market for commercial sources, and in 1978, 18%. But in 2000, it will take up 30% (I) and 25% (II) of this market and 38% (I) and 31% (II) in 2020.

Starting from a billion TOE in 1978, the quantities involved will be between 2.2 and 3.3 GTOE in 2000, and between 4 and 6.5 GTOE in 2020 (a volume, in that zone, equivalent to today's world commercial consumption).

Therein lies the paradox: the Third World becomes ineluctably a giant by the large size of its demand, but each of its inhabitants runs the statistical risk of remaining no less ineluctably an energy dwarf.

2.3 EVOLUTION OF SUPPLIES

The contribution of the different sources of energy to the satisfaction of world demand has evolved a great deal over the course of time, and will go on diversifying in the future. Such is the conclusion to be drawn from an examination of regional consumption patterns, both from the standpoint of the global situation and from a North-South analysis by region.

2.3.1. World equilibrium

(A) If one compares, first of all, the supply patterns of the period 1960-1978 (cf. A5 Tables), a number of findings are inescapable:

(i)   A clear decline of the role played by coal (from 36% to 25%)
(ii) a corresponding growth for oil (29% to 39%)
(iii) the reinforcement of the role played by natural gas (13% to 17%)
(iv) the emergence of nuclear power (0% to 2%)

Will those trends continue in the future? Did the oil shocks of 1973 and 1979 have a determining influence on the long term evolution of supply patterns? (B) It should be noted from the start that the overall situations for 2000 and 2020 only differ from one scenario to the other very marginally. This can be seen in the detailed analysis of these scenarios. The scenarios are more the result of two similar energy variants introduced into a more or less troubled overall context, than the result of the exploration of two radically contrasted futures.

Consequently, it is significant that the part played by *coal* will not only cease to decline in the world total but will instead grow regularly again: 28% in 2000 and 32% in 2020. No-one doubts that the coal alternative is unanimously hoped for as the most desirable substitute for oil because of the size and the range of its resources. Nevertheless, it is necessary to measure the quantities involved: whereas consumption had progressed from 1.2 GTOE to 1.7 GTOE from 1960 to 1978, it is anticipated that between 2.8 and 3.2 GTOE will be consumed in 2000, a time when coal will again be the most used energy

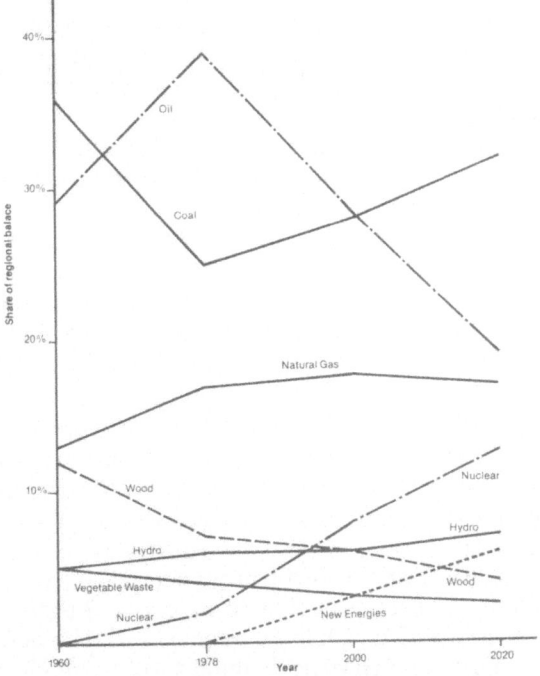

World energy consumption: evolution of supplies

source alongside oil. In 2020, the amounts consumed will be no less than between 4.4 and 5.7 GTOE. As from the turn of the century, coal once more becomes the unrivalled world leader which it was during the 1960s. Is this a caprice of history? Or is it improvidence? Or is it a return to secular tendencies after the historical parenthesis of oil?

(C) The other massive substitute for oil is *nuclear power,* even if its development is currently experiencing some delay and if its contribution remains small compared to that of coal, at least during the period analysed. Its contribution will be 8% of the world balance in 2000, then overtaking hydropower, and it will reach some 12-13% in 2020. The quantities involved are, here again, considerable: 0.8 to 1 GTOE in 2000 (or 3600-4400 TWh as against 650 TWh in 1978); and, in 2020, 1.7 to 2.3 GTOE (or 7400 to 10000 TWh, the former figure representing the global amount of electricity production from all sources in 1978).

Besides these two massive sources, hydropower will be called upon to play a not inconsiderable support role, since it will maintain its share in the context of a world balance undergoing great expansion. From 0.4 GTOE in 1978, its contribution will be 0.6 to 0.7 GTOE in 2000 and 1 to 1.3 GTOE in 2020 (or 4500-6000 TWh), thus representing then a multiplication by 2.3-3.5 of its present-day production capacities.

New energies progressively emerge by a parallel motion. Their contribution, which is insignificant today, becomes significant in 2000 (3% of the balance) and even more so in 2020 (6%), contributing quantities of the order of 300 MTOE in 2000 and of 800-1000 MTOE in 2020 (thus equal to the estimated nuclear production in 2000 and nearly at the level of hydropower in 2020).

As for traditional energies, they will experience a constant decline.

In 1960 and again in 1978, *wood* appeared as the world's 4th supplier after fossil fuels. In 2000, it will be overtaken by nuclear power and will be caught up by hydropower. It will lose its place to new energies in 2020, where it will be confined to a marginal role on the world scene.

Vegetable and *animal waste* will not succeed in leaving the bottom rank they occupied as from 1960, despite rather favourable estimates created by the play of energy equivalences (cf. A5).

Turning finally to hydrocarbons, the considerable development efforts in *natural gas* should, it would seem, at least allow it to maintain in the long-term its market share in the world balance (17%). This statement is made in the knowledge that the consumption of natural gas will grow continually, from 1.2 GTOE in 1978, to 1.8-2.2 GTOE in 2000, and 2.4-3.2 GTOE in 2020 (or a 2-2.5 multiplication of its 1978 contribution).

Consequently, energy saving efforts will primarily concentrate on *oil* under the aegis of industrialised countries. The simultaneous success of all the alternative policies referred to above will succeed in reversing past tendencies and in reducing oil's contribution to the world balance. From 39% in 1978, it will regress to 28-29% in 2000 and to 18-20% in 2020. One will thus witness a quasi-static ceiling for world oil consumption at its present level, at least in Scenario (II): 2.8 GTOE in 2000 as against 2.7 today, with prospects of a decrease to 2.4 GTOE in 2020. In Scenario (I), the action of the other sources will not be sufficient to stifle demand totally (which will still be at 3.4 GTOE in 2000 and 3.6 GTOE in 2020).

However, within the context of this overall world evolution, the situation of

North and South reveal themselves to be very different.

2.3.2. The industrialised nations

The tendencies observed at the global level are even stronger here. The decline of coal is spectacular: from 40% of the balance of these countries in 1960, it falls to 25% in 1978. The rise of oil is no less vigorous: from 33% to 43%. It is accompanied by a sustained rise in natural gas (16–21%), while hydropower maintains itself at 6% and while nuclear power begins to make its mark (3%).

Non-commercial sources nearly disappear (5% in 1960 and 2% only in 1978). They will nevertheless experience a reprieve in that forecasts give them some 3% both in 2000 and in 2020, to the extent that they will again develop from a consumption of 116 MTOE in 1978 (and nearly 130 MTOE in 1960) to about 200 MTOE in 2000 and 2020.

The recovery of *coal* will also be substantial in this zone since its contribution will rise to 31% in 2000 and to 38% in 2020, thus returning for all intents and purposes to its 1960 percentage but contributing far greater quantities: 2.2–2.4 GTOE in 2000 (1.3 in 1978 and 1 GTOE in 1960), 3.4–4.1 GTOE in 2020.

Hydropower will find it difficult to maintain its contribution, which will fall towards 5% in 2020 by reason of the relative scarcity of its reserves. By contrast, nuclear power will makes its presence felt in large proportions since

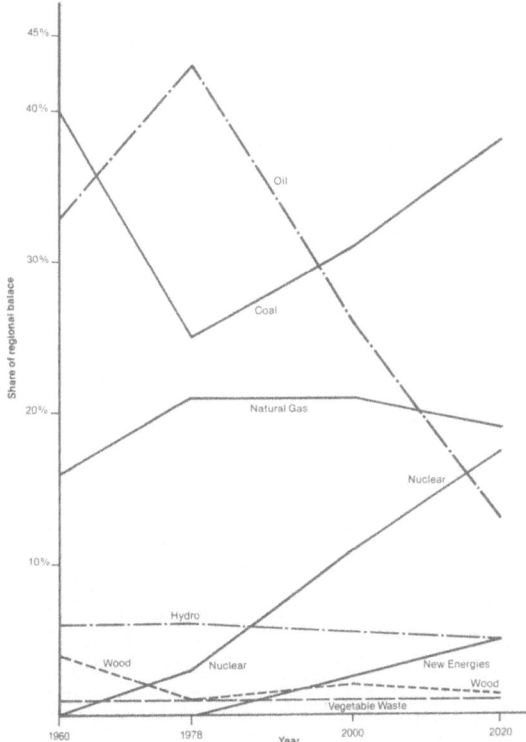

Energy consumption in industrialised countries: evolution of supplies

it will supply 11% of the zone's demand in 2000 (or 0.8–0.9 GTOE) and 17–18% in 2020, (or 1.5–1.9 GTOE) as against 3% in 1978. *New energies* will follow the median curve, arriving to 2–3% in 2000 and 5% in 2020, providing at that later date some 500–600 MTOE.

Despite considerable progress in Eastern Europe, *natural gas* will suffer from a shortage of regional reserves and its growth will be correspondingly hampered. Up to 2000, it could maintain its contribution at 21%, but in 2020 its decline will already have begun (19%).

The drawing together of these various favourable hypotheses will lead to a spectacular decline of oil, a decline which will be, in part, relative: from 43% in 1978 to 26% in 2000 and 13% in 2020. The same is true in terms of absolute quantities since demand will fall from 2.2 GTOE in 1978 to 1.9–2 GTOE in 2000 and 1.2–1.4 GTOE in 2020. Oil will then be overtaken by coal, natural gas and nuclear energy, and will only be in fourth place among the supplies of this region. This is an ambitious forecast, and one which is made with an objective in mind; moreover, it is one which is possible if there occurs a successful combination of economic substitution policies.

2.3.3. The Third World

For its part, the Third World presents a supply picture which is largely different from that of the industrialised countries. This is because non-commercial sources continue to play a major role (37% in 1978), even if less preponderantly than in the past (54% in 1960). Moreover, the commercial consumption picture which is very influenced by the weight of the Centrally Planned Asian Countries (Region 9) confers a much larger role to coal.

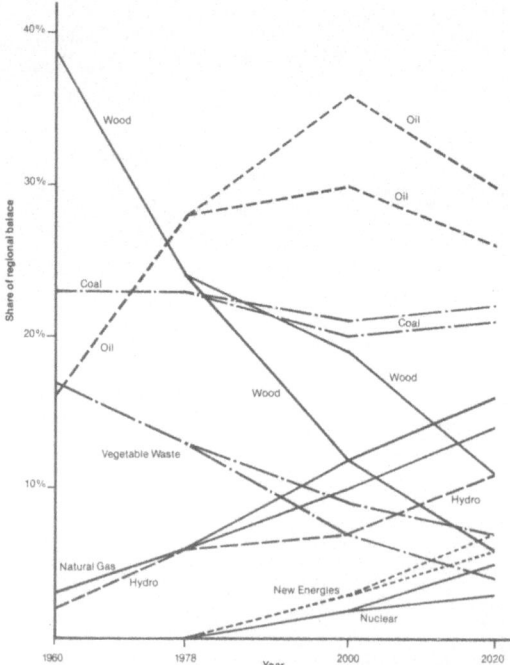

Third World energy consumption: evolution of supplies

Coal has not suffered any appreciable regression and covered 23% of the zone's overall demand in 1978 as in 1960, while the growth of oil, which is even vigorous, is much more limited than in the North in its extent: 16% in 1960, 28% in 1978. Finally, natural gas and hydropower remain marginal, despite a certain progress (from 2–3% in 1960 to 6% in 1978).

Future trends also turn out to be very different from those of the North. The growth of oil cannot be immediately checked, at least before 2000, so large is the unfulfilled demand. Thus oil's contribution continues to grow from 28% in 1978 to 30–36% in 2000 to return to levels close to those of the present day in 2020: 26–30%.

However, the translation of these shares in the terms of quantities is a cause for concern. Contrary to industrialised countries, the quantities involved rapidly grow: from 0.5 GTOE in 1978 to 0.9–1.4 GTOE in 2000 and 1.2–2.2 GTOE in 2020.

The share of the Third World on the global consumer market consequently rises spectacularly. It was only 13% in 1960 and 18% in 1978. In 2000, it rises to 41% according to Scenario (I) and 33% in (II), and in 2020 to 61% in (I), and 50% in (II). Beyond 2000, the Third World becomes a gradually dominant buyer on the world market. This cannot be without strategic consequences.

After a fashion, *coal* maintains itself between 20 and 22% during the entire period, without much difference in size between the Scenarios and without succeeding in gaining ground. Unlike the North, coal is not the substitute energy source of choice for oil.

However, it should be noted that the quantities involved are greatly increased: from 0.4 GTOE in 1978 to 0.6–0.8 GTOE in 2000 and 1–1.6 GTOE in 2020, which is not far (at lease in Scenario I) from contemporary global production.

On the other hand, *natural gas* and *hydropower* will be more and more sought after as substitute agents. Their responsibility for the supply of the Third World increases from 6% at the present day to 10–12% for gas in 2000 and 14–16% in 2020. As for hydropower, after a laborious start (7% in 2000 as against 6% in 1978), it takes off after that date and will provide 11% of supply in 2020, providing annual production quantities of up to 500–800 MTOE or 2300–3700 TWh, then accounting for 50–60% of world hydro capacity.

Nuclear power will not be sought after in the Third World in anything like the quantities of the North. Indeed, its contribution will be limited to 2% in 2000 and to 3–5% in 2020, which confines it to the last rank among supply after vegetable waste. At least until 2020, its development prospects are therefore very limited in the South. The 5% in 2020 (I) nevertheless represents an annual production of nearly 400 MTOE, to 1700 TWh; in 2020 (II), this only rises to 160 MTOE, which, however, remains superior to the total world production of 1978.

Under the impetus of Latin America in particular, *new energies* find a favourable field of application in the Third World and could contribute 3% of demand in 2000, and 6–7% in 2020, for quantities of the order of 300–400 MTOE (thus approaching the quantities involved in the North).

Non-commercial energy sources must be stressed especially here. Although they are destined to yield ground gradually to modern energy sources, they are nevertheless sources of energy on which the Third World will rely over the entire period, and especially according to Scenario (II). *Wood* retreats from 24% in 1978 to 12–19% in 2000 to 6–11% in 2020.

Vegetable waste decreases from 13% to 7-9% in 2000 and 4-7% in 2020.

However, the quantities involved continue to grow: from 0.4 GTOE of wood in 1978 to 0.5-0.6 in 2000, and back to 0.4-0.5 in 2020. An increase of consumption of vegetable waste to about 300 MTOE in 2000-2020 as against a little more than 200 MTOE in 1978 is recorded even. Consequently, the pressure on soil and environment will not easily subside, especially in the zones with high population densities.

It will be noted that the patterns relating to the two Scenarios are much more contrasted than for industrialised countries. The two development trajectories differentiate even more the types of energy resources which are involved. One of the reasons for this is, no doubt, that the range of options remains much wider for a zone which is far from having wholeheartedly entered into the age of the industrial society.

In addition, it will be remembered that if one examines the Third World as a whole, except for Region 9 (essentially by omitting China), the scale of penetration of energy souces remains very near to those recorded above. The only significant difference lies in a much reduced role for coal (8% in 1978 as against 23%) and more uncertain prospects for coal, although these remain steadily growing: 11-12% in 2000 and 14-15% in 2020. This works to the benefit of oil which, in this zone, plays a preponderant role which will only be slowly challenged during this period. In 2020, oil clearly remains the main source of supply, with a contribution of 27-33% as against 37% today. It is followed a long way off by natural gas with 16-17%.

The oil constraint which the Third World (and especially the Third World omitting China) experiences, will thus be overcome only very gradually, despite a diversification of substitution which is apparently better balanced than the North's. Even if this zone can hope to rely, by preference, on the hydrocarbon deposits of the OPEC countries, this diversification appears to be won at the price of a perhaps disproportionate reliance on traditional sources, and in fact seems to be induced by a lack of coal reserves.

Table 4, taken from Annex 11, recapitulates the respective evolution of the shares of the consumption market between the North and the South.

Table 4 EVOLUTION OF NORTH/SOUTH MARKET SHARES

Shares of PEC %	Coal N S	Oil N S	Gas N S	Hydro N S	Nuclear N S	New Energies N S	Wood N S	Waste N S	CEC N S	NCEC N S	PEC N S
1960	86 14	87 13	95 5	89 11	— —	— —	26 74	14 86	88 12	23 77	77 23
1978	77 23	82 18	91 9	75 25	98 2	— —	16 84	15 85	82 18	16 84	76 24
2000 I	74 26	59 41	78 22	59 41	90 10	63 37	22 78	16 84	70 30	20 80	66 34
2000 II	77 23	67 33	83 17	64 36	93 7	63 37	19 81	17 83	75 25	18 82	69 31
2020 I	72 28	39 61	64 36	40 60	83 17	56 44	25 75	21 79	62 38	23 77	60 40
2020 II	77 23	50 50	71 39	49 51	90 10	59 41	21 79	17 83	69 31	20 80	65 35

It will be noted that, for all commercial sources, the Third World is more successful in making its presence felt in the context of the rapid growth Scenario I than in Scenario II, the total gap growing to 5 points in 2020 (40% as against 35% between I and II). A faster growth favours the emergence of the Third World on the international market.

2.4 WORLD ENERGY PRODUCTION

2.4.1. Total Production

Total production differs from consumption within the overall economic context because of variations in the level of stocks. It also differs from consumption in terms of its structure by virtue of the very definition of demand. In the sense used in this study, demand particularly excludes those *bunkers* and *non-energy products*, the raw materials of which (essentially oil) must nevertheless be produced at the primary stage of the transformation cycle.

(A) In our forecasts, we have ignored the movement of stocks. However, production must always exceed consumption, by virtue of intermediary factors (bunkers and non-energy products) which must be integrated with it. This may be observed in the results: the world's production is thus forecasted to grow from 7.2 GTOE in 1978 to 12.4–10.6 GTOE in 2000 and to 18.5–14.4 GTOE in 2020.

However, there is another essential factor which intervenes at this stage, but one which at the same time is a factor causing discrepancy. It is the product of the consultation procedure which was adopted. The regions' forecasts for production and consumption were recorded separately, without further adjustment made to reconcile them together. Everything else being equal, it is thus normal to find a *residual discrepancy* between the forecasting prognoses of *production* levels and those for *consumption*.

For this study to have meaning, the important thing is that this discrepancy should remain unimportant. Happily, this is what may be ascertained.

The 1978 production/consumption discrepancy is c. +5% (or of the order of 360 MTOE), and 95% of this is attributable to oil. Apart from one single miniscule exception (in the localised case of a possible use of Brazilian alcohol in the basic chemistry), it has been postulated that world production would be equal to world consumption for hydropower, nuclear energy, the new energies and the non-commercial sources, without transfer to chemical uses. As a result, the only discrepancies which subsist at a global level, lie in the area of fossil fuels, and these can be explained on the one hand by the existence of bunkers and non-energy uses which are systematically excluded from energy consumption, and, on the other hand, by the failure to reconcile supply and demand forecasts.

(B) The total discrepancies will be c. +5% in 2000 and remain at a level of +3 to +5% in 2020. In relative value, the discrepancy remains at the level recorded in 1978 (+5.3%). In terms of quantities, the discrepancies are more important, stabilising at a level between 500 and 640 MTOE for the entire period.

Included in the total discrepancy, there exists a part which is not accounted for, but which is attributable to the forecasting supply/demand discrepancy and which remains very weak yet positive. This is, however, not so in the case of 2020 (I), where this residual discrepancy is a little higher and negative, revealing a slight shortfall in demand compared to supply (since in order to equal demand, production must be raised to 200 MTOE worldwide).

We will return to this more specific analysis in 2.5.2.

(C) Whereas the uses of coal and natural gas for chemical purposes remain very limited, the promised development of a brilliant future for international

Table 5 WORLD CONSUMPTION DIFFERENTIALS

MTOE	1978	2000		2020	
		I	II	I	II
Total differential, of which bunkers	363	640	507	57	627
+ non-energy users	331	561	452	784	579
(of which oil)	(321)	(519)	(419)	(668)	(508)
Forecasting differential MTOE	32	79	55	− 207	48
(% of production)	(0.4)	(0.6)	(0.5)	(1.1)	(0.3)

air and maritime transport on the one hand, and of petrochemistry on the other, leads *oil production* to a substantially higher level than that created by purely energy-producing consumption.

For this reason, the path of the graph curve for oil production is longer drawn out and, in particular, is higher than the curve representing consumption: 3 GTOE in 1978, but 3.2–3.9 in 2000 and again 3–4.1 GTOE in 2020. The following diagram illustrates the problem. The efforts agreed upon to lower the purely energetic oil curve are not translated into an equally substantial effect on the production curves. In Scenario (I), one should envisage an

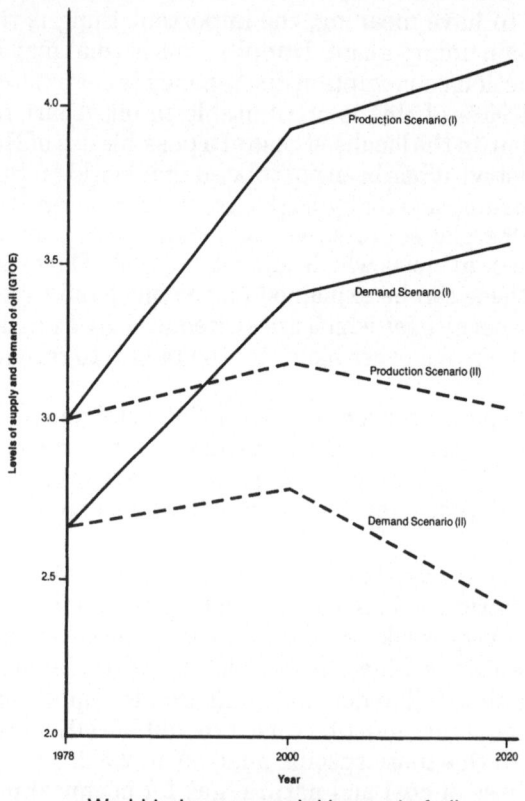

World balance: supply/demand of oil

increase in production of at least 1 billion TOE before 2020 (+1.2 GTOE if one takes account of the global energy production shortfall of 200 MTOE revealed above). Even in Scenario (II) where energy demand drops noticeably after 2000 to go far below the 1978 level in 2020, production at best only succeeds in stabilising itself on a long-term basis at its present level of 3 GTOE. Transposed in terms of total production, the oil problem remains more acute than it appeared on the basis of the forecasts of purely energy consumption alone.

2.4.2. North–South production structures

Beyond the world production structure, an analysis of the respective contributions of North and South reveals significant asymmetries.

A. PRODUCTION SHARES

58% of total production was contributed by the North in 1978 (4.2 GTOE as against 3 GTOE in the South). Despite what was observed with regard to demand, the North will keep its share for the duration of the forecasting period at a level of between 55% and 60% in 2000 and in 2020.

The Third World as a whole will therefore not be able to increase its presence any further on the supply market.

From the start the industrialised nations dominated coal production by an overwhelming majority (77% of the production market in 1978); they do the same for natural gas (87%), hydro (76%) and nuclear power (98%). The area of non-commercial energies above eludes them (16%), as well as oil production, of which 58% is produced by the South.

How will this situation evolve? In the final analysis, it will change very little for *coal*, where the North permanently will hold 73% to 78% of world production for the entire period. In other words, coal development should operate in a highly parallel fashion in the North and in the South. By contrast, the weakening of the North's position is very substantial for *natural gas*, especially before 2000. Indeed, the industrialised countries will only account for 68-75% of production in 2000 and 56-64% in 2020.

Whereas *nuclear power* remains very much the long-term preserve of the industrialised nations (90-93% in 2000, and 83-90% in 2020), they will yield ground as regards *hydropower* on which the Third World is primarily devoting its resources. The North's share falls to 59-64% in 2000, and below the 50% mark in 2020 (40-49%). The long-term production share of new energies tends towards a certain equilibrium; 54-56% for the North in 2020. Of course, traditional sources, even if they are in relative decline, remain connected to the Third World which continues to provide 77-80% of their global production in 2020.

It is interesting to note that the South strengthens its position on the oil supply market. Its contribution (58% in 1978), indeed, rises to 60-65% in 2000, and to 67-72% in 2020.

B. COMPARISON OF PRODUCTION AND CONSUMPTION PATTERNS

The comparison of North and South production and consumption patterns according to energy sources allows the tensions creating currents of trade between the two zones to appear. These physical tensions, resulting from the failure of matching offer and supply within each zone can lie at the origin of political tensions. By definition, they do not exist regionally for the sources of energy where production is apparently equal to consumption (hydro, nuclear, new energies and traditional energies). They can only subsist for fossil fuels.

On the *coal market,* at the level of this very wide North–South survey, such a comparison is of no consequence: the shares of the market, as much from the point of view of supply as that of demand, practically coincide. In 1978, the North produced 77% of world coal and absorbed 77% of it. Even in forecasting, very little trade appears necessary between the two zones: on average, 1–2% of world production, or some 100 MTOE in 2000 and 20 MTOE in 2020 in the North, as against 20 MTOE and 100 MTOE in the South.

As far as *natural gas* is concerned, the situation is already fuller of contrasts. In 1978, the North produced about 87% of world gas and consumed 91% (a deficit of c.30 MTOE). The situation deteriorates in the future: the gap increases to 8–10% in 2000 (130–190 MTOE), and diminishes later on to 7–8% in 2020; but it nevertheless represents at that time a North–South deficit of the order of 160–205 MTOE.

Finally, one knows that the situation as far as *oil* is concerned is much more unbalanced. The North only produced 42% of world production in 1978 and accounts for 82% of consumption (a shortfall of −40%). In the future, the rise of Third World demand tends to reduce this initial asymmetry. Of course, in 2000 the North will only produce 35–40% of world oil, but, at the same time, its consumption is reduced to 59–67%. The deficit is only 24–27%. This trend continues beyond 2000: the North will only satisfy 28–32% of world production then, but only 39–50% of demand; the remaining deficit accounts for −11/−17% therefore.

In terms of quantities, one goes from a 1978 North–South deficit of 1.2 GTOE to one of c. 0.9 GTOE in 2000 and 0.4–0.5 GTOE in 2020. Globally, the North–South oil trading currents therefore tend to retract themselves very much in the long-term. The oil market will therefore become globally less extroverted.

2.5 INTERREGIONAL TRADE AT A GLOBAL LEVEL

(A) It should be recalled at the start that we are only concerned with the analysis of the trade in fossil fuels made between the 10 regions of this study: this excludes intraregional trade between neighbours. The world market is only partially covered by this analysis, mainly in its intercontinental components and long-distance sales.

The overall picture will consequently be more or less complete, depending on the fuel in question. In analysing the 1978 situation (which serves as a point of reference), one indeed discovers that interregional sales represent about 33% of all natural gas sales but 53% of coal and 88% of oil (which is by itself responsible for 85% of the world total fuel trade measured in terms of TOE).*

(B) The forecasting method of choice has enabled each region to be free to determine the level of production on the one hand, and that of consumption on the other, for each of the fossil fuels which it uses. The differences between the two creates an export surplus or an import deficit depending on the case. The ten regional results were then aggregated without iteration to obtain a

*We are here concerned with crude oil, excluding trade in refined products.

picture of world trade. There is therefore no *a priori* reason for the total import demand to coincide exactly with the export supply total for each of the resources studied. It is only to be hoped that the total resulting figure will not be too high, so that it will be possible to consider the adjustment between demand and supply as almost made for each product.

2.5.1. The world market

(A) Let us first turn to the global size of interregional trade. In *1978,* it covered a total of c. 1.5–1.6 GTOE worldwide, the comparison of export and import statistics leaving an unexplained deficit of c.90 MTOE or 5.6% of the total. This world trade was supplied as to 92% by crude oil, as to 5% by coal and as to 3% by natural gas.

(B) According to the scenarios, global interregional trade ("GIRT") will either slightly increase or diminish.

If one bases oneself on import demands which provisionally account for forecasts of consumption, Scenario (I) provides a picture of a slight expansion: from 1.6 GTOE in 1978, the GIRT could rise to 1.85 GTOE in 2000 and exceed 2 GTOE in 2020, or an average rate of increase of 0.6% per annum.

By contrast, Scenario (II) predicts a slow reduction of the GIRT, which will fall back to 1.5 GTOE in 2000 and to 1.35 GTOE in 2020.

We should note in this connection that the GIRT accounted for 23% of world energy demand in 1978. The progressive weakening of oil's role and the substitutions operated either by local sources (not requiring interregional

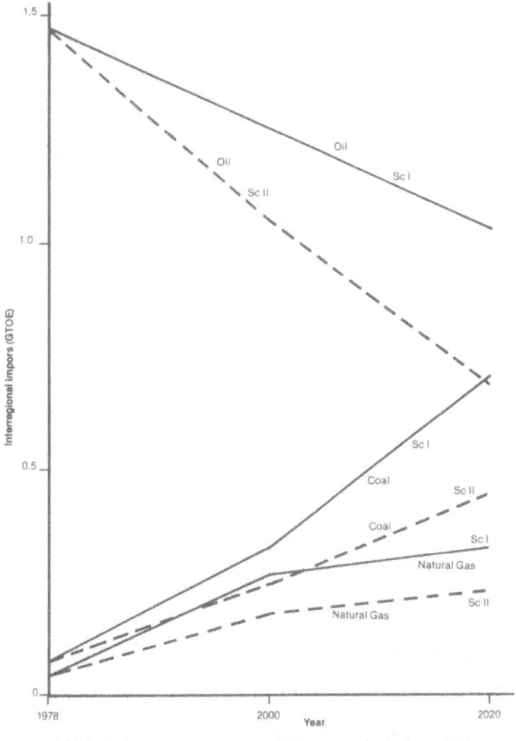

Evolution of world interregional imports

transport) such as nuclear, hydro and new energies, or by resources of a less mobile nature than oil (gas and especially coal), will lead to a reduction of the GIRT. In the future, demand will be fulfilled on the spot to a much greater extent. Indeed, one observes that the role played by the GIRT decreases until it only fulfils 15–16% of world demand in 2000 and 10–11% in 2020.

(C) The difference between the two Scenarios is less marked than one might have anticipated in view of the hypotheses which underlie them. Indeed, Scenario II implies much greater stresses between regions, and, in its economic hypotheses, a certain decrease of international trade. Measured in terms of energy, the results are less explicit — the trade differential of the GIRT compared to the total demand being only 1–1.5% between the two Scenarios.

(D) As for the adjustment of export supply and import demand, this appears to be satisfactory globally. Indeed, one notices that the residual difference (the export–import differential) diminishes in volume compared to its original 1978 level in three of the four configurations studied (except in 2020 according to I). In relative value, it also diminishes very clearly. In 1978, the residual difference represented some 5.6% of total imports; however, in 2000 it will represent 0.6–2.3% according to the scenario, and in 2020, according to (II), it will represent 3.5%. The only noteworthy deficit which is more difficult to reduce concerns 2020 (I): 10%.

However, if it is compared to the total production of fossil fuels of the same year, the percentage of residual differential becomes very small, as the following table demonstrates. And in any case, it always remains equal or greatly inferior to the initial 1978 proportion:

Table 6 WORLD RESIDUAL DIFFERENTIAL:
Global Imbalance of interregional fossil fuel trade

	1978	2000		2020	
		I	II	I	II
Residual differential (E–I) in MTOE	– 89	+ 43	– 9	– 206	– 51
% Total Imports	– 5.6	+ 2.3	– 0.6	– 10.0	– 3.5
% Total Production of fossil fuels	– 1.5	+ 0.5	– 0.1	– 1.5	– 0.5
% Total Energy Production	– 1.2	+ 0.3	– 0.1	– 1.1	– 0.4

One can thus conclude from an examination of this table that the "natural" adjustment between region imports and exports functioned well, even in the absence of any iterative procedure. This is without doubt one of the successes of this study.

2.5.2. The trade of fuels

In this context, how does the trade in the different fuels evolve? As can be ascertained from the following scheme, the evolution differs very much from case to case.

Coal will see its trade increase very rapidly, so necessary is its substitutive actions in enabling us to extricate ourselves from oil. Consequently, the total reliance on imports exceeds 320 MTOE in 2000 (I) (240 in 2000 in II) and 700 MTOE in 2020 (I) (440 in 2020 in II). Coal, which only accounted for 5% of the GIRT in 1978, experiences a spectacular progression: 17% of interregional trade in 2000 and 33% in 2020.

Natural gas also develops its channels of trade in the same proportion as coal up to 2000, but much more moderately so afterwards. This is because local needs in the producing regions will rise to such an extent that they will no longer allow large surpluses to be released for export. The trade in gas thus increases from 50 MTOE in 1978 to 180–270 MTOE in 2000, and then to 270–325 MTOE in 2020. It therefore accounts in 2000 to c. 13% of the GIRT (as against 3% in 1978), and to 16% in 2020.

It is *oil* which suffers from the rapid expansion of its rivals. Interregional trade in oil decreases continually in Scenario I: 1260 MTOE in 2000 (1470 in 1978) and then 1020 MTOE in 2020. In Scenario II, it falls rapidly: 1050 MTOE in 2000 and less than 700 MTOE in 2020. Its contribution, which was overwhelming at the start (92% without taking account of trade in refined products), regresses to 68–71% in 2000 and falls to 50% in 2020, a time when the total trade in gas and coal becomes equal to those in oil.

Chapter III

EVOLUTION OF REGIONAL ENERGY BALANCES

3.1 NORTH AMERICA

North America comprises the USA (including Puerto Rico), the world's greatest economic power, and Canada. It is at the centre of global strategies and stakes in the field of energy. It has been the main driving force of the modern industrial and consumer society. More than any other region, it has contributed to propagate this model all over the world, irrespective of any frontiers resulting from different development levels and even of political ideologies.

Consequently, it is of the utmost importance to follow closely the evolution of its energy system which cannot fail to continue to influence strongly the world energy balance.

3.1.1. Demographic and economic environment

(A) In 1978 the population of North America was 245 million.* Even if its population growth will experience the sharp slow-down common to the whole of the industrialised countries, it should, nevertheless, be comparatively more sustained than that of the other OECD countries: 1.1% from 1960 to 1978, 0.9% by 2000, and 0.8% beyond. Thus its population will grow by about 100 million by 2020.

(B) North America's economic growth, as measured by its GNP, was 4% per annum in the past, increasing its *total GNP* from nearly G$1200 in 1960 to more than G$2350 in 1978, in real terms.

Growth prospects are affected by the slow-down observed over the last few years and the structural changes induced by the successive oil shocks. According to Scenario (I), the growth rate will be 2.5% p.a. up to 2000, and 2% beyond. According to Scenario (II), which is more restrictive, it will be 2% up to 2000, and 1.5% beyond.

This would nevertheless bring its total GNP to respective levels of G$5000 and G$6000 in 2020 (as opposed to G$2350 in 1978).

*A recent re-evaluation made as a result of the 1980 US census increases this figure to 249 million, or +4 million. The RWT considers that one should retain the total estimates for GNP and PEC, since the corresponding downward adjustment would only affect the per capita GNP and PEC levels. The adoption of this new demographic basis would result in a population increase for R1 in 2020 of the order of +12 million.

(C) This slow-down in total terms, combined with population forecasts, is immediately reflected in the per capita value. North America kept the first place among the ten regions in 1978 ($9580), closely threatened, nevertheless, by the growth of the Industrialised Countries of the Pacific (R3).

North America's per capita income will be $12300–$13700, according to the scenario in 2000, and $14200-$17400 in 2020. At the end of the forecasting period, in the most favourable case, the region will have practically doubled its economic level between 1978 and 2020.

(D) It should be noted that North America, which in 1978 as well as 1960 held second place among the regions according to total GNP (after Western Europe), will continue to be the second region over the whole period, even if its relative weight on a world-wide scale of wealth will gradually decrease. It produced 31% of the total world GNP in 1960, and only 25% in 1978. According to Scenario (II), it will hold its own better: 23% in 2000, and 21% in 2020. In Scenario (I), its relative position will deteriorate more clearly: 21% as from 2000, and 17% by 2020.

3.1.2. Energy consumption

A. TOTAL CONSUMPTION

(A) Concurrently with its economic strength, North America enjoyed an even stronger primacy in the field of energy consumption; it represented, by itself, 35% of world energy demand in 1960, and 30% in 1978, and this for a population representing slightly less than 6% of the world's total population. More than 2 billion TOE are consumed by North America.

(B) The projections of the growth of North America's total consumption are lower than past trends, especially up to the year 2000: 1.2% in Scenario (I) and 0.8% in Scenario (II), compared to 3.1% from 1960–1978. If one measures the corresponding income elasticity in its most current definition (the ratio between average annual growth rates of commercial energy consumption and total GNP), one observes a very sharp drop from 0.80 for 1960–1978 to 0.44 (Scenario I) and 0.35 (Scenario II). These projections underline the size of structural adjustments of demand anticipated in North America.

It is hoped that the draconian constraints prevailing up to 2000 will be slightly less severe after that date, once the essential adjustments in the field of energy have been made. Consequently, total growth rates could again become more flexible after 2000: 1.5% in (I), and 1.2% in (II). One would then witness an increase in total elasticities in an economic context of continuous slower growth; these would then be at 0.75 (I), and 0.80 (II), the same levels as during the reference period (cf. Robert Lattes[35]).

(C) The quantities involved, therefore, would grow moderately: 2650 MTOE (I) and 2440 MTOE (II) in 2000. Nevertheless, 3 GTOE (3080 MTOE) will be reached by 2020 (II), and 3.5 GTOE (3575 MTOE) in 2020 in (I). As far as total consumption is concerned, North America will then be followed very closely by Eastern Europe.

North America's energy future will be marked by the contrast between the great constraints prevailing up to 2000 and greater flexibility after this date. This is illustrated by its lesser relative weight in the world market. The essential part of this decrease will be finished by 2000: from 30% in 1978 to

22–24% in 2000; during the next twenty years it will decrease only slightly, to reach a level of 20–22% in 2020.

B. ENERGY CONSUMPTION PER CAPITA

North America has, by far, the largest per capita energy consumption in the world: on average, it is 8.3 TOE per annum in 1978, while that of its nearest rival, Eastern Europe, is still only 3.8 TOE. This difference of +4.5 TOE is considerable. It has, in fact, grown over the past, since it was only +4 TOE in 1960.

The growth rate of North America's per capita energy demand was higher than those of three other regions in the past (+2% per year); in the future it will be the lowest of all regions, according to both scenarios and to both forecasting horizons. The combined impact of slower population growth with the extent of the gradual slow-down of energy demand will lead to minimal per capita rates: 0.3% in (I) by 2000, and even negative (−0.1%) according to (II). After 2000, growth rates will be slightly higher: 0.7% in (I), and 0.3% in (II). Nevertheless, according to Scenario (II), total per capita energy consumption will remain practically unchanged: 8.3 TOE in 1978, 8.2 TOE in 2000, and 8.8 TOE in 2020. According to Scenario (I), its growth will be very limited: 9 TOE by 2000, and 10.2 TOE by 2020. Certainly, North America will still be the world's largest energy consumer in per capita terms by 2020, but its "lead" over Eastern Europe, the next largest consumer, will have been reduced by nearly 50% in relation to 1978: 2.4 TOE to 2.8 TOE by 2020.

3.1.3. Supply pattern

North America's supply pattern changed very little between 1960 and 1978, but it will be radically changed over the whole forecasting period.
(A) From 1960–1978, the relative importance of coal and gas decreased slightly (−3% for both sources), while that of oil increased by 3%. In addition, nuclear energy's part rose from 0% in 1960 to 4% in 1978. These few data give a good idea of the small importance of the changes in North America's past supply pattern.

In 1978, North America was the world's largest consumer of oil and natural gas, as well as hydraulic and nuclear power. It was second only for coal (after Eastern Europe) and for new energies (after Latin America).

North America's impact on world demand of all commercial energy sources, without any exception, is overwhelming, but it depends for 45% of its total consumption on oil, and for 26% on gas; thus more than 70% of its energy needs are covered by hydrocarbons alone, or about 1440 MTOE (in other words 38% of the world demand of hydrocarbons).
(B) This situation will change radically by 2000. The share of coal will leap to 33–34% (as against 18% in 1978); that of nuclear energy will increase from 4% to 11%, and new energies will represent 4%. Finally, the share of natural gas will be reduced from 26% to 20%, and that of oil will drop to 21% (compared to 45% in 1978). These are indeed very deep changes in North America's supply pattern, and this development will continue on an even greater scale between 2000 and 2020.
(C) In 2020, coal will be called upon to cover practically one half (48–49%) of the region's total energy demand, as it did more or less in the 1940s. New energies will represent nearly 10%, while the growth of nuclear energy will

slow down considerably, representing only 13%. The share of natural gas will continue to decrease to 13%. Finally, oil will have become a marginal source of energy, with a share of only 7%. Oil will be only the region's fourth largest energy source, its share being about the equivalent of that of hydropower.

The above figures show clearly that the substitution by coal is vital. It should be noted that, at that period, it will no longer be only a matter of coal uniquely consumed in its solid form: a part of coal's contribution to demand (which is not ascertained here because this study is solely concerned with primary energy) will be converted into gaseous and liquid vectors, which are much more efficient competitors of hydrocarbons.

In terms of primary energy, *coal* consumption will thus increase from 360 MTOE in 1978 to 800-900 MTOE by 2000, and will reach 1500-1800 MTOE by 2020; that is, in the high Scenario (I), the equivalent of the present level of world production putting all the regions together. This gives an idea of the size of the task that has to be accomplished.

This development will make it possible to reduce slowly *natural gas* demand to a volume slightly lower than its 1978 level (528 MTOE) by 2000, and to the 400-460 MTOE range by 2020.

The growth of *nuclear energy* will be considerable: from 77 MTOE in 1978 to 270-290 MTOE by 2000. The nuclear option in 2020 will be more flexible: 400-480 MTOE, or the equivalent of 1800-2200 TWh.

New energies will represent 100 MTOE by 2000, and around 300 MTOE by 2020. Even the consumption of *non-commercial energy* sources will find a new lease of life and their contribution will increase threefold by 2020, to reach more or less 100 MTOE (that is more than present production of nuclear energy). In particular, it should be noted that North America, by one of those odd reversals of fortune, will then be the third or fourth consumer of fire-wood, representing 13-17% of the world market.

Nevertheless, it is in the field of *oil* that the reversal of fortune is most remarkable: its consumption will decrease from more than 900 MTOE in 1978 to 510-550 MTOE by 2000, and to 210-230 MTOE by 2020. About 75 years later, therefore, oil consumption will be back to its pre-1945 level in the context of a five times increased energy consumption. This is a spectacular reversal, since in 1978 oil contributed 45% of energy demand, while in 2000 it will only contribute 21% and, in 2020, 7%.

This is indeed a very ambitious oil substitution policy, and its success is far from certain. Will coal consumption really amount by 2000 to 800-900 MTOE, or the equivalent of 1.2 to 1.35 billion metric tons of hard coal? And what about the forecasts of nuclear production (1200-1300 TWh), rather jeopardized by numerous delays in the programmes? Will it really be possible to develop new energies up to the threshold of 100 MTOE? If total demand will develop according to forecasts, any deviation from these supply projections will have to be compensated by hydrocarbons and this will slow down their substitution proportionally.

In the year 2020, if the supply pattern has changed according to these projections, North America will become, by far, the world's largest coal consumer, representing one third of world demand. It would also represent one third of the world consumption of new energies. It would be in second place for hydropower and natural gas, as well as number three for nuclear energy (while it will still be number one in 2000).

Finally, North America will be only the world fifth (Scenario II) or even seventh oil consumer (Scenario I). Is this really compatible with the economic development which North America will have to sustain? At any rate, this is a challenge fit for North America.

3.1.4. Energy production

North America, the world's largest energy consumer, is also its first energy producer. It contributed 24% of world energy supply in 1978, when it represented 30% of world demand.

(A) The region's production pattern is more balanced than that of its consumptions. In fact, oil and natural gas represent one third each, compared to 45% for oil and 26% for gas in consumptions, while the share of coal is nearly 25% for production and 18% for consumption.

However, the strong growth of coal demand will change this picture. In 2000 37–38% of total energy production will be based on coal and nearly 50% in 2020. On the other hand, oil and gas production will decrease at similar rates and each contribute only 18–22% to total energy production in 2000, and only 12–13% in 2020. Hydropower will experience difficulties in maintaining its level and its share will decrease gradually from 8% in 1978 to 7% in 2000, and 5% in 2020. Nuclear power and new energies (together with coal) will progressively take over from hydrocarbons. Nuclear power's share will be 10% in 2000, and 11–12% in 2020, and that of new energies will be 3–4% and 8%.

(B) This change in production pattern will occur in the context of a total energy production which will experience a vigorous development under the impulse of the coal sector. The region will thus be able to maintain its position as the world's first energy producer over the whole forecasting period.

North America's total energy production will increase from about 1.7 GTOE in 1978 to 2.6–2.8 GTOE by 2000, and to 3.5–4.0 GTOE by 2020. It should be noted that coal's contribution alone will represent 70% of the total growth of production between 1978 and 2020, compared to 32–35% for the whole of nuclear power and new energies.

This is due to the fact that, at the same time, the level of the production of hydrocarbons will, at best, be more or less maintained up to 2000 and, whatever the circumstances, decrease after that date, to decline from 1.1 GTOE in 1978 to only 0.8–0.9 in 2020. This ineluctable downward trend of production levels is inevitable because of the relatively low level of the region's reserves, despite the great exploration efforts that are being, and which will be, made to attempt to alleviate this situation.

(C) In 1978, North America was the world's first and second producer of gas and oil, respectively, but it will be only number three for both energy sources by 2020 in Scenario I and number two in Scenario II.

Whereas North America will replace Eastern Europe as the world's first coal producer as from 2000, it will be only number three for nuclear power (whereas it was still number one in 2000) and number two for hydropower.

The size of coal development (which will occur at an average rate of 3.5–4.0% p.a. between 1978 and 2020, resulting in a fourfold or fivefold multiplication of present day production) conceals to a certain extent the general decrease of the region's relative importance in the production of all other energy sources (except for new energies whose role, however, will still be rather marginal).

3.1.5. Interregional energy trade

(A) North America, the world's first producer and consumer of energy, plays a very assymmetrical role in world energy trade. In fact, its production, which is considerable, is almost entirely absorbed by its domestic market, only a very small part of it being available for export at present (2% of world-wide interregional exports). By contrast, this production does not cover domestic needs, and in 1978 North America was the world's second importer of fossil fuels (22% of interregional imports) after Western Europe. Its total imports amounted to about 350 MTOE, practically all in the form of crude oil. It should be noted that if one were to add its imports of refined products, which are very important in this region, total fuel imports would have been of the order of 390 MTOE in 1978.* North America's only significant energy exports are based on coal (27 MTOE in 1978). This means that the total balance of its international energy trade was rather negative in 1978: − 320 MTOE (and even − 360 MTOE if refined products are included). It is well known that this balance has become much more negative in the past few years since this region's oil needs have been covered less by domestic production. Consequently, it has been forced (or has had to accept) to rely more on imports.

(B) How will this situation develop in the future? Projections for the total trade of fossil fuels are eloquent: in 2000, the negative trade balance in the field of energy will be virtually eliminated; in 2020, the situation will have become clearly one of surplus, with exports of about 135 MTOE. Two factors explain this change for the better: on the one hand, there is the increase of coal-based exports and, on the other hand, the decrease of oil imports.

In fact, coal exports will rise from 27 MTOE in 1978 to 160–180 MTOE during the period 2000–2020. The greater part of this increase will take place by 2000. Concurrently, after a slight transitory increase up to 2000, natural gas imports will be eliminated in the long run, because the region's consumption will adjust to falling production of natural gas (with the precious help of synthetic gas from coal). In addition, oil imports will be strongly reduced from 343 MTOE of crude in 1978 to 120–160 MTOE by 2000, and to only 40–50 MTOE by 2020.

On the one hand, the success of this strategy will make it possible for North America to become the world's third (Scenario II) or fifth (Scenario I) energy exporter by 2020. On the other hand, the region will be only the sixth (Scenario II) or the seventh (Scenario I) energy importer by that time. This would considerably relieve the tensions created by North America's imports in the international oil market, because, while the region represented 23% of world crude oil imports in 1978, in 2020 it will only account for 5% at a time when supply tensions will be inevitably very great owing to the gradual depletion of the most favourable reserves. North America will also no longer be an importer of natural gas. But its share in the ever more demanding world coal market will continue to be close to its present level (38%): 42% in 2000 and 32% in 2020 under Scenario I. It will be even higher under Scenario II: 51% in 2000, and 45% in 2020. In 2000, in particular, this means that North America could represent almost one half of the interregional coal trade. This is enough to clearly show the decisive part played by the region's exports at that time.

*NB. Only interregional imports are included here.

The success of this overall strategy for the reduction of the part played by oil and for a large-scale expansion of coal (accompanied by the sustained development of nuclear power and new energies), can only be greeted on the international energy scene with relief and awaited with hope.

However, more than others, this series of optimistic forecasts should, no doubt, be understood as mere possibilities. This is all the more so because the present overall economic context, marked as it is by a reduction in the prices of oil, does not encourage the continuity of austerity measures and of active oil substitution in a zone which is traditionally versatile in energy matters and which is very sensitive to the slightest market movements.

3.2 WESTERN EUROPE

3.2.1. Demographic and economic evironment

(A) As defined extensively for the purpose of this study, that is including, in particular, Yugoslavia, Turkey, and Israel, Western Europe has the highest population among the industrial regions, but it is also the region with the lowest population growth rates. Its rate between 1960 and 1978 (less than 0.9% per year) was the lowest in the world. This situation, marked by a resulting increase in the elderly part of the population, will remain critical in the future, since population growth rates will decrease to 0.5% before 2000, and to 0.3% beyond. Western Europe's population will not reach 500 million in 2020 (from 417 million in 1978), despite the fact that the area contains nations with a growth rate that is still relatively high (such as Spain, Yugoslavia and Turkey).

(B) Western Europe is one of the richest regions in economic terms. The volume of its total GNP was the world's highest in 1978 (2600 G$), higher than that of North America. These two regions have very close ties, experiencing a parallel average economic growth between 1960 and 1978: +4% per annum. Their future economic growth prospects are also very similar: 3%* by 2000, and 2% by 2020, under Scenario (I), and 2% and 1.5% under Scenario (II). Under these conditions, Western Europe's total GNP would be G$4000 (II) or G$5000 (I) by 2000, and G$5500 (II) or G$7500(I) by 2020. This means that the volume of the region's total GNP will remain the highest in the world, in spite of the marked decrease of growth rates in the future. Nevertheless, the region's share in the total world GNP will decrease from 31% in 1978 to 26% in 2000, and 22% in 2020. The world-wide economic slowdown will have a simultaneous impact on all regions, even if it will be different from one region to another.

(C) Western Europe's GNP per capita was also the third highest in the world in 1978 ($6200), after that of North America and that of the Industrialised Countries of the Pacific. Because of the mechanics of divergent economic growth rates, the volume of Western Europe's per-capita GNP will grow slightly faster than that of the North America, but it seems that it will grow slower than that of the Industrialised Countries of the Pacific. As far as the

---

*In the light of current trends, the RWT has recently admitted that this rate was no doubt high. But it did not express the same opinion with regard to future energy consumption trends, which are in any case very restrained.

growth rate of the region's per-capita GNP is concerned, the region will be number six over the whole forecasting period, just as in the past.

Under these conditions, Western Europe's GNP per capita would be $8800–10900 in 2000 and $11300–15400 in 2020, or from 2–2½ times more wealth than in 1978.

3.2.2. Energy consumption

A. TOTAL CONSUMPTION

(i)   Western Europe's total energy consumption was more than 1.2 GTOE in 1978, thus making the region the world's third energy consumer after North America and Eastern Europe (which became number two as from the late 1960s). Its annual growth rate of consumption was only 3.6% between 1960 and 1978, as against 5.1% for Eastern Europe. The region's share in world energy demand was 18% in 1978, for a population representing only 10% of world population and a GNP representing 31% of world GNP.

(ii) Western Europe's total energy consumption will experience the effect of gloomy economic prospects. Its growth rates will continuously decrease: 1.9–1.4% up to 2000, and 1.3–1.1% up to 2020. Independently of the ups and downs of the economic situation, the region's energy demand growth rate will regularly decrease over the whole forecasting period.

   If one measures the elasticity of energy development in terms of commercial energy compared to economic development, one finds that the income elasticity, which was 0.95 from 1960 to 1978, will experience a marked decrease, as in North America, from 0.63–0.70 by 2000 (Scenarios I & II), and 0.70–0.73 beyond. The essence of the structural reorganisation of consumptions will have occurred before the end of the century.

(iii) The volume of Western Europe's total consumption will increase regularly: 1.7–1.9 GTOE in 2000 and 2.1–2.4 GTOE in 2020. The region's share in world consumption will decrease gradually: from 18% in 1978 to 16–17% in 2000, and 14–15% in 2020.

B. CONSUMPTION PER CAPITA

Among the industrialised countries, Western Europe's energy consumption is the most "modest": less than 3 TOE per annum in 1978 (compared to an average of 4.3 TOE for the whole of the industrialised countries). This should not change in the future, because the growth rates of the region's per capita consumption will be lower than those of the other industrialised regions, with the exception of North America: 1.5–1.1% up to 2000 and 0.8–1.1% beyond, as it was between 1960 and 1978: 2.7% for Western Europe versus 5.8% for R3 (Industrialised Countries of the Pacific) and 4.1% for R4 (Eastern Europe). In other words, the growth rate of Western Europe's consumption per capita will be lower than the average of the industrialised regions by over half a point by 2000, and by a few tenths of a point beyond. This will be sufficient to increase the relative consumption gap. The region's per capita consumption was practically equal to that of R3 (Industrialised Countries of the Pacific) in 1978 and it was almost one TOE lower than that of Eastern Europe. Western Europe's average consumption will be near to 4 TOE in

2000 and will reach 4.5 TOE in 2020, but it will be lower by 1–2.4 TOE in Scenario (I) and by 0.60–1.7 TOE in Scenario (II) from Eastern Europe's one. Nevertheless, the gap will remain large (+2.3 TOE) with Latin America, the leading region of the Third World.

3.2.3. Supply pattern

How will this energy demand be supplied?

(i) Oil's share in Western Europe total supplies increased considerably between 1960 and 1978: from 28% to 51%. The region absorbed practically one quarter of world oil supplies in 1978. The ancillary share of gas in the region's total supplies increased during the same time from 2% to 14%. This penetration of oil and gas took place exclusively at the expense of coal, whose share in total regional supplies dropped by 34 points: from 55% to 21% in 1978. Nuclear power represented 4% of the region's 1978 energy balance, while hydropower continued to play a far from negligible part with a share of 8%. Contrary to North America, the reorganisation of Western Europe's supply pattern was very radical between 1960 and 1978: whereas hydrocarbons at present cover two thirds of total needs (800 MTOE), compared to 30% in 1960.

Western Europe was the world's second consumer of oil, hydropower, and nuclear energy in 1978, while it was its third gas and fourth coal consumer. But it should be recalled that it was the world's second coal consumer in 1960.

It is also worth noting that the region was the world's fourth consumer of vegetable waste, and that it will consolidate this position up to 2020.

(ii) Up to 2000, this pattern will again change, and Western Europe will attempt to progressively free itself from its oil dependence. This is apparent from the data available: in 1978, oil covered 51% of energy demand; in 2000, this will decrease to 34–35% and the trend continues in 2020, to fall down to 18–20%. The share of natural gas in the region's total supplies will hardly change up to 2000, but it will be gradually reduced to 11–12% by 2020. This change in the course of the region's energy policy was accelerated by the realization during the 1970s of the great danger that Western Europe's oil dependence represents. The success of this policy depends entirely on the success of simultaneous large-scale substitutions of oil by coal and, above all, nuclear power, accompanied by the general expansion of new energies; these will only represent 3% of total supplies in 2000, and 5% in 2020.

(iii) The share of *coal* in the region's total supplies, which had decreased spectacularly since 1960, should increase again to 25% in 2000, and 29–30% in 2020. This increase may seem rather small in relative terms, but it is much more impressive when expressed in volume: from 260 MTOE in 1978, the consumption of coal will reach 420–470 MTOE in 2000, thus growing at an annual rate of +2/+3 up to 2000, while it had decreased at an annual rate of about −2% between 1960 and 1978. The growth of the region's coal consumption should be sustained at an annual rate of 2% between 2000 and 2020, when its volume could be of the order of 600–700 MTOE. It is thus expected that the region's coal consumption will be multiplied by two and a half by 2020, which would then make coal Western Europe's first energy source. On the world market, however, it only just maintained its share: 13-14% of demand compared to 15% in 1978.

(iv) The other decisive component of Western Europe's oil substitution strategy depends on the success of the expansion of *nuclear power.* Its share in regional supplies will rise from 4% in 1978 to 12-13% in 2000, and, without experiencing any downturn thereafter as in North America, it will reach 25% in 2020, when it will represent the region's second supply source, only just behind coal.

It should also be noted that, compared with other regions, Western Europe will thus become the region with the highest relative penetration by nuclear energy, and, in 2020, the foremost nuclear energy producer.

The growth rates projected for nuclear power seem quite realistic: 8% per year up to 2000, and 5% per year after that date. But the realization of these growth objectives will require a very serious effort. The volume of Western Europe's nuclear power production will make available 210-240 MTOE in 2000 (compared to 40 in 1978) and 500-600 MTOE in 2020, that is the equivalent of an annual production of 2300-2800 TWh. These orders of magnitude give an idea of what is at stake, but they also give an idea of the degree of uncertainty which is inevitably implied in the region's nuclear policy. The most direct result of the success of this policy would be the stabilization of Western Europe's oil consumption at a level of 630 MTOE by 2000, that is, the 1978 level, and its gradual reduction to 400-500 MTOE by 2020. Under these conditions, Western Europe would gradually reduce its share in world oil demand to 12-17%, versus 24% in 1978, thus relieving the oil market, even if not as massively as in the case of North America.

3.2.4. Energy production

Because of its small natural resources, Western Europe has always been an energy producer disposing of relatively limited means, except when coal was the predominant energy source and during the time of the first industrial development.

(A) It is only in the field of primary electricity (hydro and nuclear power) that Western Europe was the world's second producer in 1978, but the contribution of these sources to world energy supplies was rather marginal, with a total share of 7% only for both in 1978. As far as fossil fuels are concerned, Western Europe is the world's third producer of gas, but it is only number four for coal, and number eight for oil. Its share in world energy consumption was 18% in 1978, making it the world's third consumer. Its share in world energy production was only 9%, that is, it was the world's fifth producer in that year. Thus, Western Europe's total needs were covered at a rate of 50% by imports. This very high energy deficit will hardly be reduced in the future, at least not by any increase in production. In fact, the region's share in world energy production will still be at its present level of 8-10% by 2000, as well as by 2020. Fortunately, its share in world energy consumption will be reduced from 18% in 1978 to 14-15% by 2020 and the region's energy independence will therefore improve in relative terms.

The volume of Western Europe's energy production will exceed one billion TOE by 2000, versus 630 MTOE in 1978. It will further increase to around one billion and a half TOE by 2020. Nevertheless, nuclear power excluded, its share in world energy production will decrease to 5-6% by 2020.

(B) Western Europe's present energy production pattern is dominated by coal (35%) and gas (24%), while the share of oil is only about equal to that of hydropower (15%). Nuclear power and non-commercial energies represent the rest, with similar shares.

(C) *Nuclear power* will represent Western Europe's main energy production in the long run, because it is the only source that will permit the region to maintain an acceptable energy production level. It will be the second regional supply source as from 2000 and gradually replace coal as number one after that date, before consolidating definitely its position as number one by 2020, when it will represent 36-40% of regional supplies. This large-scale penetration will make Western Europe the world's first producer of nuclear power by 2020. The volume of the growth of the region's nuclear production will be equal to that of North America or Eastern Europe by 2000, but becomes prominent between 2000 and 2020. In fact, Western Europe has hardly any alternative solution if it wants to limit its dependence on energy imports.

The shares of most other energy sources (except, of course, new energies) in Western Europe's total production will decrease regularly. Oil's share will remain at its present level (15%) up to 2000, before being reduced gradually to 7-9% between 2000 and 2020. It is worth noting that non-commercial energies will also maintain their present share up to 2000.

Coal's share in Western Europe's production will decrease gradually: from 35% in 1978 to 28-29% by 2000, and 23-24% by 2020. Because of the faster depletion of regional reserves, the share of natural gas in regional production will decrease in a more marked way: from 24% in 1978 to 11-13% by 2000, and 7% by 2020. Hydropower's share will also decline to 11-12% by 2020, versus 15% in 1978. But it should be pointed out that the volume of new energies (of the order of 100 MTOE) will furnish a contribution more or less equivalent to that of oil or gas by 2020. Therefore, the combined volume of nuclear power and new energies alone could very well represent one half of Western Europe's energy supplies by that time.

As in North America, non-commercial sources will probably remain part of Western Europe's supplies for a longer time than it was formerly thought. Their production could even double by 2020. In this case, they would stabilize their share in the region's energy production.

3.2.5. Interregional energy trade

(A) Western Europe's energy demand has been higher than its energy production for a very long time. As a matter of fact, its energy dependence had become even worse between 1960 and 1978, making it by far the world's first energy importer, with a share of 45% of interregional imports in 1978, while it was only the sixth exporter, with a share of 3%. This very pronounced energy dependence resulted in net imports representing 660 MTOE of fossil fuels. Oil alone made up 90% of this deficit. The question is if there is any chance that this situation will improve in the future.

(B) Unfortunately, forecasts at the horizons of 2000 and 2020 seem to exclude this possibility. On the contrary, Western Europe's energy trade deficit will become only worse, especially under Scenario (I): −860 MTOE by 2000, and

−950 MTOE by 2020. Prospects are slightly better under Scenario (II): −690 MTOE by 2000, and −720 MTOE by 2020.

But Western Europe's share in world-wide interregional imports will be 49% by 2000, and 47% (I) or 54% (II) by 2020. The only favourable aspect of this development is that the share of these net imports in regional demand will decrease from 54% in 1978 to 46% (I) or 41% (II) by 2000, and 39% (I) or 34% (II) by 2020. This means that the expansion of nuclear power will somewhat reduce the region's import dependence.

(C) Nevertheless, there will be very significant changes in the region's energy import pattern, because oil's share will be gradually reduced in favour of coal and gas. In fact, the region's import dependence, which used to be almost exclusively focused on oil, will become more diversified. Coal's share in Western Europe's energy imports will increase to 19% (I) or 17% (II) by 2000, and 38% (I) or 33% (II) by 2020 (compared to only 6% in 1978), while gas will represent 18% (I) or 13% (II) by 2000, and 21% (I) or 18% (II) by 2020 (compared to 3% in 1978). Oil's share will then be less than 50%, that is 41% under Scenario (I), and 47% under Scenario (II). This improved diversification will enable the volume of crude imports to decrease to 510 MTOE (II) or 565 MTOE (I) by 2000, and 350 MTOE (II) or 400 MTOE (I) by 2020, versus 656 MTOE in 1978. Inevitably, the volume of coal and gas imports will increase to 260–370 MTOE and 130–200 MTOE, respectively. This will, of course, create new tensions in the total world market of fossil fuels, because Western Europe's share in world coal and gas imports will be 50–60% and 55–60%, respectively, even if it will have arrived at stabilizing its share in world oil imports at a level of 40–50% (versus 44% in 1978), over the whole forecasting period.

(D) In reality, this policy will lead Western Europe to change its absolute import dependence on one source, that is oil, for a very pronounced import dependence on all fossil fuels, and that without changing its relative pressure on the three world markets involved. The simple statement of the above facts and figures is enough to underline the precariousness of the region's supply situation. This precariousness is already a matter of grave concern at present, but it will be worse in the future, because any delay or reduction in the expansion of nuclear power will immediately aggravate Western Europe's import dependence in the field of energy. This is no doubt the weakest point of this policy, especially as there is little room to manoeuvre, even under the best of circumstances. In addition, there is hardly any realistic and feasible alternative to nuclear power, except if Western Europe will be willing — and able — to impose even more constraints on the growth of its future energy demand.

3.3   INDUSTRIALISED COUNTRIES OF THE PACIFIC

3.3.1. Demographic and economic environment

The region "Industrialised countries of the Pacific" comprises only three countries: Japan, Australia, and New Zealand. It is mainly because of the high levels of their development, as well as their geographical isolation in relation to other industrialised countries of the West, that these nations are traditionally grouped together in studies made on the world-wide scale.

(A) This region's population is smaller than that of any of the other ten regions: 132 million, or about 3% of the world's population (1978). It seems that the growth prospects of its total population are very limited up to 2000, and almost zero after that date. In other words, this region will know little development in demographic terms, with all the sociological and economic consequences that this implies.

(B) Measured by its GNP per capita (nearly $7700 in 1978), this region is second only to North America, but its economic growth prospects are more sustained than those of the other industrial regions of the West.

Its average economic growth rate, stimulated by Japan's development, has been the highest in the world from 1960 to 1978 (7.5%), even higher than that of the Middle East. In fact, in per-capita terms, this region first caught up with, and then largely outstripped, Western Europe during that period.

(C) It is likely that its future growth will be faster than that of the other regions of the West. In this case, it will have outstripped North America by 2000 and have the highest GNP per capita of the ten regions by 2000: $13000 (II) and $15800 (I), and $17000 (II) and more than $25000 (I) by 2020.

This region's share in world-wide total GNP increased considerably from 1960 to 1978 (from 6% to nearly 12%). It should more or less maintain its share at the 1978 level up to 2000 and 2020, in spite of the handicap resulting from an almost stagnant population.

3.3.2. Energy consumption

A. TOTAL CONSUMPTION

This spectacular economic expansion was, of course, accompanied by a strong growth of energy demand. The average growth rate of energy consumption was +7% between 1960 and 1978, compared to 3.1% for North America and 3.6% for Western Europe. In fact, the region's total energy consumption increased from 117 MTOE in 1960 to nearly 400 MTOE in 1978. This consumption, from which all non-commercial sources have been practically eliminated, will, of course, slow down sharply in the future, but due to the absence of any significant population growth, demography will only play a very marginal part in this development (about 0.5%). As in the other countries of the West, the deep changes of the economic environment will be the main cause of the extremely large decrease of the growth rate of energy demand: from 7% in the past to 2.5% (I) and 2% (II) by 2000. It will be only 1.5% (I) and 1% (II) by 2020. In other words, total energy consumption will be 600 MTOE (II) and 700 MTOE (I) by 2000. It will be somewhere between 750 MTOE (II) and 920 MTOE (I) in 2020. At least up to 2000, the growth rate of total energy demand will be higher than that of the other regions of the West, where it will be limited to a ceiling of 2% per year, even under the best of circumstances. This faster growth will be mainly due to the region's higher growth rates in the past. In this context, it should be noted that the region's energy conservation efforts in the past had been equal to those made in North America or in Western Europe. In fact, measured by the income elasticity of commercial energy demand of these three regions, the values are the following for the period from 1960 to 1978: 0.81 for R3, compared to 0.80 for R1 and 0.95 for R2.

This improvement of total energy "efficiency" of the region's system should continue in the future: 0.64–0.67 by 2000, 0.60–0.67 after that date. In other words, the region's performance will be comparable to that of Western Europe (or even be rather better).

Compared to the other industrial regions, the much faster economic growth (in the past) has not resulted in wasteful or unconstrained energy use in this region, where Japan, which is by far its greatest economic power, has always been without any significant energy resources.

The region should represent 5–6% of total world energy consumption in 2020, as in 2000, 1978, and 1960.

B. CONSUMPTION PER CAPITA

In comparison to the other industrial regions, the development of energy consumption per capita has been more moderate. It is only in 1978 that consumption per capita reached Western Europe's level *(3 TOE)*, while the difference was still very great in 1960: 1.1 TOE versus 1.8 TOE. The strong increase between 1960 and 1978 is due to the region's higher economic growth rate: 5.8% versus 2.7% for Western Europe. In fact, R3 had the fastest economic growth of the ten regions between 1960 and 1978. In the future, this difference will be somewhat more marked. R3's energy consumption per capita will reach *4.1–4.6 TOE* by 2000 (with economic growth rates amounting to 1–2%), compared to 3.6–4.0 TOE for R2. It will be *5–6 TOE* by 2020, versus 4.3–5.0 TOE for R2. But Eastern Europe's per capita consumption will also grow at a similar rate: from 3.8 TOE in 1978 to 6.0–7.4 TOE by 2020, that is it will conserve its "lead". Finally, in spite of a sharp slow down, North America's per-capita consumption level will still be much higher (between 8.8 and 10.2 TOE).

Under these circumstances, R3's average consumption per capita could increase by 2–3 TOE by 2020, that is to a similar extent as during the 1960–1978 period (+2 TOE), in spite of its rather mediocre economic growth prospects compared to the past. In addition, it is worth noting that this increase would be achieved over a much less favourable period, even if it is twice as long as the 1960–1978 period. In brief, the region's energy consumption per capita, which had grown threefold between 1960–1978, would be multiplied by two under the best of circumstances by 2020.

3.3.3. Supply pattern

(i) Region R3 is very dependent on oil. In fact, its oil dependence is higher than that of any other region, and oil represented 62% of its total energy consumption in 1978. Natural gas played only a limited part (6%), while the shares of coal (21%), hydropower (7%), and nuclear energy (4%) were similar to those of the two other regions of the West.

As in Western Europe, the penetration of oil and, consequently, the concurrent retreat of coal, have been very marked in the past, because oil covered only 32% of energy needs in 1960, compared to 51% for coal.

(ii) Nevertheless, the official energy policies of the region's countries should make it possible to reduce the share of oil to one third of total supply by 2000, and even to one fifth by 2020, that is, to levels very similar to those of Western Europe.

The condition of this remarkable development is, of course, the strong growth of coal's share in total consumption, which should represent about 30% by 2000, and almost one third by 2020. In addition, the simultaneous strong growth of the consumptions of natural gas, nuclear energy, and, to a lesser degree, new energies will be necessary. Gas should represent 14-15% in the region's energy balance by 2000, and 18-19% by 2020. The share of nuclear energy should rise to 13-16% by 2000, and 17-22% by 2020. In addition, new energies should contribute 2% to total supplies by 2000, and 4% by 2020.

(iii) In relation to the development of the world pattern, this policy does not represent any major deviations. When comparing the energy supply patterns by sources in 1978 and 2020, it will be observed that there are no very significant divergences. In other words, the regional consumption's shares by sources in the world total vary only to a small degree. The only significant changes between 1978 and 2020 are the decrease of the shares of oil (from 9% to 5-7%) and hydropower (from 7% to 3%). The decrease of hydropower's share is due to the fact that available resources are already largely harnessed. There is, of course, also the increase of the shares of natural gas (from 2% to 5-6%) and new energies (from 0% to 4%).

If this strategy will succeed, oil consumption would have already reached its ceiling before 1978, because it will have decreased from 247 MTOE in 1978 to 208-225 MTOE by 2000, and to only 160-185 MTOE by 2020.

(iv) A closer review of consumption increases in absolute values between 1978 and 2020 reveals interesting changes in the energy policies of the region's countries. According to Scenario (I), coal will represent 40% of the total increase in consumption (which is estimated at +520 MTOE from 1978 to 2020), nuclear energy 36% (or nearly +200 MTOE), and gas 33%. Under Scenario (II), where the total increase of consumption is less important (+350 MTOE), coal would contribute 48%, gas 33% (as in I), and nuclear 33% to total demand growth.

For this reason, oil demand will fall by about 60 MTOE in (I), and nearly 90 MTOE in (II). It will, therefore, be observed that in volume (and this does not appear in relative terms) it is coal which will replace the larger part of the decreasing penetration of oil, followed for almost equal parts by gas and nuclear power.

Will these changes on the demand side be encouraged or discouraged by corresponding changes on the supply side?

3.3.4. Energy production

(A) Region R3 is very dependent on imports for its energy supplies. In fact, its production covered only 38% of its 1978 demand, that is, its energy dependence was by far greater than that of any other region, including Western Europe, which ran second and covered 51% of its demand.

Coal, with a share of 47% in 1978, is the main component of the region's energy production, followed by hydropower (18%), oil (17%), and nuclear (9%). Nevertheless, it should be noted that R3 is the region where the shares of hydropower and nuclear energy in total production were the world's highest at that time.

(B) But this situation will radically change in the future, because the reinforcement of their energy independence is a primary goal of the policies pursued by the region's countries: 57-60% by 2000, and 71-74% by 2020. R3 as a whole will remain a net energy importer, but it will be much less vulnerable. In fact, its energy independence will be greater than that of Western Europe by 2020.

(C) Coal should remain the region's main assets in this context. It would continue to furnish 43% (I) and 45-47% (II) of total supply, which will grow even faster than total demand: 3.1% or 3.7% per year between 1978 and 2000, according to the scenario, that is, it will represent 535-680 MTOE at the end of the forecasting period, versus 150 MTOE in 1978. Coal production (essentially in Australia) will rise from 71 MTOE in 1978 to 180 MTOE or 165 MTOE by 2000, and 290 MTOE or 240 MTOE by 2020.

But the region's production of nuclear energy will also grow considerably: from 14 MTOE in 1978 to 200 MTOE or 130 MTOE by 2020, according to the scenario. It is the same with natural gas production (from 10 MTOE to 60 MTOE or 50 MTOE), while oil production could increase from 25 MTOE in 1978 to 50 MTOE or 45MTOE by 2020.

(D) On the whole, the region's contribution to world energy supply will remain very marginal, except perhaps for nuclear power. In fact, it will not exceed 10% for any energy source by 2020. This means that the region will not be able to play a major part as an energy exporter in the world market, because its production capacity will be essentially focused on its own needs.

3.3.5. Interregional energy trade

(A) A review of the region's participation in interregional energy trade underlines the above. Its export capacity was limited to a few million TOE of Australian coal in 1978, while its total energy imports was very high: 285 MTOE (250 MTOE for crude oil alone). It is recalled that this refers only to interregional trade, excluding all intraregional exports and imports, especially Australia's coal exports to Japan. In purely national terms, Australia is, of course, a significant energy exporter.

(B) The increase in the region's energy production should make it possible to contain imports in the future. In fact, total energy imports will increase very little by 2000: from 280 MTOE in 1978 to 300 MTOE or 290 MTOE in 2000, according to the scenario, and they will even fall to 270 MTOE or 240 MTOE by 2020.

Nevertheless, the region will become the world's second oil importer, while it occupies the third place at present. Oil imports will indeed represent 200-220 MTOE by 2000, and still 140-160 MTOE by 2020, compared to 250 MTOE in 1978. This means that the region's share in world oil imports will be of the order of 15-20% over the whole forecasting period.

The considerable growth of the region's coal production will not make it a net coal exporter, in spite of its massive intervention in the international coal market. In fact, its imports will grow concurrently with its exports, resulting in an even higher negative balance in 2000 (-20 MTOE or -25 MTOE). This negative balance will, however, be practically eliminated by 2020. The region's share in world coal exports will grow from 10% in 1978 to 20-25% in 2000 as well as in 2020, but the region will hardly improve its position as a coal importer. It was already the world's second coal importer in 1978 and remains in this rather uncomfortable situation over the whole forecasting period, representing 30-40% of world imports by 2000, and still 15-25% by 2020.

The region's situation is not better in the field of natural gas where it will also remain the world's second importer up to 2020, with a share varying between 25% and 40%, compared to 31% in 1978.

(C) Nevertheless, very considerable efforts to diversify fuel imports will be made. Crude oil represented 88% of total import supplies in 1978. This extremely lopsided supply pattern will be much more balanced by 2020: 40-42% for oil, 33-31% for coal, and 27% for natural gas.

(D) Because of Australia's coal production, the region will develop a far from negligible export capacity in the future world market: from 1% in 1978, its share in world exports will increase to 4% by 2000, and 7-8% by 2020. But its vulnerability will become rather worse because its share in world energy imports will rise from 18% in 1978 to 19-25% in 2020.

In addition, it should be pointed out that this situation, whose negative consequences are, after all, limited, depends on the simultaneous development of all regional energy resources in a very large measure, as was shown above.

If this is not the case, imports would grow proportionally, and that in a region which has to live in a very open market. In fact, having diversified its supplies more than any other region (like Western Europe), it will also have diversified its energy dependence, without being able to attenuate this situation in any really significant way.

3.4 EASTERN EUROPE

Because of its great supply and demand, Eastern Europe as defined in this context* is one of the main actors in world energy. However, unlike the other great consumer regions of the West and the great producer regions of the South, it plays a relatively small part in international energy trade outside its sphere of influence, because it aims, above all, at covering its needs by its own resources.

Unlike for the other 9 regions, forecasting procedures were adapted to the specific context of Eastern Europe. This means that only data for the past were explicitly approved by the regional team, while population, economic, and energy forecasts, based essentially on official estimates up to 2000 (such as ref. 17) were established *by the central team, under its own responsibility.* The team was only kept informed in this respect.

3.4.1. Demographic and economic environment

(A) The level and the growth of rate of Eastern Europe's population are similar to that of Western Europe. In fact, it represented 372 million in 1978, and its growth rate will decrease from less than 1% between 1960 and 1978 to more or less 0.5% by 2020. In other words, the region's population will be 460

*Cf. Annex 1.

million by 2020, versus 490 million for Western Europe. Its share in world population tends to decrease continuously: from 10% in 1960 to less than 9% in 1978, 7% in 2000, and 6% in 2020.

(B) According to the World Bank's comparative estimates, Eastern Europe's *total GNP* was of the order of G$1370 in 1978 and had a very sustained growth between 1960 and 1978 (+6.1%), that is, a much higher growth rate than the average rate of the market economies of the West (+4.3%). Future growth rates have been fixed as a guide only, but they were not used as exogenous variables in establishing the energy forecasts. They would vary between 3% (II) and 4% (I)* by 2000, and 2% (II) and 3% (I) by 2020, that is, in a range higher by about half a point than that corresponding to the prospects of the whole of the countries of the West. Consequently, Eastern Europe's total GNP could be somewhere between G$2600 and G$3200 by 2000, and somewhere between G$3900 and G$5800 by 2020. This means that it would remain the world's third region in economic terms, very significantly drawing near to North America, whose total GNP would reach G$6100 by 2020, according to Scenario (I), based on faster development than Scenario (II). Eastern Europe's total GNP could, therefore, be multiplied by 3–4.2 between 1978 and 2020. Because of this growth, the region's share in world GNP, which had risen by 3 points between 1960 and 1978 to about 16%, will level off around 16–17% over the whole forecasting period, while the share of the whole of the countries of the West will decrease significantly.

(C) GNP *per-capita* growth will be similar. The difference between Eastern Europe and the countries of the West will be reduced continuously over the whole forecasting period, Eastern Europe's GNP per capita represented only 47% compared to that of the above countries in 1960, but it was estimated at $3700 for 1978, that is it represented 61%. It should be between $6200 (II) and 7700 (I) (70% and 72% respectively) by 2000, and between $8400 (II) and $12600 (I) by 2020 (75% and 82% respectively). This reduction of the differential between the two regions is due to a higher growth rate in the past (1.5%) and in the future too +0.4% (II) and +0.7% (I).

3.4.2. Energy consumption

A. TOTAL CONSUMPTION

(i) Eastern Europe took Western Europe's place as the world's second consumer region between 1960 and 1978. In fact, its share of world demand increased from 17% to 21% over this period, while Western Europe's share decreased from 20% to 18%. The region's total consumption grew at an average rate of more than 5% over this period, compared to 3.6% for the whole of the countries of the West. In other words, the region's faster economic growth resulted in a proportionally faster growth of total energy demand, for income elasticity of commercial energy demand was 0.89, versus 0.86 for the whole of the countries of the West.

(ii) The increase in volume is indeed impressive: from 0.6 GTOE in 1960 to more than 1.4 GTOE in 1978, and it will remain so in the future. In fact, the growth rate of demand is estimated at 2.5% or 2.0% by 2000,

* A value which was considered to be rather high by the RWT.

according to the scenario, and at 1.7% or 1.3% by 2020, that is, it will be, as in the past, clearly higher than that of the countries of the West (+0.8%) up to 2000, but it will be more or less equal to that of these countries after that date. Consequently, Eastern Europe's total consumption could reach *2.2–2.4 GTOE by 2000*, and *2.8–3.4 GTOE by 2020*. In other words, it could be multiplied by 2 or 2.4 between 1978 and 2020. This growth would enable the region to maintain its share in world demand at 21% in 2000 (that is, the 1978 level), and at 19% or 20% in 2020, according to the scenario, contrary to the Western countries.

(iii) Eastern Europe's GNP elasticity of commercial energy demand will be reduced significantly by 2000, as in the Western countries. In fact, it will pass from 0.89 in 1978 to $\frac{2}{3}$ in 2000, practically in line with that of Western Europe. After that date, thanks to the efforts made in the region, it should even improve in relation to the Western countries: 0.57 or 0.65, according to the scenario, compared to 0.74 or 0.77. This shows that energy conservation will be taken seriously, in the East as in the West.

B. CONSUMPTION PER CAPITA

(i)　Even if GNP per capita is much lower in Eastern than in Western Europe, the former's energy consumption per capita has grown considerably between 1960 and 1978: from 2.9 TOE to 3.8 TOE. There are many factors that may contribute to explain this very pronounced difference between Eastern and Western Europe's per-capita consumption levels, measurable by their respective energy intensity (the ratio between energy consumption and GNP): 0.5 KOE per $ in 1960 and 1978 for Western Europe, and 1.2 KOE in 1960 and 1.0 KOE per $ in 1978 for Eastern Europe. Nevertheless, the main factors explaining this discrepancy are probably to be found in the different economic structures in sectoral terms, the differences in climate as well as the diverging efficiencies in a number of respects.

(ii)　As a matter of fact, this discrepancy will be more and more marked in the future, because the growth rate of Eastern Europe's per capita energy demand (4.1% versus 2.7% in the past) will continue to be higher than that of Western Europe by half a point, at least up to 2000. This means that its average total consumption per capita could be *5.1–5.8 TOE by that date*, and *6.0–7.4 TOE by 2020*. In other words, it will exceed that of the Western Hemisphere as from 2000 under the two scenarios. This underlines both the importance of its future growth (per-capita consumption will multiply by *1.6–2 by 2000*) and the slowdown of this growth in relation to the 1960–1978 period (when it doubled) as the region's consumption per capita will also double between 1978 and 2020, but over a period that is twice as long. In brief, Eastern Europe's consumption per capita will be second only to that of North America. The difference between the per capita consumption levels of the two regions will even be gradually reduced from 4.5 TOE in 1978 to about 3.0 TOE in 2020.

3.4.3. Supply pattern

(i)　Eastern Europe's energy supply pattern deeply changed between 1960 and 1978 but less deeply than in the countries of the West, because the substitution of coal by oil was much less pronounced. In addition, the expansion of natural gas made it possible to contain considerably the penetration of oil.

The region's *coal* supply represented 340 MTOE in 1960, that is, it covered 59% of total energy demand, and Eastern Europe was the region where hard coal played proportionately the most important role. Coal still covered 41% (575 MTOE) of the region's total needs in 1978, while its share decreased concurrently in the Western countries from 35% in 1960 to 20% in 1978. As a matter of fact, Eastern Europe is now the world's first consumer of hard coal, representing 34% of the world market, versus 21% for North America.

The penetration of *natural gas* is also impressive. Eastern Europe's gas consumption represented 50 MTOE, that is, 9% of the region's and 13% of the world's supplies in 1960. It amounted to 334 MTOE, that is, 24% of the region's and 29% of the world's supplies in 1978. Eastern Europe remains the world's second natural gas consumer after North America, whose share in the world market fell from 80% in 1960 to 45% in 1978. But North America's predominance may now be considered as threatened by Eastern Europe.

The shares of hydropower and nuclear energy in total supply increased from 2% in 1960 to 4% in 1978. In both fields, Eastern Europe is the world's fourth producer. The share of non-commercial energies dropped from 8% to 3% over the same period,

Under these conditions, *oil's* share increased only from 22% in 1960 to 28% in 1978. Nevertheless, oil consumption rose considerably in volume, from 125 MTOE in 1960 to 400 MTOE in 1978, that is, it was multiplied by more than 3, at an average growth rate of 7% per year. The region is, therefore, still the world's third oil consumer.

(ii) What are future prospects? No doubt, there will be greater difficulties, because the expansion of coal will not be sufficient to enable it to maintain its relative position in the region's energy balance. And the share of natural gas will barely reach the threshold of one third of total consumption. Only nuclear energy should inject an impetus of a new type into energy growth, while hydropower and new energies only furnish marginal contributions to total needs.

(iii) The region's coal consumption will grow in particular under Scenario (I): 715 MTOE by 2000, and 930 MTOE by 2020. According to Scenario (II), it will be more difficult to achieve such a strong growth of coal consumption: 635 MTOE by 2000, and 780 MTOE by 2020. Under these conditions, coal's share in total consumption would represent only 27–30% in the region's energy balance between 2000 and 2020. Eastern Europe's share in the world market would also deteriorate: from 34% in 1978 to 22–23% by 2000, and 15–18% by 2020. However, the region would remain the world's second coal consumer up to 2000, when it would have to yield this place to North America. Indeed, it takes quite an effort to imagine what it means for an economic system already largely penetrated by coal to increase its coal consumption by around 60% (I) at the horizon of 2020, that is, over less than 40 years.

It is also true that, because of its huge reserves, Eastern Europe's prospects on the supply side are the best in the world.

(iv) *Natural gas* consumption will also grow massively: from 330 MTOE in 1978 to 770 MTOE or 670 MTOE by 2000, and even 1100 MTOE or 940 MTOE by 2020, according to the scenario. In other words, natural gas demand could be multiplied by 3, that is, this source would represent 31–34% of total regional needs. Under these conditions, the region would be the world's first natural gas consumer as from 2000, with a share of between 35% and 40%.

(v)   The expansion of *nuclear energy* will be accelerated. Its production would increase from 12 MTOE in 1978 to 240 MTOE or 180 MTOE by 2000, and 600 MTOE or 450 MTOE by 2020, according to the scenario. This means it would cover 10% or 8% of total needs by 2000, and 18% or 16% by 2020, according to the scenario. In other words, nuclear energy's share would be higher than that of oil by 2020.

Under these conditions, Eastern Europe will be the world's third producer of nuclear power by 2000. It would even be number two by 2020, after Western Europe, and represent more than one quarter of the world's nuclear power supply.

(vi) The growth of the region's *hydropower* production will be considerable in volume: from 42 MTOE in 1978 to 70 MTOE or 65 MTOE by 2000, and 130 MTOE or 100 MTOE by 2020, according to the scenario. Hydropower's share in the region's total supplies would represent 3% by 2000 and 4% by 2020, while its share in world production would remain 10% over the whole forecasting period. The region's climate is relatively unfavourable for *new energies*. Nevertheless, these could represent 2% of total supplies in 2000 and 2020, that is 80 MTOE or 50 MTOE.

(vii) As far as *oil* is concerned, all these considerable changes in the region's supply pattern will not limit its importance. In fact, oil demand will continue to rise, even if at a more moderate rate, to 595 MTOE or 545 MTOE by 2000, according to the scenario, compared to 400 MTOE in 1978, to decrease again to 530 MTOE or 420 MTOE by 2020. Its share in the region's total demand will drop to 24–25% by 2000, and 15–16% by 2020. However, as the level of oil's penetration was more moderate in the past, this gradual decrease will be less significant than, e.g., in Western Europe. Being less urgent, oil substitution will take place at a more moderate pace. In 1978 as well as in 1960, Eastern Europe was the world's third oil consumer. It will be the second by 2000 and 2020 (except under Scenario I, when it would be still number three in 2020). The process of oil substitution will not be easy in the context of an increasingly tighter situation on the oil reserve level, and this in spite of the great efforts made to develop alternative sources at a scale corresponding to the importance of the problem. As a matter of fact, between 1978 and 2020, coal consumption would be increased by a volume representing the equivalent of North America's present coal production level. The additional penetration of natural gas would be equal to the total present consumption of the countries of the West. Additional hydropower would represent Western Europe's total current production. Finally, the expansion of nuclear energy would be equal to present world production multiplied by four.

These comparisons speak for themselves and give an idea of the region's enormous future needs and problems, especially considering that the question will not only be to cover these needs, but also to develop an energy export capacity which is vital to Eastern Europe's economic development.

3.4.4. Energy production

(A) Eastern Europe is the world's second energy consumer and producer. Its share in world production was 22% in 1978, and its share in world consumption was almost of the same order. The pattern of its energy supply (more than 1.6 GTOE) did not vary much from its consumption pattern during this period. Coal represented 37%, natural gas 21%, and oil 35% of total supplies (compared to 28% in total consumption). Eastern Europe was also the world's first coal producer in 1978 (35% of world supply), and its second oil (19%) and gas (29%) producer.

(B) In the future, the growth of production will be slightly lower than that of consumption: 2.2% or 1.4% by 2000, according to the scenario (versus 2.5% or 1.9% for consumption), and 1.6% or 1.2% by 2020 (versus 1.7% or 1.3% for consumption). This means that the ratio between production and consumption will deteriorate from 115% in 1978 to 102-106% between 2000 and 2020.

Production volume will be considerable: 2.2-2.6 GTOE by 2000, and 2.8-3.6 GTOE by 2020.

(C) The application of the principle of self-sufficiency in the field of energy is well illustrated by the region's increasingly identical supply and demand pattern, which will be almost absolutely identical as from 2000. In fact, deviations per source will amount to less than 1% by 2020. It is difficult to assess the relative importance of the political and economic factors determining this development, especially as far as the parts played by the political will of independence and the subordination of demand to supply are concerned. However this may be, as far as results are concerned, the region tends to achieve a strict balance between supply and demand for all energy sources. Consequently, we shall not repeat here the comments made in respect of the region's demand pattern because they are also applicable to its supply pattern.

However, it is worth noting that Eastern Europe will yield its place as the world's first coal producer to North America even before 2000. It will also be only the third or fourth oil producer by 2020, according to the scenario. This demonstrates well the increasing depletion of its oil resources. On the other hand, the region will reinforce its position as the world's number one in the field of natural gas, and that long before 2000. And it will also be the world's third producer of nuclear power by 2000, and even the second by 2020.

3.4.5. Interregional energy trade

Because of its policy of concurrent expansion of domestic demand and supply, Eastern Europe only plays a marginal part in the interregional energy market. However, it should be noted that this study, for obvious reasons, cannot take into account a great part of the region's international trade, that is, the huge energy exports and imports inside COMECON.

The volume of the region's interregional energy trade outside COMECON was relatively small in 1978: 100 MTOE for exports (or 7% only of world exports), and 35 MTOE for imports (or 2% of world imports), with a positive balance of about 65 MTOE.

The region's fossil-fuel export pattern is well balanced: 26% for coal, 57% for oil, and 17% for natural gas. In fact, Eastern Europe's energy exports are the most diversified in the world. At a smaller scale, it is the same with region's fossil-fuel imports, even if they are more clearly focused on oil.

Future development will vary under the two scenarios.

(A) *Scenario (I),* which stresses international cooperation, favours the concurrent increase of interregional energy flows, but it also assumes a greater expansion of exports: by 155 MTOE by 2000, and 190 MTOE by 2020. However, the pattern by sources of these exports will change radically. Because of the decreasing depletion of reserves, oil exports will almost completely disappear as from 2000 (and that in both scenarios). Coal and natural gas exports will be about equal in 2000, but natural gas exports will be clearly higher than coal exports in 2020 (120 MTOE). These two sources will share export markets in the future.

The region's share in world energy exports will increase slightly to 8% in 2000, and 10% in 2020. Eastern Europe will still be the world's second gas exporter in 2000. It will be number one by 2020, with one third of interregional gas exports. It will also be the world's second coal exporter by 2000 (75 MTOE), but only the fifth in 2020 (70 MTOE).

Imports likewise increase, but in a lesser manner: 85 MTOE in 2000, 80 MTOE in 2020, with crude oil accounting for 75–80%. R4 nevertheless remains a marginal contributor to the world import market: 5% in 2000, 4% in 2020 (with 6% of the crude oil market).

(B) *Scenario (II)* (based on greater international tensions) assumes reduced export potentials: only 85 MTOE by 2000 and 90 MTOE by 2020, that is, below present exports. Oil exports will have completely disappeared, and the prospects of gas and coal exports are much worse than in Scenario (I): 55 MTOE in 2000, and 70 MTOE by 2020, for gas, compared to 30 MTOE and 20 MTOE, respectively, for coal. Eastern Europe will conserve its share in a generally smaller world gas market: 29% in 2000, and 31% in 2020, but it will only remain number two. As far as coal is concerned, it will be only the world's fourth exporter by 2000, and the fifth by 2020, while it was number two in 1978.

Energy imports will also feel the impact of the general economic slowdown (but less than exports): 75 MTOE by 2000, and 60 MTOE by 2020, compared to 35 MTOE in 1978. In addition, future imports will almost exclusively consist of crude oil, that is, Eastern Europe will be the world's fourth importer of crude oil by 2000, and even the third by 2020.

Under these conditions, the region's positive trade balance, which remained at a level of 70 MTOE in 2000, and increased even to 110 MTOE by 2020, under Scenario (I), will be reduced very considerably under Scenario (II): 10 MTOE by 2000, and 30 MTOE by 2020, versus 65 MTOE in 1978.

This means that the region's situation in the field of international energy trade will become tighter under Scenario (II). But in both scenarios, oil will be the problem, for the positive oil trade balance in 1978 will become negative in a significant way, with a deficit of 70 MTOE in 2000, and 60 MTOE or 50 MTOE in 2020, according to the scenario.

The question is if coal and natural gas exports, for which especially Western Europe will be a great market, will compensate the handicap of the inevitable oil dependence that the region will know, and whose impact it will feel just like the other regions.

3.5 NORTH AFRICA AND MIDDLE EAST

The decisive part played by this region on the world energy scene because of its oil resources was made even clearer by the successive oil shocks since 1973. In fact, it is by far the world's greatest oil producer and exporter. Therefore, its prospects are of particular interest, especially as they are here assessed by regional experts, that is, from the inside.

3.5.1. Demographic and economic environment

(A) This region, limited to the Arab World but including Iran, is rather under-populated. Its 1978 population was only 184 million, or a little over 4% of world population. Nevertheless, like the whole of the Third World, its population growth is very fast: 2.8% per year between 1960 and 1978, even faster than the Third World average (2.35%). This growth should continue in the future. In fact, the region's population growth rate is estimated at 2.5% up to 2000, and at 1.9% after that date, compared to 1.9% and 1.4%, respectively, for the Third World average. This means that R5's population will be 320 million by 2000, and 465 million by 2020, when it will represent 6% of world population.

(B) R5 is already now the richest region of the Third World, at least if measured by the *average GNP per capita:* very nearly $1600 in 1978, versus $1400 for Latin America (the second-richest Third World region), and only $500 for the whole of the Third World.

The growth of the region's GNP per capita was very fast between 1960 and 1978: +6% per year, compared to only +3.5% for the LDC average. This growth will continue in the future: +3.3% by 2000, and +3.1% by 2020 (I); +2% by 2000, and +1.7% by 2020 (in II) but it would become much more aligned on the LDC average. Under these conditions, the region's GNP level would reach $2400 (II) or $3300 (I) by 2000, and $3400 (II) or $6000 (I) by 2020.

It will be observed that the two scenarios give clearly diverging projections for the region. In fact, the 2020 forecasts vary by nearly 100%. At any rate, the region's GNP per capita would remain the highest in the Third World over the whole forecasting period.

(C) Nevertheless, because of its relatively small population, the region's position is more limited as measured by the volume of its *total GNP,* even if this also rose at a rate of more than 9% between 1960 and 1978. With a total GNP of G$300 in 1978, the region was only number six in the world, largely behind Latin America. In the future, this will not change much. R5 will remain number two in the Third World, with a total GNP of G$1600 (II) or G$2800 (I) by 2020. But its GNP growth rates will be in sharp contrast with the very sluggish growth of world GNP: +6% by 2000, and 5% by 2020 (I), and 4.5% and 3.5%, respectively, under II, compared to a world average of the order of 4% and 3%, respectively (I), and 3% and 2%, respectively (II). This means that the region's share in world GNP could represent 8.1% or 6.6% by 2020, according to the scenario, versus 3.4% in 1978.

3.5.2. Energy consumption

There is a great contrast between the region's relative importance as a producer and as a consumer in the world market. On the one hand, it is an essential factor as a producer, for it furnished in 1978 18% of the world's energy supply. On the other hand, it is only a marginal factor as consumer, for its consumption represented only 2% of world energy demand in the same year. For the time being, the region is still very much turned toward the outside world, but this will fundamentally change in the future, because of the concurrent increase of regional needs and competing energy sources. This increase will gradually balance its supply and demand but also limit its relative importance as an energy exporter.

A. TOTAL CONSUMPTION

The region's total consumption was only 134 MTOE in 1978, but it is worth noting that it was a mere 33 MTOE in 1960. Consumption grew at a phenomenal rate of more than 8% per year between 1960 and 1978. For commercial energy sources alone the rate was nearly 10%.

(i) In fact, the region has reduced the part played by non-commercial sources more than any other Third World region up to date: 13% versus 37% for the average of the Third World. But this reduction of the share of non-commercial energy has been going on for a long time already. Even as far back as 1960, non-commercial sources represented only 33% of total energy consumption, compared to 56% for the average of the Third World.

It is worth noting that *non-commercial sources* represented only 18 MTOE in 1978, that is 2% of world consumption. In addition, these sources were predominantly vegetable waste in this region, which has been traditionally poor in wood resources. Therefore, the development prospects of these consumptions are limited. Fire wood represented 8 MTOE in 1978, but it is practically excluded that its total consumption will be higher than 10–11 MTOE in the long run. Vegetable waste consumption, tied to the development of agriculture, might grow faster, especially under Scenario (II): from 10 MTOE in 1978 to 13 MTOE or 15 MTOE by 2000, according to the scenario, and to 14 MTOE or 19 MTOE, respectively, by 2020. But it is obvious that the total growth of non-commercial sources in quantitative and relative terms will not be sufficient to avoid the continuous reduction of their position in the region's energy balance. In fact, they will represent only 4% or 7% of total needs by 2000, according to the scenario, and 2% or 5% by 2020, compared to 13% in 1978. Their share in total world supply of non-commercial energy will, however, remain of the order of 2%.

(ii) Consequently, *commercial energy* sources will cover almost all the growth of demand. This means that R5 will become in this particular respect much more similar to the industrialized countries.

Growth rates will continue to be very high, especially under Scenario (I) (assuming more favourable economic conditions): 7.2% by 2000, and 4.5% by 2020. Under (II): 5.2% and 3.2%, that is, practically higher by 2 points than the Third World average by 2000 according to both scenarios. Even the growth rate between 2000 and 2020 will be one point higher.

Therefore, commercial energy consumption could be between 350 MTOE (II) or 530 MTOE (I) in 2000 and 670 MTOE (II) or 1300 MTOE (I) in 2020. These two demand forecasts reflect the economic growth differential resulting from the scenarios, which will amount to nearly 100% in 2020.

R5 will be more subject to the development of extra-regional conditions than any other region. Therefore, contrasts between the results of the two scenarios are the sharpest.

It will be also observed that the region's GNP elasticity of commercial energy demand, which was slighty higher than 1.0 in the past (1.07), would have an upward trend up to 2000 (1.20 under (I), and 1.13 under (II)), but it will decrease to around 0.90 between 2000 and 2020.

(iii) This very fast growth of demand will have a significant impact on the region's position as an energy consumer. It was only number ten in 1960 as well as 1978. Under Scenario (I), it will be number seven by 2000, and even number five by 2020. Under Scenario (II), it will be the world's ninth energy consumer by 2000, and the sixth by 2020. At the same time, its share in world demand will increase from 2% in 1978 to 5% in 2000 and to 7% in 2020 under Scenario (I), compared to 4% and 5%, respectively, under Scenario (II). This means that R5 will no longer be a marginal consumer in the future, and that its supply strategy will more and more influence the world's energy balance.

B. CONSUMPTION PER CAPITA

The above development will, of course, have an impact on the levels of consumptions per capita.

(i) First of all, there will be a marked decrease of non-commercial energy consumption, which had followed closely population growth between 1960 and 1978, to level off at around 0.10 TOE per capita. It should decrease to 0.07 TOE (I) or 0.08 TOE (II) by 2000, and even to 0.05 TOE (I) or 0.06 TOE (II) by 2020.

(ii) Total consumption per capita will grow very fast, stimulated by commercial energy demand: from 0.7 TOE in 1978 to 1.7 TOE by 2000, and 2.8 TOE by 2020 (I). In fact, under Scenario (I), R5 will replace Latin America as the first energy consumer among the Third World regions and its total per capita consumption will reach a level not far from Western Europe's present level. Under Scenario (II), R5's per capita consumption would be 1.2 TOE by 2000, and 1.5 TOE by 2020. This means that the region's per capita consumption levels will be very different under the two scenarios by 2020: four times the 1978 level (I), or twice the 1978 level (II).

The growth rates of the region's energy consumption per capita will correspond to this expansion. In fact, they will be the highest in the world over the whole forecasting period and under both scenarios: 4% by 2000, and 2.5% by 2020 under Scenario (I), and 2.2% and 1.2% respectively under (II), compared to average world growth rates of 0.9% (I), and 0.2–0.3% (II), between 1978 and 2020.

In per capita terms, R5 was only number nine of the 10 regions in 1960. It was number two of the Third World regions in 1978, but the level of its per capita consumption will almost equal that of Latin America as from 2000.

3.5.3. Supply pattern

(i) R5's supply problems will be simplified by its considerable hydrocarbon reserves, especially as these forms of energy are optimal for the consumption systems now developed in the whole world.

As from 1960, hydrocarbons represented more than 60% of the region's total supplies, with oil alone representing 55%, compared to 33% for non-commercial energy sources. By 1978, natural gas had replaced most non-commercial sources, whose share had fallen to 13% versus 22% for natural gas. But oil's share had also risen to 59%. In other words, the share of hydrocarbons in the region's balance had reached an overwhelming 81% by 1978, compared to 61% in 1960.

(ii) In the absence of any significant coal or hydraulic resources, the future share of these sources will be very marginal in relation to the fast growth of the region's energy needs. *Hydropower's* share should be limited to 3% of total supplies. *Coal's* share should not represent more than 2% up to 2000, and it might possibly increase to 5% (I) or 6% (II) by 2020. The marginal part played by non-commercial sources in the long run was already underlined above.

For the time being, *nuclear power* and *new energies* are absent from the region's energy scene. It is estimated that the share of each of these sources might represent 1% or 2% of total supplies by 2000, but their shares could be more significant by 2000: 4% (I) or 5% (II) for new energies, and 7% (I) or 5% (II) for nuclear power. These latter figures show that R5 is conscious of the fact that even its enormous hydrocarbon reserves are finite and that its long-term future, in other words the post-petroleum era, has to be prepared, very gradually.

It should also be noted that the region's shares in world demand of each of the above sources will hardly change up to 2000: 1–2% of the world market, except for vegetable waste (4%). On the other hand, R5's share in supplies of new energies will represent 5% under both scenarios by 2020. In addition, its share in world nuclear power supply will amount to 4% by 2020 under Scenario (I), that is nearly 100 MTOE, or more than North Amercia's present production.

(iii) The predominance of *hydrocarbons* will remain unchanged over the whole forecasting period. Their share will certainly reach a maximum towards the end of our century, when it could be 88% (I) or 85% (II). It will decrease to 80% (I) or 76% (II) by 2020, when coal, nuclear, and new energies will have begun to replace hydrocarbons.

It is only normal that the share of *natural gas* in the region's energy balance should continue to increase. Its transport being more difficult and more expensive than that of oil, it is subject to a tendency to be used preferentially in local consumptions in a region with large hydrocarbon reserves, reserving oil for exports. Therefore, the share of natural gas could increase from 22% in 1978 to 28% (I) or 27% (II) by 2000, and even 34% (I) or 35% (II) by 2020. The volumes involved will increase considerably: from 30 MTOE in 1978 to 155 MTOE (I) or 104 MTOE (II) by 2000, and 450 MTOE (I) or 250 MTOE (II) by 2020. The region would thus become one of the greatest natural gas consumers, whose share in the world market could rise from 2% in 1978 to 7% (I) or 6% (II) by 2000, and even 14% (I) or 10% (II) by 2020. In other words, it could be the world's third gas consumer by 2020 (versus the fifth only in 1978), with a consumption level similar to North America's present level under Scenario (I), and Western Europe's present level under Scenario (II).

Because of the increasing penetration of gas, the share of oil in regional needs will level off at about 60% by 2000. It could decrease to 46% (I) or 41% (II) by 2020, even if oil's growth in volume will remain very high, especially under Scenario (I): 335 MTOE (I) or 218 MTOE (II) by 2000 (compared to 80 MTOE only in 1978), and 600 MTOE (I) and 300 MTOE (II) by 2020. In fact, the volume under Scenario (I) would be the equivalent of Western Europe's present oil consumption. Under these condition's, R5's contribution to world oil demand will increase considerably: from 3% in 1978 (when it was number eight only) to 10% (I) or 8% (II) by 2000, and 17% (I) or 12% (II) by 2020. In particular, under Scenario (I), R5 would become in the long run the world's second oil consumer, after Latin America, but before all the industrialised regions. Even under Scenario (II), R5 would be number four by 2020. This development will take place in a world where oil will be increasingly scarce. Therefore, it will be necessary to replace oil by other sources that will be less in demand in the world market, even in R5, that is a region largely favoured by nature in the field of hydrocarbon resources. In other words, an adequate oil substitution strategy must be developed also for this region, and it must be developed in time.

3.5.4. Energy production

(A) The region's production is, of course, dominated by oil, which represented 1180 MTOE in 1978, that is 93% of R5's total supply, while gas represented only 60 MTOE, or 4%. The region was by far the world's largest oil producer, with 39% of world supply. In addition, it was already then number four for gas supply.

(B) Under Scenario (I) it is assumed that the potential oil supply of the Arab World will be of the order of 25 Mbl/day over the whole forecasting period. This value is not far from that for 1977 and represents production capacity according to present estimates. Of course, owing to the unfavourable development of the world oil market, effective production has considerably decreased for the time being. The above production capacity is also perfectly consistent with the region's proven reserves. In fact, even at a production level of 25 million bl/day over the next 40 years, the region's cumulative consumption (including Iran) over the same 40 years would be about 50 GTOE, that is 90% of the reserves classified as proven at present.

(C) Under Scenario (II), a supply level of 20 million bl/day had been envisaged as a working hypothesis. Owing to the more pessimistic demand prospects, it was decided to adopt a lower level, that is, 19 million bl/day. Including Iran, this gives 1340 MTOE (I) or 1070 MTOE (II) by 2000, and 1310 MTOE (I) or 990 MTOE (II) by 2020, compared to 1180 MTOE in 1978. These volumes seem perfectly realistic for R5, a region without extreme total supply problems. Consequently, R5 will certainly be able to remain the world's first oil producer, whose share will represent 34% (I) or 33% (II) by 2000, and 31% (I) or 32% (II) by 2020, that is around one third of world supply.

(D) It was already mentioned that the increasing share of gas in the region's local consumptions will reduce the penetration of oil and contribute to conserve its oil export capacity to a certain extent. In fact, R5's gas supply could rise from 59 MTOE in 1978 to 260 MTOE (I) or 190 MTOE (II) by 2000, and 515 MTOE (I) or 315 MTOE (II) by 2020. The share of gas in the region's total supplies would concurrently increase from 4% in 1978 to 16% (I) or 14% (II) by 2000, and 25% (I) or 24% (II) by 2020. This would limit in itself oil's share to 80% (I) or 82% (II) in 2000, and 65% (I) or 68% (II) in 2020 (93% in 1978). R5 will become the world's third gas producer by 2000, and the second (I) or third (II) by 2020.

(E) Nevertheless, in spite of this development, R5 would not be able to conserve its position as the world's third energy producer (all sources combined) in the very long run, that is after 2000, when Latin America will occupy this place and also be the Third World's first energy producer. R5's share in world energy production will decrease continuously: from 18% in 1978 to 13% (I) or 12% (II) in 2000, and only 11% (I) or 10% (II) in 2020. This will be due to the fact that the region's production of non-hydrocarbon sources will each be under 15 MTOE in 2000, representing only 1% of world supply. In addition, the total share of nuclear power and new energies in R5's total supplies will not be significant before 2020 (2-5%). It is also certain that development prospects for hydropower are very limited, because the available resources potential is well known and scarce. Finally, there is no indication that there are any significant coal resources in the region.

3.5.5. International energy trade

The development outlined above will have amplified repercussions on the region's interregional energy trade.

(A) First of all, the region's share in world energy production will decrease, while its share in world energy demand will increase. The difference between these two shares was +16% in 1978 (18% for supply and 2% for demand). But this difference will decrease to 8% by 2000, and 4% (I) or 5% (II) by 2020. Under these conditions, the region's energy independence factor, that is, the ratio between R5's total production and total consumption, will be reduced enormously: from 945 in 1978 (it was the world's highest, when production represented almost ten times consumption) to 300 (I) or 350 (II) by 2000, and 155 (I) or 210 (II) by 2020. In other words, the region will still enjoy a very comfortable situation as far as its energy independence is concerned, but production will not grow proportionally to demand.

The region's energy exports are impressive. Its share in the world's exports of fossil fuels was 68% in 1978, representing over one billion TOE. Its share of world oil and gas exports was 73% and 46%, respectively, in the same year. R5 was indeed by far the world's first oil and gas exporter in 1978.

(B) The region will conserve these dominating positions, with shares of 54% (I) and 57% (II) of the world export in *2000*. But it will maintain its export volume of over one billion TOE only under Scenario (I), versus 830 MTOE in Scenario (II).

R5 will consolidate its dominating position in world crude oil exports: 76% (I) or 78% (II) by 2000. But its dominating position in world gas exports will be menaced by Eastern Europe as from 2000, in spite of a share representing 37% (I) or 39% (II). There will also be some small-scale coal imports by 2000 (2% of world imports).

(C) By *2020*, the region's exports will only represent 38% (I) or 48% (II) of potential world exports, that is, 700 MTOE (I) or 630 MTOE (II), respectively. Under Scenario (I), oil supply available for exports will have fallen to 665 MTOE, compared to 1000 MTOE in 1978 and 930 MTOE in 2000. Under Scenario (II), it will have decreased to 760 MTOE in 2000, 550 MTOE in 2020.

In addition, gas exports, which will rise from 27 MTOE in 1978 to 95 MTOE (I) or 75 MTOE (II) in 2000, will also decrease by 2020, in spite of the considerable growth of production, because the region's needs will absorb the by far greater part of the growth of production capacity. In fact, in 2020, gas exports will represent only 40 to 80 MTOE, according to the scenario.

The result of this development will be that R5 will maintain its share in world oil exports at a level of 71% (I) or 78% (II) by 2020, but its share in world gas exports will drop sharply to 11% only under Scenario (I), making it number four only among the 10 regions. Under Scenario (II), R5 will remain number one, with a share of 37%. Concurrently, the region's coal imports will rise to 58 MTOE (I) or 35 MTOE (II), representing 8% of world coal imports and making it the world's third (I) or fourth (II) coal importer.

It should be pointed out that this deliberate diversification of supply by coal imports might perhaps not be achieved because of the long-term supply tensions in the world coal market. In addition, there will be always the temptation to choose the easy way out, that is, to increase hydrocarbon production to cover local needs.

3.6 AFRICA SOUTH OF THE SAHARA

On the one hand, any analysis of this region's energy balance is bound to be problematical because of the absence of any significant cooperation or relations between practically all countries of Black Africa and South Africa for obvious political reasons.

On the other hand, geographical and methodological reasons require one to analyse the combined data of Black and South Africa as a whole, especially as the study's terms of reference explicitly exclude purely national forecasts. Under these conditions, it was, of course, impossible for South African experts to join the regional team of experts which, consequently, had to limit its work to *Black Africa* alone. Nevertheless, independent consultations of South African sources and experts made it possible to establish additional forecasts for this country and to reconstruct *ex post* the greater region (R6) as it is presented in its table of forecasts.

However, South Africa's energy weight and specificity are such that it seemed advisable to make two separate analyses: firstly, one for "Black Africa" (excluding South Africa) which represented the forecasting area for the RWT properly speaking; and, secondly, one for the whole of the Region R6: "Sub-Saharan Africa" (including South Africa). These two analysis are presented successively in this section. For the first, reference will be made to the two "Black Africa" forecasting sheets (Annex 10, tables 6 and 17) which provided most of the material for comment.

3.6A BLACK AFRICA: AFRICA SOUTH OF THE SAHARA WITHOUT SOUTH AFRICA

3.6.1.A Demographic and economic environment

(A) Black Africa's (R6 excluding South Africa) *population* is one of its great assets and at the same time one of its great liabilities. It was 320 million in 1978, but the region's density of population was still rather low. In fact, in demographic terms, Black Africa was only number seven among the ten regions.

The growth rate (2.5%) of the region's population was very near the Third World average between 1960 and 1978, but this will change in the future: 3.1% by 2000, while the rest of the world will gradually have lower population growth rates. It will remain very high even after 2000: 2.7% versus 1.4% only for the Third World average. This means that Black Africa's population growth will be much less under control than in most other parts of the Third World. In fact, its population will be 630 million by 2000, and 1070 million by 2020. As from 2000, the region will be the world's number three in terms of population, after the two giants, South Asia and Communist Asia, but its population will approach those of these two regions by 2020.

(B) Black Africa is still a marginal factor only in economic terms. Its total GNP (G$122) was the lowest of the ten regions in 1978. In addition, the growth rate of its GNP was lower than the Third World average between 1960 and 1978: 4.3% versus 5.1%. The future growth of its total GNP will be stimulated by its population growth: 4.5% by 2000, and 3.5% by 2020, under Scenario (I), and 4% and 3%, respectively, under Scenario (II). This population growth will absorb an essential part of Black Africa's future economic growth. These growth prospects should, however, take into account that the world average will be lower by half a point under Scenario (I), and one point under Scenario (II), and that the Third-World average will be much higher under Scenario (I), and little higher even under Scenario (II).

(C) As far as Black Africa's GNP *per capita* is concerned, the situation is hardly promising. For example, while it was comparable to South East Asia's (R8) in 1960 ($298 versus $287), it was only $382 in 1978, compared to $652 for R8. In fact, the growth rate of Black Africa's per capita GNP was the world's lowest over this period: 1.4% versus 2.7% for the Third World average. Only South Asia had such a low growth rate.

Because of the combined effect of the very limited growth of its total GNP volume and its very high population growth, the prospects of Black Africa's GNP per capita growth rates are very limited. Compared to the other Third World regions, they even tend to be worse: 1.3% by 2000, and 0.8% by 2020, under Scenario (I), and 0.8% and 0.3%, respectively, under Scenario (II). This means that the volume of Black Africa's GNP per capita will hardly reach $500 (II) or $600 (I) by 2020, versus nearly $400 in 1978. As a matter of fact, as the region will become gradually one of the world's giants in terms of population, it will be, like R7 (South Asia), increasingly the victim of the chief handicap inherent in this development, that is, substantial efforts to increase total GNP seem to be nullified by any referability to effects on per capita GNP. Nevertheless, in this context, it should not be forgotten that Black Africa, which is largely a rural region, has a very active and powerful parallel economic system that does not appear in any statistics.

3.6.2.A Energy consumption

A. TOTAL CONSUMPTION

(i) The difficulties of Black Africa's economic development are reflected in its energy situation, whose main features are the great relative import- ance of non-commercial consumptions and the exceptionally *low commer- cial energy demand*. In fact, the region's total consumption of commer- cial sources was only 27 MTOE in 1978, that is, by far the lowest in the Third World. This level of commercial consumption is about equal to Norway's or Hungary's, with populations of 4 milion and 10 million respectively, and not 320 million as in Black Africa's case! Nevertheless, the growth of the region's commercial consumptions was rather fast between 1960 and 1978: 5.5%. In fact, total commercial demand repre- sented only 10 MTOE in 1960.

But the most striking aspect of Black Africa's energy situation is the *great relative importance of non-commercial sources*. The volume of these consumptions was some 96 MTOE in 1978, not far lower than South Asia's or South East Asia's level. It is obvious that any analysis not taking into account, at least in an approximate way, this phenomenon could only give a completely deformed and inadequate picture of Black Africa's energy situation and prospects. As a matter of fact, the region represented 7.5% of world population, but only 0.4% of the world commercial energy demand, versus 13% for non-commercial sources.

(ii) The growth rate of Black Africa's *commercial energy consumption* was lower than the Third World average between 1960 and 1978: 5.5% versus 6.7%. It will be slighty higher than the Third World average in the future, in spite of a more difficult world-wide context: 6.3% by 2000, and 4.9% by 2020, under Scenario (I), versus 5.2% and 3.6%, respectively, for the Third World average. Under Scenario (II): 4.4% by 2000, and 4.7% by 2020, versus 3.4% and 2.8%, respectively, for the Third World average. The level of departure of this growth is, however, so low that commercial consumption volumes will be realtively moderate, even in the long run: 103 MTOE (I) or 69 MTOE (II) by 2000, and 270 MTOE (I) or 174 MTOE (II) by 2020, when Black Africa will still represent only 1.5% of world demand of commercial energy.

The GNP elasticity of Black Africa's commercial energy demand (calculated *ex post*) was 1.38 between 1960 and 1978. It will remain practically constant over the whole forecasting period under Scenario (I) (1.40), while it will vary under Scenario (II): 1.10 by 2000, and 1.57 by 2020.

(iii) The growth rate of Black Africa's *non-commercial energy consumption* was 2.5% per year between 1960 and 1978. Contrary to the average of the rest of the world, it will remain around that level up to 2000: 2.3% (I) or 2.8% (II). But it will decrease to 1.1% (I) or 1.6% (II) by 2020, while the average of the rest of the Third World will stagnate. This shows that Black Africa will continue to use non-commercial sources for a longer time than the other Third World regions. This is, of course, also due to the fact that there is no very pronounced scarcity of these resources, except for the Sahel areas. Because of the exceptional growth of its non- commercial consumptions, Black Africa's share in world demand of non-commercial energy will increase from 13% in 1978 to 17% by 2000 and 22% by 2020.

Black Africa's share in total world energy consumption will not increase much up to 2020. The combined effect of the diverging development of commercial and non-commercial demand will be an average growth rate of total energy consumption that will hardly differ from the 1960–1978 average: around 3% per year over the whole forecasting period. It will probably be a little higher than 3% up to 2000, and a little lower than 3% between 2000 and 2020.

Black Africa's share in total world energy demand was 2% in 1978, as in 1960. It will increase to only 2.5–3% by 2020, when non-commercial sources will still represent 42% (I) or 58% (II) of total regional demand, versus 78% in 1978. By that time, non-commercial energy will represent only 10% of total Third World and 5% of world demand. Wisdom or anachronism? Only the future can tell.

B. CONSUMPTION PER CAPITA

Inevitably, Black Africa's energy consumption per capita appears to be very low: 0.35 TOE in 1960 and hardly more in 1978 (0.38 TOE). But its per capita consumption of commercial energy was only 84 KOE in 1978, that is one hundred times less than North America's! It is also worth noting that commercial consumption represented more than 50% of total consumption in only 3 (Gabon, Zambia, Reunion) of the region's 46 countries.

In spite of the relative importance of non-commercial energy, the growth rate of Black Africa's consumption per capita was very low between 1960 and 1978: 0.4% per year, compared to a Third World average of 2.1%. As it was already pointed out, the future growth rates of Black Africa's total energy consumption will be very near the Third World average, but this has very little effect on its consumption per capita: 0.39 TOE (II) or 0.41 TOE (I) by 2000, and 0.39 TOE (II) or 0.43 TOE (I) by 2020, versus 0.38 TOE in 1978. In fact, only South Asia's per-capita consumption will be lower than Black Africa's by 2020. To a great extent, this slow and difficult growth of Black Africa's total energy consumption per capita may be explained by the massive shift from non-commercial sources. However, this slow growth also indicates without any doubt a marked improval of the region's overall energy efficiency. This higher efficiency is, of course, not apparent from the data as they are presented.

Black Africa's commercial energy consumption per capita will increase from 85 KOE in 1978 to 250 KOE (I) or 162 KOE (II) by 2020, that is, it will double (II) or treble (I). Its non-commercial energy consumption will fall from 300 KOE in 1978 to 180 KOE (I) or 224 KOE (II) by 2020.

These few figures give an idea of the scale of the changes in Black Africa's basic consumption pattern, in spite of the apparent stability of total consumption levels.

3.6.3.A Supply pattern

(i) *Fire-wood* is the king-pin of Black Africa's energy supply. It still covered 69% of total needs in 1978, versus 76% in 1960, that is, its share in total demand was much higher than in any other Third World region, including South Asia (35% only).

Far behind fire-wood, *oil* was Black Africa's second supply source with a share of 14% in 1978, followed by vegetable waste (9%). Apart from oil, hydropower was the region's only other commercial supply source (5%), with a production of around 30 TWh. Natural gas and coal were practically not used in 1978.

(ii) The growth of Black Africa's *non-commercial consumptions* will be considerable. Under Scenario (I), based on faster economic development and industrialization, the substitution of non-commercial sources by commercial energy will take place at a faster rate. Under Scenario (II), assuming greater development problems, this substitution process will be slower. The region's fire-wood demand will, therefore, represent 200 MTOE by 2020, under Scenario (II), and 157 MTOE only under Scenario (I), compared to 85 MTOE in 1978. The region's vegetable waste demand will also vary according to the scenario: 40 MTOE (II) or 36 MTOE (I) by 2020, versus 11 MTOE in 1978.

In spite of these increases in volume, the share of fire-wood in Black Africa's total supplies will drop sharply by 2020: 48% (II) or 34% (I), while that of vegetable waste, whose volume depends mainly on the development of agriculture and related industries, will remain at 8–10% over the whole forecasting period. The total share of non-commercial energy in regional supplies will decrease from 78% in 1978 to 60% (I) or 72% (II) by 2000, and 42% and 58%, respectively, by 2020. This shows that the future development of commercial and non-commercial consumptions depends very much on the basic assumptions of the two scenarios. In spite of its decreasing share in total demand, fire wood will remain Black Africa's first supply source by 2020, with oil following rather far behind, and that even under Scenario (I), where its substitution will take place at a much faster rate. But Black Africa will also be the world's first consumer of fire wood by 2020 under Scenario (I), with a share of 29% in world demand (versus 17% in 1978), far ahead of South Asia (20%).

(iii) The shares of the commercial sources in total regional energy supply will increase considerably in the future. *Oil's* share in total regional demand will be 25% (116 MTOE) by 2020 under Scenario (I), while it will hardly increase under Scenario (II): 15% (63 MTOE) versus 14% in 1978. Black Africa's oil dependence will, therefore, increase under Scenario (I), but in very moderate proportions.

The share of *hydropower* in the region's total supply will also increase regularly: from 5% in 1978 to 8% (I) or 6% (II) by 2000, and 15% (I) or 11% (II) by 2020, compared to only 1% in 1960.

In fact, hydropower is one of Black Africa's great assets in the field of energy. The production corresponding to the above market shares would be the equivalent of 70 MTOE (I) or 45 MTOE (II) by 2020, that is, 315 TWh (I) or 200 TWh (II), compared to 30 TWh in 1978.

The market shares of coal and gas will also become more significant as from 2000 (between 2% and 4% of total needs). Coal alone could represent 4% of the region's total supply by 2020, with a volume of 16–20 MTOE, versus 3 MTOE in 1978 while the share of gas could be 4% to 6%, with a volume of 15 to 30 MTOE. Nuclear power will not make its appearance before 2000, but it could represent 1% to 2% of the region's total supply by 2020, that is the equivalent of 5 to 10 MTOE (or a production of 20 to

50 TWh). Finally, the prospects of *new energies* may be considered as favourable: 8 MTOE (II) or 10 MTOE (I) by 2000, and 24 MTOE (II) or 30 MTOE (I) by 2020, when they could cover 5% (II) or 7% (I) of Black Africa's total needs,

Black Africa's share in world demand of commercial sources has been insignificant up to date. It will be only marginal by 2020: 3% for oil, 3–5% for hydropower, 3–4% for new energies, and 1% for natural gas.

Under these conditions, the region's share in total world demand of commercial energy will, of course, remain extremely modest: 0.5% in 1978, 1% in 2000, 1.5% in 2020.

3.6.4.A Energy production

By convention, all data and comments concerning non-commercial consumptions are directly applicable to non-commercial productions. In other words, their consumption and production patterns are identical. It is quite different in the case of fossil fuels, especially as far as oil is concerned. Apart from the region's present production (Nigeria, Gabon, etc . . .), a number of countries in Black Africa have no doubt great potential oil resources.

(A) The region's *oil production* was 113 MTOE in 1978. This volume represented 4% of world oil production, that is, more than the whole of Asia (without the Middle East, of course). Black Africa's coal and natural gas production, however, was insignificant in 1978. This explains the small shares of these fossil fuels in its total consumptions.

In other words, oil largely dominated Black Africa's production pattern, with a share of 52% in regional supply in 1978, compared to 39% for fire wood, and 9% only for all other sources.

(B) Great oil exploration efforts, financed in part by the World Bank, are made in Black Africa. These will no doubt increase the regions production capacities in the future: 160 MTOE (I) or 150 MTOE (II) by 2000, and 250 MTOE (I) or 180 MTOE (II) by 2020. Black Africa could, therefore, very well represent 5% of world production by 2000, and 6% by 2020.

However, in spite of these impressive volumes, oil's share in total regional production will decrease to 47% (I) or 41% (II) by 2000, and 42% (I) or 34% (II) by 2020.

The share of fire-wood in Black Africa's total production will also decrease, at least under Scenario (I): from 39% in 1978 to 35% (I) or 41% (II) by 2000, and 27% (I) or 38% (II) by 2020. But the share of hydropower will increase considerably; from 3% in 1978 to 5% (I) or 4% (II) by 2000, and 12% (I) or 9% (II) by 2020. In fact, all other sources will gain a certain share by 2020: 4% (I) or 6% (II) for new energies, and 5% (I) or 3% (II) even for natural gas.

Nevertheless, apart from oil and non-commercial energy, Black Africa will only have significant shares in the future world supply of hydropower and new energies: 3% for hydropower as from 2000, and 3–4% for new energies by 2020.

3.6.5.A Interregional energy trade

(A) Black Africa's participation in interregional energy trade is, and will be, limited to oil. Its coal and gas production will, probably, be too limited to allow significant exports.* In addition, it is assumed that its coal and gas needs will be covered by the greater region comprising Black and South Africa.

(B) Black Africa's crude oil exports amounted to nearly 100 MTOE in 1978. The should increase only in a very moderate way, especially up to 2000, because the growth of regional oil demand will absorb the greater part of additional supplies. Black Africa's interregional export capacities would, therefore, be limited to 120-130 MTOE in 2000 as well as 2020 (No 2 in the world). However, the region's share in world oil exports will increase from 7% in 1978 to 10-12% by 2000, and 14-17% by 2020. Under these conditions, Black Africa would, of course, be in a very favourable position, because by that time the price of crude oil will be inevitably very high.

(C) But Black Africa will also increase its crude oil imports: from 7 MTOE in 1978 to 15 MTOE (I) or 10 MTOE (II) by 2000, and 25 MTOE (I) or 20 MTOE (II) by 2020. It is obvious that these imports would offset to a certain degree the favourable impact of the above oil exports on Black Africa's overall prospects. The increase in the region's oil imports will be due to a number of reasons: lack of intraregional transport facilities, cooperation problems, excessively great distances from regional refinery units. In the case of the East African countries, the relatively short distances to the Middle East will, of course, facilitate imports. Black Africa's net oil exports will, therefore, increase only in a very moderate way: from 90 MTOE in 1978 to 110 MTOE by 2000. They will even decrease to 100 MTOE between 2000 and 2020.

Nevertheless, Black Africa's oil export capacities are, and will be, one of the few substantial assets for its economic development, even if they are in reality limited to a few privileged countries. There are also long-term projects of natural gas exports.

3.6B AFRICA SOUTH OF THE SAHARA

The following refers to all Africa South of the Sahara, that is, Black Africa (R6) and South Africa combined.

The impact of South Africa's addition to Black Africa (R6) is far from negligible in statistical terms. First of all, South Africa's commercial energy consumption alone is twice that of all Black Africa. Also, its energy production and consumption patterns are quite different, because they are dominated by coal and marked by the almost complete absence of non-commercial sources. In addition, nuclear power and new energies will play significant parts in South Africa's supply and demand patterns. In fact, in order to get an idea of the total energy prospects of the whole of this vast region, that is, Black and South Africa combined, the additional analysis presented hereafter is required.* It is, therefore, recommended to read the comments that follow after those of 3.6A, that is, Black Africa.

---

*Even if mention was made of LNG projects in Nigeria, Congo and Cameroon, which are currently suspended and which were not taken into account by the RWT.

*We should like to stress again that we do not intend to prejudice the possibility of political changes or of improved political relations in this region. The aggregation of South Africa and Black Africa made here is solely done for technological reasons related to the geographical division used in this study.

3.6.1.B Demographic and economic environment

(A) The greater region's population has grown, and will grow, at the same rate as that of Black Africa: from 348 million in 1978 to 681 million by 2000, and 1152 million by 2020.

(B) The greater region's total GNP was G$166 in 1978, but South Africa's share alone represented 26%, compared to a share of 8% only in total population. The greater region's GNP growth prospects are also similar to those of Black Africa: G$400 (II) or G$440 (I) by 2000, and G$710 (II) or G$870(I) by 2020. This means that the greater region's growth prospects in GNP terms are definitely better than those of South Asia under both scenarios, and even better than those of Communist Asia under Scenario (II). In spite of this progress, the share of the greater R6 in world GNP will only increase from 2% in 1978 to 2.5–3% by 2020.

(C) The greater region's GNP per capita was higher by nearly $100 than that of Black Africa alone in 1978. It will grow at rates similar to those of R6 in the future: $580 (II) or 640(I) by 2000, and $620 (II) or $750 (I) by 2020, compared to $478 in 1978. This means it will be very much higher than that of South Asia under both scenarios, and exceed that of Communist Asia under Scenario (II).

3.6.2.B Energy consumption

The greater region's demand pattern is very different from that of Black Africa alone.

A. TOTAL CONSUMPTION

(i)   The greater region's total consumption was considerably higher than that of Black Africa alone: 177 MTOE in 1978, versus 123 MTOE. This means that it was higher than R5's (North Africa and Middle East).

The greater region's future growth rates will be in the range of 3.2–3.9% over the whole forecasting period, that is, slightly higher than in the past (3.2%). This will increase the volume of total demand to 400 MTOE (I) or 360 MTOE (II) by 2000, and 870 MTOE (I) or 680 MTOE (II) by 2020. The greater region's share in world demand will, therefore, increase very moderately from 3% in 1978 (as well as in 1960) to 4–5% by 2020. In other words, the greater region will not become a key factor in world demand.

(ii) The share of non-commercial sources in total consumption was still 56% in 1978, for a total volume of nearly 100 MTOE. The region's share in world consumption of non-commercial energy was 14% in 1978, that is, almost the same as that of Black Africa alone, because South Africa non-commercial consumptions were negligible (3 MTOE).

But the share of commercial sources in the greater region's total consumption was twice that of Black Africa alone: 44% versus 22%, representing 1% of world demand of commercial energy.

(iii) The growth rates of the greater region's commercial and non-commercial energy consumption were very different in the past: 4.4% versus 2.5%. They will differ even more in the future under Scenario (I): 5.3% versus 2.2% by 2000, and 5.2% versus 1% by 2020. The relative importance of non-commercial sources will remain much greater under Scenario (II), where the difference between growth rates will become smaller, at least up to 2000: 3.9% for commercial energy versus 2.7% for non-commercial energy, compared to 4.5% and 1.5%, respectively, by 2020.

At any rate, the result of these growth rates will be that commercial energy will represent more than 50% of the greater region's total consumption in the future. This will come about sooner under Scenario (I) than under Scenario (II): 61% (I) or 50% (II) by 2000, and 78% (I) or 64% (II) by 2020. It will happen in spite of the high growth rates of non-commercial consumptions, exceptional in the Third World: 2% or 3% (II) by 2000, and 1% (I) or 1.5% (II) by 2020.

The income elasticity of the greater region's energy demand will have a rising trend: from 1.02 between 1960 and 1978 to 1.18 (I) or 0.98 (II) by 2000, and 1.48 (I) or 1.5 (II).

B. CONSUMPTION PER CAPITA

The greater region's energy consumption per capita was much more comparable to those of other Third World regions in 1978 than that of Black Africa alone. In fact, with a level of 0.51 TOE, it was much closer to the levels of 0.62 TOE for R9 and 0.65 TOE for R8. It also was far higher than the level of 0.25 TOE of R7.

The future growth rates of the greater region's per-capita consumption remain, however, much lower than the Third-World averages, especially up to 2000: 0.7% (I) or 0.2% (II) versus 2% (I) or 0.9% (II). The difference between growth rates will be reduced by 2020: 1.3% (I) or 0.5% (II) versus 1.6% (I) or 0.8% (II). Nevertheless, the greater region's consumption per-capita will be lower than South-East Asia's (R8) and even Communist Asia's (R9) by 2020. In fact, with 0.76 TOE (I) or 0.59 TOE (II), it will be the world's lowest, excepting South Asia.

But the deep changes in the greater region's demand pattern, especially the increasing substitution of non-commercial sources by commercial sources over the whole forecasting period, will no doubt be of even greater importance than the development of total per-capita consumption, as it will also be the case for Black Africa alone. The greater region's per-capita consumption of commercial energy amounted to 224 KOE in 1978, versus 284 KOE for non-commercial energy. By 2020, these consumptions will represent 568 KOE and 169 KOE, respectively, under Scenario (I) and 381 KOE and 211 KOE under Scenario (II). In other words, non-commercial consumptions will drop sharply under both scenarios. Concurrently, an increasing share of total consumption will be covered by new energies. These changes will take place in the context of a low growth of total energy consumption. The above aspects will certainly be the essential factors on the demand side of the greater region's — and Black Africa's — future energy development.

3.6.3.B Supply pattern

(i)   On the one hand, the greater region's energy balance, as compared to that of Black Africa (R6) alone, is primarily marked by the integration of South Africa's relatively massive coal production.

On the other hand, there is hardly any difference between the two 1978 supply patterns as far as the shares of oil (15%), gas (1%), and hydro-power (4%) are concerned. Indeed, the share of coal in the greater region's total supplies was 24% in 1978, while firewood represented 49%, and vegetable waste 7%. As compared to the supply pattern of Black Africa alone, this massive entry of coal decreases, therefore, almost exclusively the shares of these two latter sources, even if they still represented over one half of the greater region's total supply in 1978,

(ii) The decrease of the total share of firewood and vegetable waste in the greater region's energy supplies will also be reinforced by the gradual emergence of nuclear power and new energies, as well as gas and hydropower, over the whole forecasting period. In fact, coal will represent 29% (II) or 32% (I) by 2000, and 38% (II) or 44% (I) by 2020, while wood will represent 33% (I) or 42% (II) by 2000, and 18% (I) or 30% (II) by 2020 (although the quantities involved will be clearly greater: 160–200 MTOE).

(iii) The net volume of the greater region's future coal supplies will increase considerably: from 43 MTOE in 1978 to 100 MTOE (II) or 130 MTOE (I) by 2000, and 260 MTOE (II) or 380 MTOE (I) by 2020. The greater region's share in world consumption will also increase from 3% in 1978 to 4% in 2000, and 6–7% in 2020.

But the greater region will also have a significant share in the world demand of each of the other sources by 2020: 4% (I) or 3% (II) for oil, 5% for hydropower, 4% (I) or 5% (II) for new energies, 1% for gas, and even 1% for nuclear power. In addition, it will still be by far the world's first firewood consumer by 2020, with a share of 29% of world demand. Finally, it will still be one of the world's major consumers of vegetable waste by that time, with a share of 10–11% in total world demand.

(iv) One of the most positive results of the expansion of coal and the maintenance of a high firewood consumption will be the limited growth of the greater region's *oil* consumption: from 27 MTOE in 1978 to 126 MTOE (I) or 63 MTOE (II) by 2020. Under these conditions, the share of oil in total demand would remain practically stable at 15–16% up to 2020 (versus 15% in 1978) under Scenario (I), and it would even decrease considerably under Scenario (II) (9–10%). This means that the greater region would have a rather well balanced supply of alternative sources in the long run: natural gas 4% (I) or 2% (II), hydropower 8% (I) or 7% (II), nuclear energy 3% (I) or 2% (II), vegetable waste 4% (I) or 6% (II), and new energies 4% (I) or 6% (II). In this context, it is worth noting that nuclear power production would be 70 TWh (II) or 110 TWh (I) by 2020.

3.6.4.B Energy production

Apart from its contribution to world supplies of non-commercial energy, because of its oil production, Black Africa is also represented on the supply side of the world market of commercial energy. Owing to the integration of South Africa's coal production, the greater region's contribution to world energy supply is more diversified.

(A) The greater region's coal production was 53 MTOE in 1978, with a share of 3% in world coal production. Coal represented 19% of regional energy production in the same year, compared to 41% for oil (versus 52% for Black Africa alone), and 32% for firewood (versus 39% for Black Africa alone). As from 1978, commercial sources represented 63% of the greater region's total energy production, compared to 37% only for non-commercial energy.

(B) *Coal* and oil are the greater region's greatest assets for its future energy development, but its hydropower potential is also far from negligible. Coal production will be 150 MTOE (II) or 190 MTOE (I) by 2000, and 310 MTOE (II) or 470 MTOE (I) by 2020, when its level will be comparable to that of present US or Chinese production (380 MTOE and 310 MTOE, respectively). South Africa having hardly any prospects to become a significant crude oil or natural gas (as distinct from synfuels) producer, the data for the greater region's natural hydrocarbon supplies are the same as for Black Africa (R6) alone. But the increase of the greater region's coal production should, of course, take into account South Africa's plans for the expansion of its present coal gasification and liquefaction capacities.

Under these conditions, the share of coal in the greater region's production pattern will increase considerably: 29% (II) or 32% (I) by 2000, that is, about the same shares as for oil. By 2020, coal's share in total supply will rise to 37% (II) or 43% (I), versus 21% (II) or 23% (I) for oil, 15% (II) or 23% (I) for firewood (versus 32% in 1978), 7% for hydropower (versus 3% in 1978) 4% (II) or 5% (I) for new energies, and 2–3% each for natural gas and nuclear energy.

The greater region's shares in world production of commercial sources will also increase considerably by 2020: from 3% in 1978 to 7–8% for coal, from 4% to 6% for oil, and from 2% to 5% for hydropower. In addition, the greater region's production of new energies could represent 4–5% of world production.

In brief, with a population representing 15% of world population, the greater region's shares in total world supply and demand would be 6% and 5%, respectively, by 2020. This means that the greater region will be a net energy exporter by that time.

3.6.5.B Interregional energy trade

This can be ascertained in an analysis of the trade patterns which affect the region.

(A) The greater region was a rather important net energy exporter in 1978. Mainly because of its oil exports, its net exports represented +84 MTOE in that year. Total energy exports were composed of 97 MTOE of oil and 9 MTOE of coal in 1978, while South Africa imported 22 MTOE of crude oil. As far as natural gas is concerned, the greater region has not had, and will not have, any significant interregional trade.*

(B) The greater region's available coal export capacities will increase regularly in the future: 55 MTOE (I) or 40 MTOE (II) by 2000, and 80 MTOE(I) or 50 MTOE (II) by 2020. Its oil exports will increase sharply up to 2000, but level off after that date: 120 MTOE (II) or 130 MTOE (I) in 2000 as well as 2020. Oil imports will rise regularly: 30 MTOE by 2000, and 40 MTOE by 2020, under Scenario (I), while they will not exceed 20 MTOE under Scenario (II). In addition, the greater region will also become a small coal importer during the forecasting period. The result of these trade flows will be a more balanced export pattern as far as oil and coal are concerned, as well as a higher total energy trade surplus: 149 MTOE (I) or 138 MTOE (II) by 2000, and 164 MTOE (I) or 148 MTOE (II) by 2020. The two scenarios do not diverge very much in this respect. This means that the trade surplus will be proportion-

*Unless one assumes (unlike the RWT) a long-term development of LNG exports from Nigeria, Congo and Cameroon.

ately higher under Scenario (II), consequently leading to a better export situation than under Scenario (I).

In fact, the greater region's share in world energy exports will increase from 7% in 1978 to 10% by 2000, and 11% by 2020, under Scenario (I), and 11% and 13%, respectively, under Scenario (II).

Therefore, the greater region will be the *world's second energy exporter,* with an export volume that will be about equal to North America's. This simple statement is enough to underline the importance of the part that it will play in future world energy exports, especially as its share in interregional energy imports will remain around 2% over the whole forecasting period.

It is also worth noting that in 2020 the greater region will be the world's third coal exporter, after North America and the Industrialised Countries of the Pacific, but before Eastern Europe and Communist Asia.

In addition, the greater region will consolidate its position as the world's second crude oil exporter over the whole forecasting period.

In other words, the reinforcement of its interregional cooperation and the expansion of its energy exports will be vital for the greater region's future, because its economic development will depend to a very great extent on its export receipts.

3.7 SOUTH ASIA

3.7.1. Demographic and economic environment

(A) The subcontinent of India is one of the world's poorest and most populated regions. Population grew from 560 million in 1960 to 850 million in 1978. In fact, after Communist Asia, the region is the world's greatest concentration of human beings. Its population growth rate has been, and probably will be, exactly equal to the Third World average, in spite of all birth-control efforts: 2.3% from 1960 to 1978, 1.9% up to 2000, and 1.2% after that date. In other words, the region's population will represent nearly 1.3 billion by 2000, and 1.65 billion by 2020, when it will replace Communist Asia as the world's first power in demographic terms, with a share of 21% in world population, compared to 20% in 1978, and 19% in 1960.

(B) Unfortunately, the region's GNP is far from reflecting this demographic position. In fact, its GNP per capita was the lowest in the world in 1978: $177. Under both scenarios, this situation will not change by 2000 and 2020.

The growth of the region's GNP per capita has been also the lowest in the world between 1960 and 1978: 1.4%, compared to a world average of 2.8%. The region's GNP per capita represented only 9% of the world average in 1978, compared to 11% in 1960. In brief, its relative position became even worse.

Even under the best conditions, the future growth of the region's GNP per capita will be only comparable to that of a few other Third-World regions, but it will remain lower than the LDC average. As in 1978, it will only represent 9% of the world average by 2000, and 10% by 2020, when it will amount to $310 (II) or $440 (I). In other words, it will barely double between 1978 and 2020.

(C) In view of the region's enormous population growth, any increase in total GNP is hardly reflected in per-capita terms. Total GNP will increase from

G\$150 in 1978 to G\$500–720 by 2020, but its share in world GNP will still be only about 2% at that time, compared to a share of 21% in world population. In addition, the region's economic prospects are hardly favourable and will not contribute to improve GNP per capita in any significant way.

3.7.2. Energy consumption

The region's energy consumption per capita is very low. In addition, demand grows at a very low rate and is still largely made up of non-commercial sources. On the other hand, the balance between local supply and demand is quite satisfactory. These are the main features of the region's energy situation.

A. TOTAL CONSUMPTION

(i)   Total consumption rose from 119 MTOE in 1960 to 214 MTOE in 1978, at an average rate of +3.3% per year, compared to a Third World average of 4.5%. The growth rates of commercial and non-commercial consumptions were quite different: 5.1% and 2.2%, respectively.

(ii) Very much like in R6 (Sub-Saharan Africa), non-commercial sources represent still the greater part of the region's total energy consumption: 55% in 1978, versus 67% in 1960. The region was the world's second consumer of non-commercial sources in 1960 and 1978 (16% of world consumption).

Contrary to Africa, animal and vegetable waste is the region's most important non-commercial source. This is mainly due to the massive use of cow dung in the rural areas. The region is number three for fire wood, and number two for waste.

(iii) It is only a marginal consumer of commercial sources, representing hardly 2% of the world market in 1960 and 1978, with a volume of 100 MTOE in 1978, versus 40 MTOE in 1960.

(iv) This situation will not change fundamentally in the future, in spite of a number of inevitable developments in the long run. The region's share in total world demand was 3% in 1960 and 1978, and it will be only 4% in 2000 and 2020. Its share in world demand of commercial sources alone will increase also from only 2% to 3% over the whole forecasting period. In the field of non-commercial sources, the region will represent 17% (I) or 19% (II) of world demand by 2000, and 21% (I) or 24% (II) by 2020, that is, its share will be higher than Africa's. This is not a very positive development of the region's energy sector, which will remain more dependent on traditional sources than in other parts of the world.

(v)   Nevertheless, the total volumes involved will grow considerably: from 214 MTOE in 1978 to 460 MTOE (I) or 390 MTOE (II) by 2000, and 780 MTOE (I) or 570 MTOE (II) by 2020, but R7 will have the world's lowest total energy consumption in 2020, just as in 1978. Under these conditions, non-commercial sources will no longer be able to meet demand, that is, their share in total supplies will decrease to 38% by 2000, and 28% by 2020, under Scenario (I). It will decrease less under Scenario (II): 46% by 2000, and 40% by 2020 (with quantities that are double those of 1978: 228 MTOE).

It should be pointed out that the GNP elasticity of the region's commercial energy demand will decrease to 1.24 (I) or 1.16 (II) by 2000, and 1.0 (I) or 0.96 (II) by 2020, compared to an average of 1.34 between 1960 and 1978.

B. CONSUMPTION PER CAPITA

(i) The region's energy situation in per-capita terms may hardly be called rosy. In fact, its consumption per capita is the world's lowest: 0.25 TOE in 1978 and 0.21 in 1960, compared to a world average of 1.6 TOE, and a LDC average of 0.55 TOE, in 1978.

The growth prospects of the region's per-capita consumption are worse than those of the Third World average: 1.7% by 2000, and 1.4% by 2020 (I), and even only 0.8% under Scenario (II). This means that South Asia's consumption per capita will merely reach 0.36 TOE (I) or 0.30 TOE (II) by 2000, and 0.48 TOE (I) or 0.35 TOE (II) by 2020, that is a total increase of between + 0.10 TOE or + 0.23 TOE over 42 years. In this context, it should be recalled that the average per-capita consumption of the industrial countries will rise by + 1.5 TOE (II) and + 2.7 TOE (I).

(ii) Nevertheless, as in Africa, in spite of this rather poor performance of the growth of total per-capita consumption, there will be a concurrent and massive substitution of traditional sources by commercial sources. The respective shares of commercial and non-commercial sources in the region's total per-capita consumption were 0.11 TOE and 0.14 TOE, respectively, in 1978. Under Scenario (II), per-capita consumption of non-commercial sources will remain at a level of 0.14 TOE up to 2020, while per-capita consumption of commercial sources will increase to 0.16 TOE by 2000, and 0.21 TOE by 2020. Under Scenario (I), non-commercial consumption will be 0.13 TOE in 2000 and 2020, but commercial consumption will rise to 0.22 TOE by 2000, and 0.35 TOE by 2020.

3.7.3. Supply pattern

(i) The region has important coal reserves, even if this coal is not very high grade. Therefore, coal remained the region's predominant source of commercial energy between 1960 and 1978, with a share of 23%. This means that it was mainly oil whose share increased from 7% in 1960 to 14% in 1978, that compensated the relative decline in non-commercial consumptions. But hydropower, whose share rose from 2% in 1960 to 6% in 1978, also contributed in a significant way to this development. Nevertheless, fire-wood and vegetable waste, with shares of 33% and 22%, respectively, conserved their positions as the region's first and second energy source over this period.

(ii) The region's *coal* reserves represent its greatest asset. In fact, coal will cover 28% (I) or 26% (II) of needs by 2000, and 31% (I) or 26% (II) by 2020. The volumes involved will become significant, even if they will remain marginal in relation to world consumption (3% to 4% of the world market): 100–130 MTOE by 2000, and 150–250 MTOE by 2020, coming exclusively from regional production.

(iii) The rise of the share of hydrocarbons will remain moderate: from 16% in 1978 (1960: 8%) to 19% (II) or 24% (I), in 2000 and 2020. The share of gas in total energy supplies will rise faster than that of oil: from 2% in 1978 to 6% (I) or 4% (II) by 2000, and 8% (I) or 6% (II) by 2020.

Consequently, the increase of oil needs will be limited from 30 MTOE in 1978 to 85 MTOE (I) or 60 MTOE (II) by 2000, and 130 MTOE (I) or 75 MTOE (II) by 2020.

(iv) The volume of hydropower could grow considerably: from 13 MTOE in 1978 to 27 MTOE (I) or 20 MTOE (II) by 2000, and 54 MTOE (I) or 40 MTOE (II) by 2020, that is, about the equivalent of natural gas supplies. But this expansion will be barely sufficient to maintain hydropower's share at a level of 6-7% in the region's energy balance.

Nuclear energy will make its appearance in a very modest way by 2020, when it might represent 2–3% of total needs, or 24 MTOE under Scenario (I), and 12 MTOE under Scenario (II). The penetration of new energies will be more pronounced by 2020: 60% (II) or 7% (I), versus 2% in 2000 under both scenarios.

(v) The relative decline of total non-commercial consumptions has already been pointed out. In fact, fire-wood will be replaced as number one by coal as from 2000, but its total consumption could rise nevertheless from 70 MTOE in 1978 to 100 MTOE by 2000, and around 110 MTOE by 2020. It is easy to imagine the challenge represented by this growth in volume in a region threatened by an accelerated deforestation. As far as animal and vegetable waste is concerned, this source will practically conserve its share in the region's energy balance under Scenario (II): 22% in 1978, 21% in 2000, and 20% in 2020. Under Scenario (I), this share will drop more sharply: 17% in 2000, and 14% in 2020.

However, the volume of the total consumption of animal and vegetable waste will rise considerably: from some 50 MTOE in 1978 to 80 MTOE by 2000, and about 110 MTOE by 2020. For reasons of policy or necessity, non-commercial sources will, therefore, play an important part in the region's energy supply, even in the long run, exactly as in the other poor regions of the Third World.

The share of the region in the world consumption of commercial sources will be only marginal. In fact, it will be under 5% for each commercial energy source. But the region's share in world consumption of animal and vegetable waste will be equal to that of Communist Asia as from 2000, and it will be higher by 2020. As far as fire-wood is concerned, South Asia will become the world's second consumer by 2020, after Black Africa.

3.7.4. Energy production

The persistently great part played by non-commercial sources, as well as its relatively abundant coal reserves, should make it possible for South Asia to develop its total energy production at a rate corresponding almost to its total needs.

(i) Coal, whose predominant share in commercial supplies was already pointed out, will, of course, also dominate commercial production. The expansion of the region's coal production will result in a corresponding expansion of its coal demand. Therefore, no coal should be available for exports, because the regional market's absorption capacity will be far from saturation.

The future development of the region's coal production will, therefore, correspond very much to the future development of coal consumption from 50 MTOE in 1978 to 104 MTOE (II) or 132 MTOE (I) by 2000, and 151 MTOE (II) or 254 MTOE (I) by 2020. Consequently, coal production will represent 29% by 2000 and 33% by 2020 of regional supply under Scenario (I), and 27% under Scenario (II), compared to 24% in 1978,

But it should be pointed out that R7's share in world coal supply will remain marginal only, even in the long run: 3-5%.

(ii) The region's prospects in the field of natural gas are similar, but conditions in this respect are more favourable. R7 may not only reasonably expect to cover its own needs without any great problem, but it will no doubt also develop a certain export capacity. In fact, natural gas production could increase from 8 MTOE in 1978 to 40 MTOE (I) or 27 MTOE (II) by 2000, and 80 MTOE (I) or 54 MTOE (II) by 2020. This increase will not change the region's share in world production, which will remain at a level of 2% by 2000 as well as 2020. Nevertheless, natural gas will increase its share in total regional energy supplies from 4% in 1978 to about 10% by 2020.

(iii) By the nature of things, supply and demand are balanced for hydropower, nuclear power, new energies, and non-commercial sources. Consequently, only oil will remain a problem for R7. In fact, in order to reduce costly oil imports, considerable efforts will be made to limit its future share in the region's energy balance, even if regional oil production will rise from 12 MTOE in 1978 to 55 MTOE (I) or 40 MTOE (II) by 2000, and 85 MTOE (I) or 50 MTOE (II) by 2020. This means that oil's share in regional production will be higher than 10% as from 2000. However, R7's share in world oil supply will remain under 1% or 2% over the whole forecasting period.

(iv) The overall result of the developments outlined above will be that R7 will practically achieve a continuous balance between its total energy supply and demand. The massive penetration of non-commercial sources will certainly be an essential factor in this context, at least to the extent that they will continue to represent a large part of regional energy production: 59% in 1978, 39% or 47% (I and II) by 2000, and 28% (I) or 40% (II) by 2020, with a total volume that will practically double by 2020: 118 MTOE in 1978, 214 MTOE (I) or 228 MTOE (II) in 2020. This also means that energy independence will increase from 94% in 1978 to 96% (I) or 98% (II) in 2000, and 98% (I) or 100% (II) by 2020. However, this balance between supply and demand depends on the maintenance of a demand level which, in fine, is very low, as well as the success of the fast expansion of the region's coal production capacity, and the possibility to maintain the high level of non-commercial energy use for a very long time. In view of all these uncertainties, this fragile balance might easily be upset.

3.7.5. Interregional energy trade

(i) The region's self-sufficiency in the field of coal, as well as the potential increase of natural gas exports from 3 MTOE in 1978 to 10 MTOE by 2000, and even 20 MTOE by 2020, were already underlined above. On the negative side, there is the necessity to import a far from negligible part of the region's oil needs. R7's crude imports (22 MTOE) represented three fourths of its energy-related oil consumption in 1978. It may be reasonably expected that the expansion of the region's oil production will make it possible to contain this import dependence to a level of 50% up to 2000 and even 2020, with import volumes representing 42 MTOE (I) or 30 MTOE (II) in 2000, and 61 MTOE (I) or 37 MTOE (II) in 2020.

(ii) The balance of the region's total interregional fuel trade will remain negative, with a tendency to deteriorate in terms of volume: from −19 MTOE in 1978 to −36 MTOE (I) or −25 MTOE (II) by 2000 and −47 MTOE (I) and −27 MTOE by 2020. However, the share of this deficit in total regional energy demand will be less important: 9% in 1978, 8% (I) or 6% (II) in 2000, and 6% (I) or 5% (II) in 2020. It may, therefore, be assumed that South Asia's energy situation will not be too precarious, at least as far as its interregional energy trade balance is concerned. In addition, the share of its total energy imports in world energy imports will be very marginal: 1% in 1978, 2% in 2000, and 3% in 2020, compared to a share in world energy exports of hardly 1% in 2020.

As a matter of fact, South Asia should remain relatively protected from possible upheavals in the world energy market.

3.8 SOUTH-EAST ASIA

This region, which comprises a number of new industrialized countries, has been able to achieve high economic growth rates. In addition, its growth prospects are relatively promising compared to those of other Third-World regions. This section will analyse the conditions and consequences of this situation in the field of energy.

3.8.1. Demographic and economic environment

(A) South-East Asia's population has grown at a very fast rate: +3% between 1960 and 1978. In fact, its population growth rate was the highest in the world, and its total population increased from 220 million to 338 million over this period. However, the region's future population growth rates should be much lower, even lower than the Third World average (unlike those of Sub-Saharan Africa): 1.8% by 2000, and 1.1% by 2020, compared to a Third World average of 2.35% between 1960 and 1978, 1.9 by 2000, and 1.4% by 2020. Under these conditions, R8's total population could reach 500 million by 2000, and 630 million by 2020. At these rates, the region should maintain its share in world population of 8% over the whole forecasting period.

(B) The region's share in world GNP was 2.5% in 1978, versus 1.7% in 1960. Its total GNP represented G$220 in 1978, versus G$63 in 1960, but the growth rate of its GNP was the second highest in the Third World (after that of R5 (North Africa and Middle East) between 1960 and 1978: 7.2%, compared to a Third-World average of 5.9%.

The region's future GNP growth rates will remain high: 6.2% by 2000, and 4.2% by 2020 under Scenario (I), and 4.5% by 2000, and 3% by 2020 under Scenario (II). This means that the region's GNP would represent G$1850 (I) or G$1050 (II) by 2020. In other words, it would be multiplied by 5 (II) or 8 (I) between 1978 and 2020. Under these conditions, R8's share in world GNP would increase from 2.5% in 1978 to 4.2% (I) or 3.7% (II) by 2000, and 5.3% (I) or 4.4% (II) by 2020, for a population representing a steady 8% of world population between 1978 and 2020. This means that the situation of R8's population will continuously improve, in the medium as well as the long term. (C) This is also shown by the development of the region's GNP per capita. Again, its growth rate (4.7%) was the second highest in the Third World, after that of R5 (North Africa and Middle East) between 1960 and 1978. But R8 will definitely have the highest growth of per-capita GNP under both scenarios in 2000 and 2020. In fact, its GNP per capita will increase from some $650 in 1978 to $1640 (I) or $1150 (II) by 2000, and $3000 (I) or $1700 (II) by 2020. The increase of R8's per capita GNP is also impressive in comparison with the industrial countries. It represented only 9% of the average of these countries in 1960, and 11% in 1978, but it will represent 15% (I) or 13% (II) by 2000, and 19% (I) or 15% (II) by 2020. Compared to the Third-World average, the increase is not less impressive: 108% in 1960, 132% in 1978, and 160% as from 2000.

South-East Asia should, therefore conserve a relatively favourable situation in demographic and economic terms, as compared to other Third-World regions.

3.8.2. Energy consumption

A. TOTAL CONSUMPTION

(i) The region's economic growth is well reflected by its energy development. Again, the growth rate of the region's energy demand was higher than that of any other Third-World region, except R5 (North Africa and Middle East): over 5% per year between 1960 and 1978. But the growth rates of commercial and non-commercial consumptions were very different: +8.8% and +2.7%, respectively.

The total volume of the region's energy demand had been the lowest in the world in 1960, except for R5. The region was already number seven in 1978, before Sub-Saharan Africa and South Asia, with an increase from some 90 MTOE in 1960 to some 220 MTOE in 1978.

(ii) Future growth rates of total energy consumption will be closer to the Third-World average up to 2000: 4.6% (I) or 3.2% (II). They will be even lower than the Third World average between 2000 and 2020: 2.9% (I) or 1.9% (II). Total demand in volume will rise to 584 MTOE (I) or 436 MTOE (II) by 2000, and 1030 MTOE (I) or 640 MTOE (II) by 2020. The region's share in world energy consumpton will rise accordingly: from 3% in 1978 to 5% (I) or 4% (II) by 2000, and 6% (I) or 5% (II) by 2020. Under Scenario (I), R8 will preserve its position as the world's seventh energy consumer up to 2020, while it will be only number nine under Scenario (II).

The GNP elasticity of the region's commercial energy demand will decrease continuously and sharply: 1.22 between 1960 and 1978, 1.05 (I) or 1.0 (II) between 1978 and 2000, 0.80 between 2000 and 2020.

B. CONSUMPTION PER CAPITA

South-East Asia has the third highest energy consumption per capita among the Third-World regions as from 1978, after Latin America and R5 (North Africa and Middle East). It will preserve this position up to 2020. In fact, the growth rate of its per-capita consumption will be clearly higher than the Third-World average up to 2000, and more or less the same after that date.

The region's consumption per capita increased from 0.40 TOE in 1960 to 0.65 TOE in 1978. It will reach 1.16 TOE (I) or 0.87 TOE (II) by 2000, and even 1.6 TOE (I) or 1.0 TOE by 2020. This means that its 2020 level will not be far from Europe's 1960 level, but 60 years later.

3.8.3. Supply pattern

(i)   As in most Third-World regions, non-commercial sources play still a great part in South-East Asia's supplies: 101 MTOE in 1978, or 46% of total supplies, but this must be compared to the 1960 level of 63 MTOE, or 71% of total supplies at that time, when R8 was the world's number one for its dependence on non-commercial sources.

This will change in the future. Under Scenario (I), which is based on a rather higher industrialization rate, the volume of R8's traditional consumptions should reach a ceiling of 112 MTOE. Under Scenario (II), the volume of non-commercial sources will increase from 101 MTOE in 1978 to 125 MTOE by 2000, and even 138 MTOE by 2020, but at a growth rate of only 0.7% per year.

R8's per-capita consumption of non-commercial sources will drop sharply: from 0.30 TOE in 1978 to 0.22 TOE (I) or 0.25 TOE (II) by 2000, and 0.18 TOE (I) or 0.22 TOE (II) by 2020. The share of non-commercial sources in total supply will decrease to 19% by 2000, and 11% by 2020, under Scenario (I), and to 29% by 2000, and 21% by 2020, under Scenario (II). But it should be noted that R8 had still the world's highest per-capita consumptionof non-commercial sources in 1978.

There are already strong supply pressures in the field of fire wood. The region's demand increased from 43 MTOE in 1960 to 63 MTOE in 1978, but any significant demand growth seems excluded in the future: 65 MTOE (I) or 70 MTOE (II) by 2000, 55 MTOE (I) or 65 MTOE (II) by 2020. This shows that demand volume will decrease as from 2000.

The region's supplies of animal and vegetable waste will grow moderately but continuously: from 38 MTOE in 1978 to 47 MTOE (I) or 55 MTOE (II) by 2000, and 57 MTOE (I) or 73 MTOE (II) by 2020. The growth of the available volume of these sources will depend on to the development of agricultural production, which, fortunately, will not be subject to the same constraints as fire wood.

R8's future share in the world supply of fire-wood will decrease to 9–11% (13% in 1978), and its share in animal and vegetable waste will be 15–18% (15% in 1978). But its share in total world supply of non-commercial sources will vary little: 14% in 1978,12% in 2000, 13% in 2020.

(ii)   It will not be the same with commercial sources, where South-East Asia's share in world supply will increase from 2% in 1978 (as in 1960) to 4% in 2000, and 5% in 2020.

As in the two other Third World regions with high per-capita demand levels, the share of *oil* in R8's total consumption increased considerably: from 17% in 1960 to 42% in 1978. R8 is the Third-World region where the increasing penetration of oil was by far the most pronounced in the past. In fact, oil's share in the region's energy balance increased by +25% between 1960 and 1978.

R8's oil consumption increased from 15 MTOE in 1960 to 92 MTOE in 1978. Under Scenario (I), it will rise dramatically to 247 MTOE by 2000. This growth will slow down considerably after 2000: 310 MTOE by 2020.

Under Scenario (II), the part played by oil will be much more modest in the future: 132 MTOE by 2000, and only 80 MTOE by 2020. At least under Scenario (I), oil substitution will take place at a moderate rate. Oil's share in the region's energy balance will represent 42% by 2000, and still 30% by 2020. In addition, oil will remain the region's first source of supply, even by 2020. Under Scenario (II), the region's efforts to contain oil's penetration will be more successful. Oil's share will be 30% in 2000, and only 13% in 2020. In addition, oil will be only the third source of supply by 2020. Under Scenario (II), the region's share in world oil consumption will remain marginal: 5% by 2000, and 3% by 2020, compared to 4% in 1978. But this share will be significant under Scenario (I): 7% and 8%, respectively. As will be shown later on, the region's main problem in the field of energy will be its insufficient energy production in the long run.

(iii) In order to contain as far as possible the inevitable increase of the penetration of oil, South-East Asia, like other regions, will turn to *coal,* which covered 7% of needs 1978, with a volume of only 14 MTOE. The volume of R8's coal consumption will rise to 110 MTOE (I) or 100 MTOE (II) by 2000, and 260 MTOE (I) or 170 MTOE (II) by 2020. Coal's share in the region's energy balance will represent 19% (I) or 23% (II) by 2000, and 25% (I) or 27% (II) by 2020. In fact, it will the region's second source of supply by 2020, and even the first under Scenario (II).

Nevertheless, the region's future share in the world coal market will be modest: 3% by 2000, and 5% (I) or 4% (II) by 2020, but this should be compared to a share of merely 1% in 1978.

(iv) *Natural gas* will also be used to contain the penetration of oil, but this will begin at a later date. In fact, the region's natural gas consumption will increase from only 7 MTOE in 1978 to 50 MTOE (I) or 30 MTOE (II) by 2000, and even 150 MTOE (I) or 100 MTOE (II) by 2020. Its part in R8's total supplies will be 9% (I) or 7% (II) by 2000, and 14% (I) or 16% (II) by 2020, compared to 3% in 1978. This will make gas the region's third source of supply. R8's share in the world gas market will be 2% by 2000, and 5% (I) or 4% (II) by 2020, compared to 1% in 1978.

This shows that R8 tends to arrive at a more balanced supply pattern as far as the three fossil fuels are concerned.

(v) *Hydropower's* share in the region's energy balance will increase significantly, especially after 2000. It was 2% in 1978 as well as 1960, but could increase to 3% (I) or 4% (II) by 2000, and even 8% (I) or 9% (II) by 2020. Under Scenario (II), South-East Asia would then be the world's number two for the share of hydropower in total energy supplies, and number three under Scenario (I). R8's hydropower production would rise from 5 MTOE in 1978 to 20 MTOE (I) or 17 MTOE (II) by 2000, and even to 80 MTOE (I) or 60 MTOE (II) by 2020. Concurrently, R8's share in world hydropower production would rise from 1% in 1978 to 3% by 2000, and 6% by 2020.

(vi) The region produced no *nuclear power* in 1978, but it will develop nuclear energy at a faster rate than any other Third World region: 35

MTOE (I) or 20 MTOE (II) by 2000, and 80 MTOE (I) or 40 MTOE (II) by 2020. R8's share in world nuclear energy would be 3% by 2000, and 4% by 2020, under Scenario (I), while it would be 2% under Scenario (II).

(vii) Finally, *new energies* could furnish a far from negligible contribution, especially after 2000. Their consumption would exceed 10 MTOE by 2000, and would reach 40 MTOE (I) or 50 MTOE (II) by 2020, representing 4% or 8% of the regional energy balance. The region's share in world demand of new energies would be 3% (I) or 5% (II) by 2000, and 6% (I) or 5% (II) by 2020.

It is remarkable that South-East Asia will arrive at a more balanced supply pattern than that of any other region, mainly because of the differences between the future growth rates of its supply sources. In fact, if these forecasts prove just, the share of no single source will represent less than 4% in the region's energy balance by 2020 under Scenario (I), and even at least 6% under Scenario (II).

The relatively small difference between the shares of the first and the last source in the region's energy balance also shows that South East Asia will diversify its supplies more than any other region. Of course, the drawback of this policy is that the region cannot do without any of these sources, because any interruption of supplies would threaten the balance between total energy supply and demand.

3.8.4. Energy production

If the demand side of the region's energy balance is positive, it is not so with the production side. In fact, the region's supply and demand patterns were well balanced in 1978, but they will have diverging trends in the future. This will create serious problems for the region's energy independence, especially under Scenario (I).

(A) South-East Asia's oil producton of 100 MTOE is at present its best asset in the field of international energy trade. Unfortunately, development prospects of the region's oil production are not very favourable. Under Scenario (I), oil production will be 150 MTOE by 2000, but it seems inevitable that it will decrease to only 120 MTOE by 2020. Under Scenario (II), oil production will begin to decline even before 2000. In fact, it would represent only 85 MTOE by 2000, and even only 50 MTOE by 2020. Under these conditions, oil's share in total regional supply would fall from 43% in 1978 to 30% (I) or 21% (II) by 2000, and 16% (I) or 8% (II) only by 2020. The region's share in world oil supply would be barely maintained at 2% (II) or 3% (I) by 2020 (3% in 1978).

(B) This negative development of the region's oil production makes its prospects in the field of coal production even more important. South-East Asia's coal production will increase from 12 MTOE in 1978 to 70 MTOE (I) or 80 MTOE (II) by 2000, and 130 MTOE (I) or 150 MTOE (II) by 2020. It is worth noting that, exceptionally, prospects under Scenario (II) are better than under Scenario (I) as far as the compensation of the sharp decline of oil production is concerned. This means that coal's share in total regional energy production will increase from 5% in 1978 to 14% (I) or 20% (II) by 2000, and 17% (I) or 24% (II) by 2020. Nevertheless, the region's share in world coal production will be only around 3% over the whole forecasting period, but coal will replace oil to a large extent in the future.

(C) It is in the field of natural gas that production is expected to increase fastest: from 18 MTOE in 1978 to 100 MTOE (I) or 65 MTOE (II) by 2000, and 200 MTOE (I) or 130 MTOE (II) by 2020. This means that the region's natural gas production would double between 2000 and 2020 under both scenarios. Under these conditions, the share of natural gas in South-East Asia's total energy supply would increase from 7% in 1978 to 20% (I) or 16% (II) by 2000, and 26% (I) or 21% (II) by 2020. The region's share in world supply of natural gas would rise from 2% in 1978 to 4% (I) or 3% (II) by 2000, and 6% (I) or 5% (II) by 2020.

(D) The development of the supply volumes of all other sources will correspond to demand forecasts. However, the share of each of these sources in total regional supply will be slightly higher than in the demand pattern by one to two points on the average, especially under Scenario (I), where total supply will become more and more insufficient in relation to total demand.

3.8.5. Interregional energy trade

The increasing divergence between the region's supply and demand will inevitably create supply tensions. The following two observations,based on Annexes 15 and 16, well illustrate this problem.

(A) On the one hand, South-East Asia's share in world energy consumption will increase from 3% in 1978 to 5% (I) or 4% (II) by 2000, and 6% (I) or 5% (II) by 2020. On the other hand, the region's share in world energy production (3% in 1978) will level off at 4% as from 2000. The difference between supply and demand will, therefore, increase constantly.

(B) In addition, total supply represented 108% of total demand in 1978, but will represent only 93% by 2000, and 97% by 2020, under Scenario (II), and only 85% by 2000, and 74% by 2020, under Scenario (I). The energy independence of other regions (R4, R5, R6, R10) will also decrease, but their supply and demand will always be at least balanced. In fact, South-East Asia is the only region whose energy independence will decrease to the point of having a durable deficit in its interregional energy trade.

(C) The region's total trade deficit for fossil fuels was only 3 MTOE in 1978, but it will rise to 117 MTOE by 2000, and even 305 MTOE by 2020, under Scenario (I). It will be much lower under Scenario (II): 52 MTOE by 2000, and 35 MTOE by 2020.

(D) Nevertheless, the 1978 deficit, though it was very small, concealed strong negative energy trade trends. In fact, the region's high crude exports (73 MTOE) were compensated already then by even higher oil imports (85 MTOE). This situation was reflected in the region's share in world oil exports and imports (5% and 6%, respectively). In fact, gas exports (over 10 MTOE in 1978) compensated more or less the region's trade deficit in oil already at that time.

(E) Three trends will strongly mark South-East Asia's future international energy trade. Coal imports will increase substantially under Scenario (I): 40 MTOE by 2000, and 130 MTOE by 2020. They will increase only moderately under Scenario (II): 20 MTOE in 2000 and 2020. Oil exports will completely disappear under both scenarios, while oil imports will increase substantially under Scenario (I): 127 MTOE by 2000, and 225 by 2020. Oil imports will increase much less under Scenario (II): 67 MTOE by 2000, and 45 MTOE by 2020. These negative developments will be only slightly compensated by the rise of gas exports up to 2000: 50 MTOE (I) or 35 MTOE (II). But even gas exports will level off at 50 MTOE (I) or 30 MTOE (II) by 2020.

(F) This means that South-East Asia will be the world's fourth oil importer by 2000, with a share of 10% in world oil imports, and the second (after Western Europe) with a share of 22% by 2020, under Scenario (I). Its share in world oil imports will be much lower under Scenario (II): 6% to 7% (as in 1978) over the whole period.

At the same time, the region's share in world coal imports will increase from 3% in 1978 to 12% by 2000, and 18% by 2020 under Scenario (I), that is, it will be number three in 2000, and even number two in 2020, after Western Europe. Under Scenario (II), its share in world coal imports will be only 8% by 2000, and 6% by 2020.

In addition, the region will not be able to maintain its share in world gas exports: 18% in 1978, 20% (I) or 18% (II) by 2000, 14% by 2020. Nevertheless, it will remain the world's third gas exporter (which it was already in 1978) over the whole forecasting period.

The above developments show that the balance between the region's supply and demand will become more and more precarious, in spite of an economic environment which would normally be considered as favourable because of high growth. The simple fact is that South-East Asia has not the energy resources necessary to cover the growth of its demand. Consequently, it will have to adapt to an increasing energy dependence, and that at a time when world-wide conditions will be more and more unfavourable. Nevertheless, in this context it should be recalled that Scenario (I) is based on a high degree of international cooperation as well as *open markets*. This implies inevitably a higher degree of uncertainty, while Scenario (II) is based on much more conservative assumptions. The region's implicit policies under each scenario are, therefore, well in line with their diverging trends. As a matter of fact, South-East Asia's acceptance of its precarious energy balance under Scenario (I), which is similar to that of Western Europe, at least in relative if not in quantitative terms, may well be the key to its greater success in economic terms.

3.9 CENTRALLY PLANNED ASIAN COUNTRIES

R9, or Communist Asia, is one of the world's most striking regions because of the density of its population, the specificity of its development mode, and the enormous relative weight of China, the third superpower. In the field of energy, Communist Asia's most striking feature is represented by its coal resources, which have deeply marked its history and will have a far-reaching impact on the future development of this very particular region.

3.9.1. Demographic and economic environment

(A) Communist Asia is the only region whose total population exceeds one billion. This means that one human being in four is an inhabitant of this region (24% of world population).

But the population growth rate of Communist Asia is already slowing down: 2% between 1960 and 1978, versus 2.5% for the rest of the Third World. This slow-down should continue in the future: 1.3% by 2000, and 0.7% only by 2020 (that is the level of the Western Europe countries at that time), compared to 2.2% and 1.7%, respectively, for the rest of the Third World. This means that the region's population would represent 1360 million by 2000, and 1580 million by 2020, when its share in world population would be only 20%.

(B) Communist Asia's population is still very poor, at least as measured by its GNP per capita:* hardly $240 in 1978, compared to over $600 for the rest of the Third World. In fact, the region's GNP per capita was only slightly higher than that of South Asia in1978. Nevertheless, the growth of Communist Asia's GNP per capita was far from negligible between 1960 and 1978: 3.2%, that is, it was practically in line with the rest of the Third World.

But in the medium term this growth may well slow down considerably, in particular under Scenario (II), where the growth rate of the region's GNP per capita will be near to zero by 2000, compared to a growth rate of 2.5% under Scenario (I), which arrives at much more optimistic results. The constraints imposed upon the growth of the region's GNP per capita will be reduced gradually between 2000 and 2020, when rates will be 3.2% (I) or 1.6% (II). But the level of per capita will still be very low by 2020: $335 under Scenario (II), that is, hardly $100 more than in 1978, and $772 under Scenario (I), that is, a multiplication by three relative to 1978. In other words, Communist Asia's GNP per capita would reach that of Black Africa by 2020 under Scenario (I).

Communist Asia's GNP per capita would only represent 35% (I) or 25% (II) of the average of the rest of the Third World, compared to 38% in 1978 and 40% in 1960. In other words, the situation of the region's population will comparatively deteriorate in the long run.

(C) Because of its enormous population, the total volume of Communist Asia's GNP tends to conceal this. In fact, with a total GNP of G$250 in 1978 Communist Asia was number three among the Third World regions. In addition, the growth rate of its GNP volume was 5.3% between 1960 and 1978. The region will be number four by 2020 under Scenario (I), with a volume of G$1200, and a growth rate of around 4%. Under Scenario (II), it will be number five by a narrow margin, with a total GNP only a little higher than that of South Asia (G$528), and an average growth rate of 1.8% over the whole forecasting period. In other words, Communist Asia's GNP will be multiplied by 5 between 1978 and 2020 under Scenario (I), but only doubled under Scenario (II). Its share in world GNP will not change significantly up to 2020: 3.5% (I) and 2.2% (II), compared to 2.9% in 1978. Like South Asia and, to a lesser degree, Black Africa, this giant in demographic terms will remain a dwarf in terms of economic development.

3.9.2. Energy consumption

A. TOTAL CONSUMPTION

Communist Asia was the world's fourth energy consumer, after the three main industrial regions (R1,R2, and R4). It will preserve this position by a narrow margin up to 2000 but it will be replaced by Latin America (R10) shortly after 2000.

The region's total energy consumption increased from 290 MTOE in 1960

*As estimated by the World Bank and thus adopted in this study. However, the estimates made by the CIA, which are based on the commercial value of goods and services at international prices, produces much higher figures.

to 638 MTOE in 1978, at a rate (+4.5% by year) comparable to the LDC average. Of course, commercial consumptions increased much faster than non-commercial consumptions: 6.2% and 2.2%, respectively. Communist Asia's share in world energy consumption was 9% in 1978.

Communist Asia's future development in the field of energy will be slightly different.

The region's total energy consumption will grow at a slower rate than that of any other Third-World region between 1978 and 2000, as well as between 2000 and 2020, and this under both scenarios. In fact, the growth rate of the region's total consumption will decrease from 4.5% before 1978 to 2.3% by 2000, and 1.5% by 2020, under Scenario (I). It will even decrease to 1.2% by 2000, and 0.9% by 2020, under Scenario (II). This unfavourable development will be due to economic difficulties.

Under these conditions, the volume of Communist Asia's total consumption will be 1060 MTOE (I) or 840 MTOE (II) by 2000, and 1440 MTOE (I) or 1000 MTOE (II) by 2020. In addition, the region's share in world energy demand will slightly decrease by 2020: from 9% in 1978 to 8% (I) or 7% (II).

Communist Asia will be the first Third-World region to have negative growth rates in the field of non-commercial consumptions. In addition, the growth rates of its commercial consumptions will be much lower than the average of the rest of the Third World. In fact, they will be lower by 3 points up to 2000, and 1–1.5 points after that date.

This means that Communist Asia's situation in terms of energy growth will deteriorate as compared to that of the other Third-World regions, and this without any explanation resulting from material differences in economic fundamentals.

Nevertheless, the volume of Communist Asia's commercial consumptions will double by 2000 (860 MTOE, versus 420 MTOE in 1978) and treble by 2020 (1360 MTOE) under Scenario (I), but it will only represent 850 MTOE by 2020 under Scenario (II), that is, the volume in 2000 under Scenario (I).

The region's non-commercial consumptions will fall sharply in the long run from 220 MTOE in 1978 to 204 MTOE (I) or 272 MTOE (II) by 2000, and only 80 MTOE (I) or 160 MTOE (II) by 2020, when they will have been replaced massively by commercial sources. This means that their large-scale substitution will only begin after 2000, that is, at a time when the region's economic growth will have become relatively faster in an more constrained world-wide context. Of course, this substitution will be easier in a period of more sustained economic growth.

The GNP elasticity of Communist Asia's commercial energy demand was 1.17 between 1960 and 1978, exactly in line with the Third World average. Future elasticity will differ according to the scenario. It will be much lower than that of the rest of the Third World under Scenario (I): 0.87 by 2000, and 0.58 by 2020, compared to 1.07 and 0.91, respectively, for the rest of the Third World. It will be very close to that of the rest of the Third World under Scenario (II): 1.04 by 2000, and 0.85 by 2020.

It is also worth noting that Communist Asia's total energy intensity, that is, the ratio between total consumption and total GNP, was exceptionally high (nearly 2.6 KOE per $) in 1978, compared to an average of 0.8 KOE per $ for the rest of the Third World, and 0.7 KOE per $ for the average of the industrial countries. The region's energy intensity will decrease in the future, but faster under Scenario (I) than under Scenario (II), as in most other regions. But it will remain much higher than the world average, even if the difference will be smaller than in 1978: 1.2 KOE per $ (I) or 1.9 KOE $ (II) by 2020, compared to 0.5 KOE per $ (I) or 0.6 KOE per $ (II) for the world average.

B. CONSUMPTION PER CAPITA

The development of Communist Asia's per-capita energy consumptions reflects its overall economic and energy development.

Communist Asia has been, and will remain, number four among the ThirdWorld regions as far as per-capita consumption is concerned, before South Asia and Black Africa. The region's consumption per capita represented 0.62 TOE in 1978, versus 0.41 TOE in 1960. It will represent 0.78 TOE by 2000, and 0.91 TOE by 2020 under Scenario (I). It will represent 0.62 TOE by 2000, and 0.64 TOE by 2020, under Scenario (II). In other words, it will increase by 50% only over 42 years under (I), and it will practically be stagnant under (II). The total volume of Communist Asia's non-commercial consumptions in 1978 was impressive, but, in per-capita terms, it was lower than that of other Third World regions: 0.21 TOE versus 0.30 and 0.29 TOE for R8 and R6.

This means that Communist Asia's population, like that of South Asia and Black Africa, is still far from getting out of its relative destitution in terms of energy, in spite of a far from negligible growth of total energy consumption and the efforts made in the field of birth-control. Such is the tragic fate of these overpopulated and underdeveloped nations, which are practically continents: any increase in the volume of energy consumption or GNP, even if it is on a massive scale, is enormously diluted because of their vast population. In other words, it is hardly reflected in per-capita terms expressed in the form of statistical averages.

3.9.3. Supply pattern

(i)   Because of the exceptionally great part played by *coal,* Communist Asia's supply pattern is unlike that of any other region. The region became the world's third producer and consumer of coal between 1960 and 1978, before Western Europe. Hard coal covered almost one half (48%) of the region's total needs in that year. Communist Asia's coal consumption grew at an average rate of +5% during the above period, that is, faster than that of any other region of the world. The region represented 10% of world coal demand in 1960, but 18% in 1978, with 306 MTOE.

(ii)   The growth of Communist Asia's oil demand was also the highest in the world between 1960 and 1978: nearly 14% per year. The volume of its oil demand rose from barely 10 MTOE in 1960 to almost 90 MTOE in 1978. Under these conditions, the share of *oil* in the region's energy balance increased from only 3% in 1960 to 14% in 1978. This means that Communist Asia, despite its geographical isolation and its strategy of giving priority to coal, was not spared from the general penetration of oil that marked the energy growth of all regions, whatever the degree and the type of their development or the nature of their economic system.

(iii)   *Non-commercial energy* has been, and will be, the region's great energy resource besides coal and oil. In fact, Communist Asia was the world's first consumer of non-commercial sources in 1978. Its part in total world consumption was 30% in that year, while its shares in world consumption of fire-wood and vegetable waste alone were 26% and 38%, respectively. The region's share in world consumption of non-commercial sources even rose between 1960 and 1978: from 27% to 30%. The volume of Communist Asia's non-commercial energy consumptions was impressive: about 220 MTOE in 1978, versus 150 MTOE in 1960. Nevertheless, the share of non-commercial sources in the region's energy balance decreased sharply between 1960 and 1978: from 51% to 34%. But fire wood and vegetable waste were still Communist Asia's second and third energy sources in 1978 (20% and 14%, respectively).

Natural gas and hydropower were only complementary energy sources for Communist Asia in 1978, each covering 2% of total needs. Nevertheless, the region's hydropower production (14 MTOE) represented 4% of world supply in 1978, that is, it was by no means negligible.

Communist Asia's future energy development will consolidate past trends, but it will also be marked by the diversification of supplies.

(iv) Coal's predominant position will not change. In fact, its share in the region's energy balance will practically remain at its 1978 level: 48% (I) or 45% (II) by 2000, and 49% (I) or 48% (II) by 2020, compared to 48% in 1978. This shows to which extent the region's energy future will depend on its capacity to develop its vast coal resources. Of course, under Scenario (I), coal demand will grow faster: 500 MTOE by 2000, and 700 MTOE by 2020, versus 300 MTOE in 1978. In spite of this growth of coal demand, Communist Asia's share in the world coal demand will gradually decrease: 16% by 2000, and 12% by 2020. It is worth noting that the 2020 level under Scenario (I), that is, 700 MTOE, corresponds to the total present coal production of the OECD countries. However, these forecasts should not be considered as too optimistic, because they are based on an average growth rate of only 2% for coal demand over the whole forecasting period, compared to 5% between 1960 and 1978.* Coal demand levels are much lower under Scenario (II): 375 MTOE by 2000, and 480 MTOE by 2020. This lower growth is due to the scenario's assumption that it will be much more difficult to expand production capacities.

(v) The sharp fall of non-commercial energy consumptions after 2000 was already pointed out. In fact, by 2000, pressures on resources and environment will reach an extent that makes it difficult to imagine that they could be borne after the turn of the century unless in the case of dire necessity. Consequently, the share of non-commercial sources in Communist Asia's energy balance will gradually tend to become marginal only by 2020, especially under Scenario (I): 19% (I) or 33% (II) by 2000, and 6% (I) or 16% (II) by 2020.

(vi) The share of oil in the region's total supplies should increase, but at a much lower rate than in the past: 20% (I) or 15% (II) by 2000, and 20%

---

*In its final remarks the RWT stressed that production might even exceed these forecasts in (I).

(I) or 19%(II) by 2020. This means that the bulk of oil's additional penetration will take place by 2000 under Scenario (I), while this development will take place only after 2000 under Scenario (II). The volume of Communist Asia's oil demand will represent 210 MTOE by 2000, and 290 MTOE by 2020, under Scenario (I), compared to 130 MTOE and 200 MTOE, respectively, under Scenario (II).

The region's share in world oil demand will also become more significant: 3% in 1978, 6% (I) or 5% (II) by 2000, and 8% by 2020, when it will be more or less equal to that of North America at that time.

(vii) It is hoped that the growth of gas and hydropower production will be very high over the whole forecasting period.

The share of gas in Communist Asia's energy balance could represent up to 11% (I) or 9% (II) by 2020, compared to 7% (I) or 3% (II) by 2000, and 2% in 1978. The corresponding volumes could be of the order of 160 MTOE (I) or 90 MTOE (II) by 2020, versus 10 MTOE only in 1978. With these volumes, the region would represent 5% (I) or 4% (II) of the world gas market by 2020.

Communist Asia's hydropower production should regularly increase over the whole forecasting period. Its share in regional supplies will grow from 2% in 1978 to 5% (I) or 4% (II) by 2000, and 10% (I) or 6% (II) by 2020. Under these conditions, hydropower production, (which is, of course, equal to demand) could arrive at 150 MTOE (I) or 60 MTOE (II) by 2020, compared to 50 MTOE (I) or 29 MTOE (II) by 2000. This shows that hydropower forecasts are very different under the two scenarios, because production would treble between 2000 and 2020 according to Scenario (I) and only double according to Scenario (II).

Communist Asia's share in the world hydropower production would be 7% (I) or 5% (II) by 2000, and 11% (I) or 6% (II) by 2020, that is, the region would become a very significant factor in world hydropower supply, even more than Eastern Europe in (I).

(viii) The shares of nuclear power and new energies will grow very gradually. Each of these sources would represent 2% of regional supplies by 2020 according to Scenario (I), versus 1% in Scenario (II). Their volumes would be limited to 30 MTOE (I) or 10 MTOE (II) in 2020. In fact, it seems that Communist Asia has not yet developed a real nuclear option up to date, because nuclear production forecasts for 2000 are situated between 14 TWh and 32 TWh only. It seems that it is the same as far as new energies are concerned.

3.9.4. Energy production

Communist Asia, just like Eastern Europe, has pursued deliberately a policy of energy independence; that is, it has tried to avoid any risk of becoming vulnerable to outside pressures that could have endangered its specific political system. In other words, the region's energy supply and demand will have been, and will be, more or less balanced. Therefore, the comments made on the region's energy demand (see 3.9.2. above) are also, *mutatis mutandis*, applicable to its energy supply.

In accordance with this policy, energy production has always been more than 100% of demand: 103% in 1978. This rate should be even higher under Scenario (I), which is based on international cooperation: 113% by 2000, and

123% by 2020, when it will be the second highest in the world after R5, while it was only the sixth highest in 1978. Under Scenario (II), it will also increase: 105% by 2000, and 104% by 2020.

The following salient features should be pointed out in this context:

(A) In order to make export capacities available, the growth of Communist Asia's coal production will be kept continuously higher than that of its coal consumption. Nevertheless, coal production's share in the region's total energy production will always be lower than coal's share in total consumption.

Communist Asia will remain the world's third coal producer over the whole forecasting period, with a volume of 780 MTOE (I) or 510 MTOE (II) by 2020.

(B) The region will also make great efforts to increase its oil production beyond the volume absolutely necessary to cover its own needs, but this volume will vary enormously from one scenario to the other: 400 MTOE (I) or 200 MTOE (II) by 2020. Therefore, Communist Asia could well become the world's fifth oil producer in the long run.

(C) Finally, Communist Asia may also become the world's fifth gas producer by 2020 under Scenario (I), with a volume of 300 MTOE, which would represent 9% of world gas production at that time. Under these conditions, the region will become a gas exporter between 2000 and 2020.

Communist Asia's share in world energy production will increase to 10% by 2000 and 2020 under Scenario (I), while it would be slightly lower under Scenario (II): 8% by 2000, and 7% by 2020, compared to 9% in 1978.

Nevertheless, the difference between the region's respective shares in world energy production and consumption will clearly increase under Scenario (I): from practically 0% in 1978 (9% less 9%) to 1% by 2000 (10% less 9%), and 2% by 2020 (10% less 8%), while this difference will remain pratically zero under Scenario (II). In other words, large quantities of energy will be available for export in the long run in (I).

3.9.5. Interregional energy trade

The policy pursued by Communist Asia in the field of energy is very well illustrated by its energy trade tables. Priority has been, and will be, given to self-sufficiency and, if possible, to the development of export capacities (which are indispensable to avoid deficits of the region's general balance of payments), whenever the international environment is favourable, as under Scenario (I). This policy has been, and will be, pursued, even if it implies very severe restrictions of domestic demand. The sharp decrease of the GNP elasticity of the region's commercial energy demand already mentioned above illustrates this point very well. In fact, if it was exactly in line with the average of the rest of the Third World in the past (1.17), Communist Asia's GNP elasticity will sharply decrease: 0.87 by 2000, and 0.58 by 2020, compared to 1.07 and 0.91, respectively, for the Third World average.

If this policy based on very constraining planning will succeed, it is certain that Communist Asia will be a significant energy exporter in the future. Its share in world energy exports may, therefore, grow from 1% in 1978 to 8% by 2000, and 15% by 2020, under Scenario (I), that is, the region would be the world's second energy exporter after R5 (North Africa and Middle East) by 2020, while it was only the fifth in 2000, and the eighth in 1978. Under these conditions, the volume of Communist Asia's energy exports would increase considerably: from 14 MTOE in 1978 to 120 MTOE by 2000, and 280 MTOE by 2020, while its energy imports would be only marginal (a few million TOE).

However, the region's exports would be much lower under Scenario (II): 35 MTOE by 2000, and 30 MTOE by 2020, when they would represent only 2% of world energy exports and consist almost exclusively of coal, representing 8% of world coal exports.

Communist Asia's increasing importance as an energy exporter under Scenario (I) is all the more remarkable because its export pattern by sources would be very diversified, contrary to that of R5 (North Africa and Middle East), whose energy export pattern will be dominated by oil, even in the long run.

Communist Asia's export pattern will be particularly well balanced by 2020, when the region's share in world energy exports would be 15% for coal (or number three), 11% for crude oil (or number three after R5, North America and Middle East and Africa), and 28% for gas (number two, after Eastern Europe). It is obvious that the successful diversification of its energy export pattern as projected would make Communist Asia a major factor in the world fuel market for each of these sources. In other words, it would have several irons in the fire.

Under these conditions, Communist Asia would fully profit from the opportunities available because of the climate of international cooperation and open frontiers on which Scenario (I) is based.

Moreover, if its economic development accelerates beyond the forecasts of (I), through the influence of favourable circumstances, Communist Asia could hope for a higher level of consumption. This level would also be reached if the drastic reduction of income elasticities cannot be faultlessly effected. For example, if the GNP elasticity of the region's commercial energy demand were to remain merely in line with the rest of the Third World (as in the past), Communist Asia's consumption of commercial energy would be 1 GTOE in 2000 and 2 GTOE in 2020 (with the GNP growth rates applicable in (I)) rather than 860 MTOE in 2000 and 1360 MTOE in 2020.

The demand differential in 2020 would be therefore be of more than 600 MTOE. It remains to be seen how such a differential could be supplied.

3.10 LATIN AMERICA

Latin America is the Third World's first economic power as measured by the volume of its GNP. It is also the Third World's number one as far as energy consumption per capita is concerned. Latin America's economic and energy growth has been high in the past and should remain so in the future. In fact, its development prospects may even be considered as the most promising in the Third World, comparable to those of R5 (North Africa and Middle East), even if the assets and the constraints of the two regions are very different.

3.10.1. Demographic and economic environment

(A) Latin America's demographic characteristics are similar to those of Black Africa and South-East Asia. Its population (345 million in 1978) is a far cry from the orders of magnitude of South Asia and R9 (Asian countries with centrally planned economies). The growth rate of the region's population has been well in line with the LDC average: 2.8% between 1960 and 1978, versus 2.5%. This situation should not change in the future: 2.2% by 2000, and 1.8% by 2020, compared to 2.2% and 1.7%, respectively, for the LDC average. Under these conditions, the region's share in world population will slighty increase: from 8% in 1978 to just over 9% by 2000 (564 million), and 10% by 2020 (802 million).

(B) Latin America's GNP per capita ($1400) was the Third World's second highest in 1978, after that of R5. In other words, the average level of the region's economic development was much higher than that of most other Third World regions. In addition, Latin America's per capita GNP was already very much higher than the LDC average in 1960: $860 versus $266. The growth rate of the region's GNP per capita was 2.8% between 1960 and 1978. Therefore, the region's favourable situation was only confirmed over this period. It seems that all the conditions necessary to ensure the future growth of Latin America's GNP per capita exist. In fact, future growth rates will be practically equal to the LDC average: 3.5% by 2000, and 2.7% by 2020 under Scenario (I), and 2.0% by 2000, and 1.7% by 2020 under Scenario (II). It is worth noting that these high growth rates will be achieved during a time when world averages will be much lower.

Under these conditions, Latin America's GNP per capita could be $3000 by 2000 and $5000 by 2020 under Scenario (I). Even under Scenario (II), in spite of its unfavourable assumptions, it could reach about $2200 by 2000, and $3000 by 2020. This means that it might be very close to that of R5 by 2000 and 2020.

(C) The volume of Latin America's GNP was by far the highest in the Third World in 1978. This should not change over the whole forecasting period, because it will grow at rates similar to the Third World average: 5.8% by 2000, and 4.5% by 2020 under Scenario (I),* and 4.3% and 3.5%, respectively, under Scenario (II), compared to 5.6% between 1960 and 1978.

Latin America should be able to maintain its economic growth better than any other region. In fact, its GNP could increase from nearly G$500 in 1978 to G$1700 by 2000, and more than G$4000 by 2020, under Scenario (I). Of course, its growth will be more limited under Scenario (II): G$1200 by 2000, and G$2400 by 2020.

Latin America, with R5 (North Africa and Middle East), will increase its share in world GNP faster than any other region, and that in a very sluggish world-wide context: from 5.5% in 1978 to 8.6% (I) or 7.7% (II) by 2000, and even 11.6% (I) or 10.2% (II) by 2020, when it will be practically equal to the share of R3 (Industrialised Countries of the Pacific).

3.10.2. Energy consumption

A. TOTAL CONSUMPTION

Latin America's total energy consumption grew at a sustained rate (4.6%) between 1960 and 1978, but growth of non-commercial consumptions was equal to zero. In fact, regional demand of non-commercial sources remained at a level of 67 MTOE over this period. This is unique among the Third-World

*Because of recent and foreseeable difficulties of the large nations of the zone, this estimate appears to be rather optimistic in the opinion of certain members of the RWT. The same applies to the corresponding rates of energy demand growth.

regions. The total growth of demand concerned, therefore, commercial sources only, whose total consumption grew at a rate of 6.6% per year, which is very close to the Third-World average. The volume of commercial demand increased from 92 MTOE to 289 MTOE between 1960 and 1978. Under these conditions, the share of non-commercial sources, which represented still almost half of the region's supplies in 1960 (42%), fell to 19% by 1978. This decrease indicates a trend comparable to that prevalent in the industrialised countries.

The region's total energy consumption stimulated by economic growth, will continue to grow at rather fast rates: 5.0% by 2000, and 3.8% by 2020 under Scenario (I), and 3.8% and 3.0%, respectively, under Scenario (II), that is, its growth rates will always be higher than the Third World averages. Its volume will increase from 356 MTOE in 1978 to 0.8–1.0 GTOE by 2000, and 1.5–2.2 GTOE by 2020. This means that the region's total consumption will be multiplied by a factor of 4 to 6 between 1978 and 2020. Latin America's share in world demand will increase constantly: from 5% in 1978 as well as 1960 to 9% (I) or 8% (II) by 2000, and 12% (I) or 11% (II) by 2020. The region will be the world's fourth energy consumer, after the three great industrialised regions, shortly after 2000, while it was only number six in 1978.

The GNP elasticity of Latin America's commercial energy demand will decrease from 1.18 in the past to 0.89 (I) or 0.94(II) between 2000 and 2020, that is, it will follow the general Third-World trend, without any particular performance.

B. ENERGY CONSUMPTION PER CAPITA

Latin America's per-capita consumption was, by far, the highest in the Third World in 1960 as well as 1978. In fact, it was the only Third-World region whose per-capita consumption was higher than one TOE in 1978, that is, twice the LDC average. This consumption's growth rate will be slighty higher than the Third World average in the future: 2.7% by 2000, and 2.0% by 2020 under Scenario (I), and 1.6% and 1.2%, respectively, under Scenario (II), compared to 1.7% between 1960 and 1978. The volume of the region's per-capita consumption will be 1.8 TOE (I) or 1.4 TOE (II) by 2000, and 2.7 TOE (I) or 1.9 TOE (II) by 2020, that is, it will be multiplied by 2 to 3 times between 1978 and 2020. In other words, the level of Latin America's 2020 per-capita consumption will be comparable to Western Europe's 1978 level under Scenario (I), and this latter region's 1960 level under Scenario (II). Consequently, the development of Latin America's, as R5's, energy consumption per capita will be somewhere between that of the industrial countries and that of the poor regions of the Third World.

3.10.3. Supply pattern

How has this fast-growing demand developed in the past, and how will it be covered in the future?

(i) *Oil* has played an essential part in Latin America's supply pattern for a very long time. Oil's share in total supplies increased from 42% in 1960 to 47% in 1978, when it represented a volume of 168 MTOE. In fact, Latin America was the Third World's first oil consumer in 1960 as well as 1978, with a share of 6% in world oil demand.

(ii) The share of non-commercial sources in total supplies fell from 42% in

1960 to 19% in 1978, while the shares of gas and hydropower increased considerably during the same period.

The share of *gas* in the region's total supplies rose from 7% in 1960 to 14% in 1978, and that of *hydropower* from 5% to 14%. The total consumption of gas and hydropower represented 50 MTOE each in 1978, when Latin America was the Third World's first consumer of these two sources, just as for oil. But Latin America was also the world's fourth gas consumer even before R3, and its third hydropower consumer before R3 and R4 in 1978, when its hydropower production alone represented 230 MTWh. Its shares in the world's gas and hydropower consumption were 4% and 13%, respectively, in the same year.

(iii) It is also worth noting that Latin America was the only Third-World region having a significant production of nuclear power and new energies (owing to Brazil's methanol programme and of production of charcoal for steel making) in 1978, when these two sources contributed 1% and 2%, respectively, to total supplies.

In brief, Latin America's development in terms of energy is more "advanced" than that of any other region of the Third World.

(iv) The region's supply pattern will not significantly change up to *2000*, at least as far as the relative importance of each energy source is concerned: 46% (I) or 44% (II) for oil, 15% for gas, 15% (I) or 16% (II) for hydropower. The share of non-commercial sources will decrease to 7% (I) or 10% (II) by 2000, but this decline will be set off by the increase of other sources: 8% for new energies, 6% (I) or 5% (II) for coal, and 3% (I) or 2% (II) for nuclear power.

These variations, which are, after all, relatively minor, should not conceal the strong growth of the volumes involved: 470 MTOE (I) or 360 MTOE (II) for oil, 160 MTOE (I) or 120 MTOE (II) for gas, 155 MTOE (I) or 128 MTOE (II) for hydropower. In addition, the volume of coal demand will be 62 MTOE (I) or 44 MTOE (II), compared to only 12 MTOE in 1978, while nuclear power will represent 32 MTOE (I) or 19 MTOE (II). Finally, and above all, the volume of new energies will be 85 MTOE (I) or 62 MTOE (II) by 2000 and nuclear energy will reach 32 MTOE (I) or 19 MTOE (II).

Under these conditions, Latin America will remain the Third World's first consumer of oil, gas, and hydropower, but it will also become the world's fourth consumer of oil and gas, before R3. In addition, it will be the world's second producer of hydropower and new energies by 2000, after North America, under Scenario (I). Finally, it will be the Third World's number two for nuclear power and also begin to be a significant coal consumer.

Latin America's shares in the world market will be 14% (I) or 13% (II) for oil (versus 7% in 1978), 7% (I) or 6% (II) for gas (versus 4% in 1978), 22% (I) or 20% (II) for hydropower (versus 13% in 1978), and 25% (I) or 23% (II) for new energies.

This means that Latin America, as from 2000, will be a major factor in world energy consumptions.

(v) Latin America's relative importance on the world energy scene will have grown enormously by 2020, because it will then be, by far, the world's first oil consumer and its first hydropower producer.

The region's oil demand will represent 730 MTOE (I) or 520 MTOE (II), that is, 20% and 21%, respectively, of world oil demand. Nevertheless in spite of this enormous growth, oil's share in the region's total consumption will decrease to 34% (I) or 35% (II). In other words, it will be 10 points lower than in 2000.

Latin America's hydropower production will increase enormously between 2000 and 2020, when it will represent 437 MTOE (I) or 288 MTOE (II), that is, 2000 TWh and 1300 TWh, respectively. Hydropower's share in total regional supply will then amount to 20% (I) or 19% (II). In fact, Latin America will represent 32% (I) or 29% (II) of world hydropower production by 2020.

The share of gas in regional demand will be 13% (I) or 14% (II) by 2020. This will represent 280 MTOE (I) or 200 MTOE (II) in volume, that is, 9% and 8%, respectively, of the world gas market.

The share of new energies in total regional supplies will rise to 10% (I) or 11% (II). This will represent 230 MTOE (I) or 160 MTOE (II) in volume, that is 23% (I) or 20% (II) of the world market. Latin America will, therefore, still be the world's second consumer of new energies, as in 2000. Finally, nuclear power's share in total regional energy supply will represent 7% (I) or 4% (II), that is, 145 MTOE and 60 MTOE, respectively. These volumes will represent 6% (I) or 4% (II) of world nuclear production.

But there will also be a very strong growth of coal demand between 2000 and 2020, when this source will cover 14% (I) or 11% (II) of total regional needs, while it was still a very minor factor in 2000. With a volume of 300 MTOE (I) or 170 MTOE (II), Latin America will in fact represent 5% (I) or 4% (II) of the world coal market in 2020, compared to 2% only in 2000.

In view of this growth of commercial energy, non-commercial sources will be practically eliminated from the region's energy balance by 2020 under Scenario (I), while they will preserve higher shares under Scenario (II), with shares in regional supply of 2% and 6%, respectively.

The simple statement of the above facts and figures gives an idea of the enormous growth in all directions that is implied in the forecasts resulting from both scenarios. These forecasts may be considered as optimistic, but they have a solid foundation in Latin America's considerable mineral resources, which to a large extent have still to be developed or even explored in a more efficient way.

Under these conditions, as in R5 (North Afrcia and Middle East), Latin America's very fast energy development should not lead to any dangerous energy dependence, as it will almost inevitably happen in the case of South-East Asia.

3.10.4. Energy production

It is certainly one of the most remarkable features of Latin America's energy development that it could, and can hope to continue to, cover the growth of its demand by its own resources. In fact, it seems that its situation in this respect is similar to that of Eastern Europe and Communist Asia, that is, the regions with centrally planned economies.

Latin America's energy supply represented 120% of its demand in 1978. Without any particular efforts, its supply should continue to exceed its demand in the future: 109% (I) or 108% (II) by 2000, and 107% (I) or 106% (II) by 2020.

As the share of non-commercial sources in regional supply will diminish, Latin America's supply and demand patterns will become progressively almost identical under both scenarios as from 2000. In fact, the difference between the supply and the demand of each source will be less than 1%. Consequently, the comments made on the region's demand pattern are also applicable to its supply pattern.

It is the same with Latin America's relative importance in world supply by sources, which is practically identical on the demand side, with a few minor exceptions in the field of fossil fuels. The world's first oil consumer by 2020, the region will be only the second producer by that date, after R5 (North Africa and Middle East), whose share will be higher by 11 points to 14 points. The world's fifth gas consumer by 2000, it will be only number four as from 2000.

Latin America's energy production will keep up with the fast growth of its demand, whose share in the world supply market will increase from 6% in 1978 to 9% (I) or 8% (II) in 2000, and even 12% (I) or 11% (II) in 2020. With the same shares, it will be the world's fourth energy consumer by 2020, while it was only the sixth in 1978. It will also be the world's third energy producer by 2020, after North America and Eastern Europe, but before Western Europe, R5 (North Africa and Middle East), and Communist Asia, while it was only number six in 1978.

The simple statement of these facts and figures is enough to underline the key part that Latin America will play in world supply and demand in the long run.

3.10.5. Interregional energy trade

Paradoxically, Latin America, which will play such an essential part in the future world energy scene, will not become a very significant factor in the interregional trade of fossil fuels. In fact, because of deliberate policy or inherent constraints, there will be a well adjusted dynamic balance between the growth of demand and supplies capacities, even if the latter will probably be very strained. This means that Latin America will hardly be able to become a significant energy exporter. In fact, because of delays in the development of regional supplies, it runs rather the risk of becoming an energy importer.

Latin America's present and future prospects in the field of interregional trade are very dim indeed, but the region should preserve an increasingly positive energy trade balance over the whole forecasting period: 4 MTOE in 1978, 16 MTOE in 2000, and 30 MTOE in 2020.

As far as crude oil is concerned, Latin America's exports and imports (75 MTOE and 66 MTOE, respectively, in 1978) will certainly be reduced without any great problems, with net exports of 40 MTOE (I) or 30 MTOE (II) by 2000, and 20 MTOE (I) or 15 MTOE (II) by 2020. Latin America represented 6% and 4%, respectively, of world exports and imports in 1978, but its share in world trade will decrease to 3% by 2000, and 2% by 2020.

Latin America's share in world coal trade will remain limited: 2% by 2000 (10 MTOE) and 3% by 2020 (20 MTOE).

But Latin America will become a significant exporter of natural gas: 25

MTOE (I) or 20 MTOE (II) by 2000, and 40 MTOE (I) or 30 MTOE (II) by 2020. It may even become the world's fourth exporter of natural gas as from 2000. In fact, its share in world exports could increase to 10% (I) or 11% (II) by 2000, and 11% (I) or 14% (II) by 2020.

At any rate, because of these developments, Latin America should arrive at a better balance of its imports and exports of fossil fuels and a reduction of total quantities in both fields.

Finally, it should be noted that Latin America's energy dependence will be reduced considerably. Imports covered 19% of its total needs in 1978. This should be reduced to 7% by 2000, and 3-4% by 2020. In addition, the total volume of the region's energy imports will decrease continously in absolute terms over the whole forecasting period.

Chapter IV

GENERAL GUIDELINES

At the end of a study which is as abundant on its data as it is plentiful in its analyses, one should not risk the formulation of definite conclusions which might over-crystallize what are only indications of tendencies or degrees of magnitude.

In addition, it would be contrary to the very essence of the enterprise, where absolute priority was given to decentralized regional analysis, to give pride of place to global syntheses.

It is, no doubt, better and, moreover, more in conformity with the reality of forecasting work, to leave to each reader the task of drawing personal conclusions from the results given at the level of each region, according to his area of interest and also according to his geographical origin.

However, one can, nevertheless, underline a *few salient traits* which emerge globally at the end of this study and which might be of assistance for forecasting thoughts.

After all, was this not the aim at the beginning of this study — to draw the future's outlines more accurately and "to highlight the principal constraints which the world and the regions will, in the future, confront?"

To produce these final analyses, we will choose as a starting point the present-day economic climate in which the energy system operates in the aftermath of the two oil shocks of 1973 and 1979. As a second step, we will analyse the long-term adaptation courses which are open to this system, using complementary indicators.

However, we will not, for all that, neglect the central aspect of this study which has been to bring together essential quotations and to place in context the variations in the picture created by the ten regional analyses which will lead one to reflect upon a tripolar geopolitical world division.

Finally, we will provisionally bring these reflexions to a close by extracting a few of the perspectives which will dominate the future of the world energy system.

4.1 OVERALL ENERGY CONTEXT AND LONG-TERM PERSPECTIVES

Undoubtedly, the present climate in which the energy system operates is not conducive to prospective thinking. Unexpected jolts to quantities produced, and especially to prices, highlight the fragility of forecasts and put into question their feasibility, even their usefulness. Such unexpected and disordered fluctuations destabilise and discourage forecasting work. Two elements nevertheless clearly emerge amidst this generalised confusion.

(A) The energy market remains largely conditioned by economic evolution, both in times of economical upturn as in times of economic slowdown. It is more difficult than one might think to separate energy growth from economic growth. The present economic depression proves this negatively by causing a notable slowdown in energy demand, even in those Third World countries which would most need to encourage it.

(B) The lessening of physical stresses on energy demand, provoked by the recession, automatically favours a lowering of pressures on supply by deferring them over time. From this fact, it is only a small step to conclude that the recession does not have only negative aspects, in that it postpones the moment for sounding alarm bells and especially that of taking Draconian oil saving and substitution measures. It is therefore to be feared that the slippery slope of facile solutions might be fatal to indispensable reconsiderations of policy. The financial difficulties of the economic climate ally with temporary, and therefore misleading, alleviations of pressures to invite laxity and myopia: the example of North American energy policy serves as a contemporary proof that these are not vain words. The appeasement of today may make the future even more difficult.

The fact remains that the accumulation of years of low growth and the persistence of uncertainty concerning the possibility of a substantial economic upturn have their effect on perceptions of the future, and have led to a clear scaling-down of long-term perspectives. The comparison of forecasts of Istanbul (1977), Munich (1980) and of those formulated here is significant in this respect (cf. A.21).

(C) Measured globally on the world's consumption, the decline since Munich is of about 1.5 GTOE in 2000, and of 2–4 GTOE in 2020. Everything is happening as if there has been a scaling-down between the two conferences:

(i) the forecasts of the present Scenario I ("desirable") are in line with those of Scenario C of Munich (then described as "pessimistic").

(ii) those of Scenario II, on the other hand, are much under what was considered to be unfavourable in 1981.

Although the Munich Conference confirmed the forecasts made in Istanbul for 2000, it also substantially reduced those for 2020. The trend is clearly reinforced by New Delhi for 2020, and this time affects the landmark year of 2000.

The change in population forecasts amounts for about 10% of the forecasting reduction between Munich and New Delhi in 2000, and for about one third of that for 2020. For their part, depressed economic forecasts account for about 15% of the reduction of per capita consumption in 2000 as in 2020. The two factors combined thus make up some 25% of the total forecasting discrepancy in 2000 and 45% (or nearly half) of that for 2020. The remainder is attributable globally to a reduced growth relationship between energy and GNP, that is to say, to an income elasticity which is significantly weaker than that envisaged in the past.

In this sense, this diminution of future consumption levels could be

Table 7 EVOLUTION OF WORLD ENERGY CONSUMPTION FORECASTS:
STUDIES OF THE CONSERVATION COMMISSION

GTOE	Base Years	2000		2020	
		SCENARIOS			
Istanbul 1977	1972	AS*	L4	AS*	L4
	6.1	12.8	11.7	22.7 (2.8%)	19.0 (2.4%)
Munich 1980	1976	B	C	B	C
	6.7	12.9	11.8	20.1 (2.5%)	17.9 (2.3%)
New Delhi 1983	1978	I	II	I	II
	6.8	11.8	10.1	18.0 (2.3%)	13.8 (1.7%)

*Alternative scenario.
()Average annual growth rates during the period Base Year–2020.

reassuring. However, it cannot alleviate two of the main problems which had already been raised in Munich and to which we will return later:

(i) the low probability of containing the long-term rise in oil demand, even under these new conditions of growth.
(ii) the all the more tragic fate of poor regions, since their development will consequently be slowed down even more (namely in that it follows trend II rather than I).

This is because the energy system is not remarkable for its suppleness: it can only readapt itself slowly, so great is its structural inertia.

4.2 THE READAPTATION OF THE WORLD ENERGY SYSTEM

A reading of some global indicators will allow one better to discern the readaptation avenues open to the world energy system.
(A) If one takes, for instance, the *ex post* size of the *income elasticities* implied by the respective commercial consumption growth and GNP hypotheses (cf. A.17), one observes a continuous decrease of this elasticity for most of the regions of the Third World. However, for the industrialised nations, one notices a certain recovery beyond 2000 (especially in North America) after a period of clear reduction up to the end of the century. These developments translate themselves globally by a substantial fall up to 2000 (from 0.9 between 1960 and 1978, to c. 0.7, followed by a return towards 0.8 after 2000). This is as if the bulk of restructurisation operates before 2000, at least as far as the North is concerned.
(B) If one examines, at present, another indicator of the same type, namely *energy intensity* (that is to say, the link between energy consumption and GNP — or energy consumption by dollar of GNP), one observes that this

ratio is destined to fall continually in industrialised countries — from 0.76 in 1960, to 0.68 in 1978, with about 0.53 in 2000 and 0.48 in 2020 (for commercial consumptions). In the past, by contrast, it rose from 0.62 to 0.70 for the Third World, to reach about 0.65 in 2000 and 0.58 in 2020. The principal effort at readaptation which is translated by the fall of this indicator appears to occur under the pressure of the industrialised countries, here again before 2000; the figures for the entire world are: 0.74 in 1960, 0.68 in 1978, 0.56 in 2000 and 0.51 in 2020.

From a rapid examination of these two complementary criteria, one can deduce, according to the experts consulted, that the world system should experience its most fundamental restructuring before the end of the century, under the urgency of the economic crisis, and that essentially thanks to the measures taken by the North. Beyond that date, one has no option but to wait for an improvement of external conditions making the effort less indispensable, or one hopes that such a stage of adjustment will have been reached that the necessity of such an effort will not make itself felt as much.

(C) Energy *dependence* (the relation between productions with total energy consumptions) also evolve very differently (cf. A. 16): rather favourably for all the industrialised countries under the impulsion of North America and the Industrialised Countries of the Pacific, and rather unfavourably for the Third World, even if overall it remains in a state of surplus. (The situation will worsen enormously for South-East Asia, it will degenerate for R5 and R10 but it will, conversely, improve for R9.)

(D) Another very simple criterion allows one to appreciate very schematically the *diversifying* effect of the regional supply structure. It will be noted that in 1960, 5 out of 10 regions depended on one single energy source for more than half their requirements. In 1978, only three are so dependent; in 2000 one remains, but there is none in 2020.

(E) Besides, one will be able to observe this diversification effort in a more global manner. If one measures the ratio between the sum of the 10 greatest consumption centres (a "centre" being defined by the consumption of one resource in one region e.g. coal in R1 or oil in R10) and global consumption for that same year, one notices that this proportion declines continuously over time: 71% in 1960, 67% in 1978 but 52% in 2000 (I) and 46% in 2020 (I). The range of consumers tends to widen and to stabilise itself. This can be corroborated at the level of production centres (the same criteria giving as a result: 69% in 1978, 57% in 2000 (I) and 56% in 2020 (I).

(F) One will also note that global interregional trade has the tendency to contract over the long-term, if only in terms of volume in (I), at least in that part referring to world consumption (23% in 1978, 15% in 2000, and 10% in 2020).

This development, associated with the evolution of regional dependencies and with the concern to diversify supply, has the effect of causing an unavoidable *retrenchment* of the *regions on their own resources.* In a more difficult general context, exporters will find it more difficult to find buyers throughout the world; importers will give priority to efforts at reducing their deficit by developing their local resources. As a result, the system's peripheral stresses which are reflected in interregional trade will tend to decrease.

(G) At the same time, it will be noticed that the evolution of the share of each region in world energy consumption closely follows the respective evolution of the shares of GNP, and reflects them very faithfully. It will also be noticed that the *structure* of world *production* is much more *stable* and remains more independent of the fluctuations of economic activity. Indeed, the share of industrialised nations in world production establishes itself for all the period between 55 and 60% (58% in 1978), while it declines significantly in terms of consumption (from 76% in 1978 to 60–65% in 2020). Likewise their GNP regresses from 82% to 67–72%.

There is much less reorganisation of supply patterns, a fact which is to the detriment of the Third World (especially for North Africa/Middle East, even if, conversely, Latin America does very well, and becomes the world's third energy producer in 2020, while it was only the 6th in 1978). With regard to demand, restructuring exclusively benefits the best placed regions of the Third World. This will be clearly demonstrated by the analysis which follows.

4.3 THE ESSENTIAL REGIONAL QUESTIONS

These questions will be understood by attempting to formulate, on the basis of the separate analysis of each of the 10 regions studied, the essential questions asked by the forecasts made (cf. the list of questions appearing at the end of each regions analysis in the summary of the study):

R1   North America: production and consumption of coal.
R2   Western Europe: coal consumption and nuclear production; trade deficit.
R3   Industrialised Nations of the Pacific: trade deficit.
R4   Eastern Europe: coal, gas and nuclear production.
R5   North Africa/Middle East: vital dependence on the international oil market.
R6   Sub-Saharan Africa: low demand; reliance on wood and coal.
R7   South Asia: low demand; massive reliance on non-commercial energy, and on coal.
R8   South East Asia: trade deficit.
R9   Centrally-Planned Asian Countries: coal production; oil exports.
R10 Latin America: strong demand; hydro, new energies, coal production.

If one regroups the 23 essential questions according to main theme, one can draw up the following table:

DEMAND:	weak	:	2	(R6–R7)
	strong	:	1	(R10)
	coal	:	$\frac{2}{5}$	(R1–R2)
SUPPLY:	coal	:	6	(R1–R4–R6–R7–R9–R10)
	nuclear	:	2	(R2–R4)
	non-commercial	:	2	(R6–R7)
	gas	:	1	(R4)
	hydro	:	1	(R10)
	new energies	:	$\frac{1}{13}$	(R10)
TRADE:	total deficit	:	3	(R2–R3–R8)
	oil exports	:	$\frac{2}{5}$	(R5–R9)

For about half, the questions concentrate on production; for the other half, they divided equally between demand and trade.

(A) A number of factors dominate discussion regarding demand; these are: concerns as to the low level of consumption in the poor regions of the Third World and uncertainties as to the coal absorption capacity of Western nations.

(B) With regard to supply, the main concern is the real development of coal capacities (a concern which by itself accounts for a quarter of all questions). This is followed by doubts about nuclear power and doubts as to the possible growth of non-commercial sources.

(C) Finally, with regard to trade, the seriousness of chronic deficits goes hand in hand with the risks involved in oil export possibilities.

One can classify the essential questions according to their frequency:

1st : coal : 8 times (or a third of questions: 6 relating to supply and 2 to demand)
2nd : trade deficit : 3 times
3rd equal : low demand : 2 times
            nuclear power : 2 times
            non-commercial energies : 2 times
            oil exports : 2 times

The main question in the future very clearly appears to be that of *coal,* as much from the point of view of its production as from that of its consumption. The problem of *trade deficit* comes second among general pre-occupations.

The other range of questions concerns *nuclear power* (the second weapon in the substitution of oil in the developed countries) and *non-commercial* sources (which are still indispensable for a long time for the poor regions of the Third World). Finally, persistently *low demand* is not without being a cause for concern to the poor regions, while even the richest countries, the *oil exporters,* are themselves concerned.

One notices that the uncertainties linked to the future of coal appear to be common to both the Industrialised Countries (4 questions for 4 regions) and to the Third World (4 questions for 6 regions). Questions regarding trade deficits are rather well distributed (2 questions in the North, 1 in the South). By contrast, the other questions are much more divergently spread out: nuclear energy, in the final analysis, really only concerns the North (2); an entire impoverished part of the South remains dominated by problems of consumption insufficiency (2), and by the continuous reliance on non-commercial sources (2); another part of the South remains a tributary of *oil,* either for export (R5) or with regard to its production in large enough quantities to create exports (R9).

4.4 DIFFERENTIATED EVOLUTIONS OF THE REGIONS

These will essentially be studied on the basis of Annex 15.

(A) With regard to demand, three principal *consumer centres* can easily be identified at the start of the period: North America (35%) in 1960, Western Europe (20%) and Eastern Europe (17%). By themselves, these three then absorbed 72% of world demand. By 1978, the situation had changed: North America's consumption diminished by 5 points (30%), Western Europe's by 2 (18%) while, by contrast, Eastern Europe's increased by +4 (21%). Thus the three centres together still account for 69% of world demand.

Let us project ourselves to 2020. North America has descended to 20-22%; it is to be noted that the bulk of its retreat occurs before 2000, when it only accounts for 22-24% of the world market. Western Europe falls back to 14-15%. Finally, after having maintained itself until 2000, Eastern Europe also experiences a slight decrease to 19-20%. The three centres only take up 53-57% of consumption as opposed to 59-62% in 2000.

Who has benefited from the relative decline of the position of the three great consumers?

Between 1960 and 1978, the effect is very insignificant: R3 is the chief beneficiary (growing from 4 to 6%).

Between 1978 and 2020, the total retreat of the three "greats" is of 16-12 points. Three quarters of that, in I as in II, is taken up by two regions: R10, *Latin America* growing by 6-7 points; and R5 *North Africa/Middle East* (5-3 points); followed by *South-East Asia* (2-3 points). These three new regions would then account for 25-21% of world demand as opposed to 10% in 1978 (9% in 1960).

(B) From the point of view of *supply*, it has already been said that changes are much more muted. The two great production centres in 1978 (R1 with 24% and R4 with 22%) together cover 46% of world production. In 2020, they continue to provide 41-44% of that production.

By contrast, the third production centre, R5 (North Africa/Middle East) undergoes a severe decline: from 18%, it falls sharply to 11-10% (or $-7/-8$ points). In total, the decline of the three centres is thus $-12/-10$ points: 64% in 1978 to 52-54% in 2020.

For half, in (I) as in (II), the decline in supply of the three "greats" was compensated by the growth of *Latin America's* share (+6/+5 points). Further behind, one finds two other marginal beneficiaries: R3 (Industrialised Countries of the Pacific) +2/+2 and R6 (Sub-Saharan Africa) +2/+2.

(C) *Trade* can be understood in terms of the difference between production and consumption. If it is positive, it signifies a surplus situation (that is to say, a generally export-inclined situation).

In 1978, the three great deficit areas were (at the beginning of the period): R2, *Western Europe* (Production-Consumption = -9 points); R1 North America (-6) and R3, the Industrialised Countries of the Pacific (-4). Conversely, only one great surplus centre existed: R5, North Africa/Middle East, with +16 points by itself.

At the end of the period, the changes are profound: as from 2000, R1 has not only redressed its balance but has become an exporter (+2), and R2 and R3 have improved their precarious positions in that they are less vulnerable: R2 $(-6/-5)$, and R3 $(-1/-2)$. The great export centre remains R5, but in considerably reduced proportions (+4/+5 as against +16). Henceforth, it is R1 which follows it with +2. Western Europe remains the main importer. But R3 has been joined in second place by R8, South-East Asia $(-2/-1)$.

Table 8 graphically sums up the principal movements, arrows indicating the large-scale movements, bold characters indicating new regions with significant ranks.

Table 8 EVOLUTION OF LARGE CENTRES

	1978		2020	
(%)	R1	30		
	R4	24	R1	20/22
			R4	19/20
Demand	R2	18	R2	14/15
			R10	12/11
	(R10	5)	R5	7/5
	(R5	2)		
Supply	R1	24		
	R4	22	R1	22/24
			R4	19/20
	R5	18	R10	12/11
			R5	11/10
	(R10	6)		
Exports	R5	+ 16	R5	+ 4/ + 5
			R1	+ 2/ + 2
Imports	(R8	0)	R8	− 2/ − 1
			R3	− 1/ − 1
	R3	− 4		
	R1	− 4	R2	− 6/ − 5
	R2	− 9		

4.5 TOWARDS A TRIPOLAR WORLD

(A) All these considerations permit one to propose a more general, and admittedly arbitrary, regrouping of the ten regions in *three great zones*. That of the industrialised countries on the one part, the common points of which are evident (zone 1). However, within the Third World, on the other hand, one can usefully distinguish between two zones with widely different profiles and futures: that of the regions with rapid development (zone 2) which comprises the main beneficiaries of growth, namely R5 (North Africa/Middle East), R8 (South-East Asia), and R10 (Latin America); and, secondly, that of the regions in transition (zone 3), which experience persistent difficulties: R6 (Sub-Saharan Africa), R7 (South Asia), and R9 (Centrally Planned Asian Countries). This tripolar global approach has been quantitatively translated in Annex 19. The reader should refer to it for a detailed analysis of the respective situation of each zone.

(B) The essential element of this annex will be retained; only zone 2 has benefited, and will benefit, from the weakening of zone 1. By contrast, zone 3 was and, it seems, will remain for all instances and purposes excluded from the benefit of growth. This is measured by the shares of the world total with regard to each of the different factors analysed. Between 1960 and 2020:

(i)   the population share of the Industralised Countries (zone 1) is reduced by 13 points (33% to 20%). Zones 2 and 3 share equally the corresponding gain (+6 points for zone 2; +7 points for zone 3).

(ii) Zone 1's share of GNP diminishes by 19-14 points (86% to 67-72%). But if that of zone 2 increases by that fact by 17-13 points, zone 3 only gains 2-1 points.

(iii) The same pattern occurs in the shares of energy consumption: Zone 1 falters by 17-12 points (77% to 60-65%); zone 2 develops by +17/+13 points and zone 3 stagnates at 0/−1 point.

(iv) Energy production, with much more restained movements, gives much less significant results. From 1978 to 2020, zone 1 evolves by −3/+2 points according to the scenario; zone 2 by +1/−2 points and zone 3 by +2/0 points.

(C) The examination of the volume of *increases* brings a complementary light, particularly in the final total increases table for 1960-2020.

The leading positions of the industrialised countries cannot be denied: whereas their population only contributes 11% of world population growth, they account for 65-70% of global GNP growth, for 56-62% of consumption, and for 54-61% of production (since 1978). This is so even if their share of the "market" is reduced.

Zone 2's progress clearly expresses itself: it occurs at about 24-29% of the growth of all factors, including population (29%).

By contrast, zone 3 contributes 60% of the increase of world population, but only 6-8% of that of GNP, and between 14 and 18% of that of supply and demand of energy.

(D) Per-capita figures are instructive. Compared to world averages, fixed at 100 for each of the factors (GNP, energy consumption and production), the gaps between the 3 zones have a tendency to widen between 1960 and 2020. In 2020, zone 1 goes from a level of about 260 in terms of GNP to 340-360; from a level of energy consumption of 230 to 300-330; from a (1978) level of energy production of 200 to 280-300. Zone 2 goes from 46 to 102-87 in GNP; from 47 to 102-83 in consumption; from 133 (1978) to 113-104 in production. Zone 3 goes from 13 to 13-12 in GNP; from 30 to 27-26 in consumption; from 29 (1978) to 30-27 in production.

Zone 1, therefore, increases its predominance continuously and does so on each factor. From 50% in terms of GNP and energy consumption, Zone 2 succeeds according to I in raising itself up to the world average, and in coming near it according to Scenario II. There are no indicators of improvements for Zone 3: for each of the three factors, it remains as far from world averages as ever. The picture which presents itself is thus a world with injustices that are more and more blatant, with a pattern of development that favours even more the wealthy (Zone 1) and placates Zone 2, but one which does nothing to alter the relative fate of Zone 3 (which will have 56% of the world's population at the end of the period).

(E) Another striking aspect will be noted in all the regional analyses (both the overall analyses and those per capita): Scenario I, which forecasts sustained growth, leads to the most favourable restructurings from the point of view of the Third World. By contrast, Scenario II, which forecasts slowed-down growth, condemns the Third World always to be subject to the rule of the industrialised countries. In these conditions, it will not come as a surprise that all the countries of the South unanimously aspire to high growth rates which will allow them to better meet their handicap *vis-à-vis* the North (of course, very incompletely, but nevertheless more substantially).

4.6 THE ENERGY FUTURE IN PERSPECTIVE

At the end of all this work, there emerge a number of fundamental questions which we will here attempt to synthesise, even at the inevitable risk of abusive simplifications. One can extract them according to two main analytical directions: on the one hand, the world and the regions, and, on the other hand, the sources of energy.

(A) *Globally,* first of all, one cannot help thinking that, with the years flowing by and uncertainties persisting, the future is developing towards more precarious situations. Between two Conferences, the *permanent slide* from "rosy" scenarios to "grey" scenarios has been noted but even today, compared to the beginning of work on this study two years ago, Scenario II which then appeared pessimistic, now appears to be "probable". And Scenario I becomes more an objective than a forecast, its "normative" aspect being stressed by that very fact.

A slower pattern of growth, and thus one less thirsty for energy: this would appear to be a good thing for those who rightly are concerned with pressures on supply. However, as has been seen, a depressed rate of growth maintains, and even worsens, the income and consumption gaps. The strong become relatively stronger, and the weak weaker.

New sources of tensions, more political than technological, are brewing in the diminution of global volumes. The Third World stands to bear the brunt of these tensions, especially those areas with the most deprived large human populations.

Moreover, even if forecasted figures are clearly lower than previous forecasts, the *quantities* involved remain *considerable.* At the global level, some 400 to 500 GTOE (cf. A.22) might be consumed before 2020, according to the scenarios (as against 85 GTOE between 1960 and 1978).

Even on the basis of a zero PEC growth from 1978 levels, the world would require some 285 GTOE up to 2020 (more than 3 times what it consumed between 1960 and 1978).

The examination of all the non-renewable energy reserves (cf. A.23) leaves a globally favourable impression. Indeed, the total of proven reserves (that is to say, available at costs near to present day costs) are of the order of 650 GTOE, without including the contribution of renewable energies and of non-conventional oils (calculated as 80 GTOE). Proven reserves easily cover the accumulative PEC up to 2020 (400 to 500 GTOE), provided that the pattern of supply coincided with that of proven reserves. Unfortunately, this is not the case because, for example, oil today accounts for 39% of world demand and its reserves account for 13% of proven non-renewable reserves (25% with non-conventional oil). By contrast, coal only provides 25% of world demand and accounts for 70% of proven reserves. This structural inadequacy will lie at the origin of additional pressures on certain energy sources.

Moreover, if the forecasting period is lengthened to a more remote future (such as 2050, cf. A. 22), the PEC volumes which are involved are of much more worrying proportions: between 2020 and 2050, some 500–700 GTOE will be required (or 900 GTOE since 1978 according to II, and 1200 GTOE to I). Whatever proven reserves there are will, therefore, be exhausted, and additional reserves will be substantially used up.

In a gloomier economic climate the putting into action of production capacities destined to meet globally with such demand will become more difficult. Room for manoeuvre is narrow.

From that, an inevitable tendency towards retrenchment is observable in interregional trade: the world, overall, will trade less energy in terms of consumption percentage, and even, according to II, in terms of global quantities.

(B) However, it is at the *regional* level that the future differentiates itself even more obviously. Putting it schematically, there are two camps: the well-off and the badly-off.

The well-off include all the industrialised regions, of the East and of the West, with, of course, nuances between them and more or less well distributed advantages. Nevertheless, North America and Western Europe reduce their place in world demand, without adverse consequences for them. Benefiting from its more than comfortable situation, North America even succeeds in re-establishing its trade balance and turning itself into an exporter in about 2020. In the general turmoil, Eastern Europe remains serene and stable on all markets. As for the Industrialised Countries of the Pacific, they continue to occupy the fringes of the system, by efficiently adapting themselves to the developments in progress.

In the final analysis, only one other region seems apt to cross the demarcation line between the two camps up to 2000: Latin America; a prediction made knowing that the forecasts formulated in its regard are perhaps a little too optimistic. All the other regions (and that means the Third World) seem to be destined to remain in the badly-off camp. Even North Africa/Middle East, which starts from an almost brilliant situation, will experience growing difficulties due to the increase of its internal needs and to the uncertainties concerning its exports. Threats overshadow the long-term equilibrium of South-East Asia due to the insufficiency of its production. As for the three human "anthills": Sub-Saharan Africa, South Asia, and Centrally Planned Asian Countries (or $\frac{1}{2}$ of humanity), everything leads one to believe that they are destined to more or less stagnate, energetically speaking.

That is to say that the relative positions of each will hardly change, even over 40 years and that acquired advantages are preponderant: this is good for some and unfortunate for the others, we might conclude, with all the more ease as we live among the well-off.

(C) The comparative analysis of energy *sources* is also worrying. While clouds gather over nuclear energy (which had been understood to be one of the two best substitutes for oil, at least as far as the North was concerned), the unanimity which surrounds *coal* (even if it is *a priori* sensible), is not reassuring for all that. Indeed, a quick calculation shows that it is forecast to be the leading supplier in 6 out of 10 regions in 2020, in second place in 2 other regions and in third place in yet another. However, in 1978, coal appeared in first place in only 2 regions, in second place in 5. This groundswell in favour of the despised source of the 1960s and 1970s, by its very size, is pregnant with questions.

On paper, coal is sought after by planners. But what about the reality of the behaviour of consumers, used as they are mainly to the convenience of hydrocarbons? Will not this inevitable pressure on prices which should result from this generalised infatuation have a "boomerang" effect on demand? Will the coal industry know how to adapt itself to this unprecedented growth? What about staff, man-power? Will one at last succeed in inventing the competitive transformation to synthetic hydrocarbons, which is the only one which could open much vaster markets for coal? What of environment? What local (and global) consequences will be engendered by a 2.5–3 multiplication of production by 2020, and by a cumulative consumption of 120–140 GTOE? What of internal transport problems, of harbour installations? What of the deficit which appears, albeit marginally but, nevertheless, clearly, between global exports and imports at the end of the period? And what of the hesitations between producer and consumer, e.g. the situation where each side waits for the other to commit itself to the coal option, before committing themselves in turn (cf. the round table on coal of Munich)? So many uncertainties which are far from being alleviated, and which condition, each for its own part, the effective development of the coal option.

Of course, it is on *oil* that not only the news limelight converges, but also most regional policies. The resource which was adulated when it had the good taste to be managed by "reliable" people, and to be available at low prices is treated with approbrium from now on. It is true that it was allowed to invade the energy balances of both West and East, and of both North and South without proper consideration and in various degrees, by basing oneself on short term views and in putting up with its social impact (via the automobile civilisation).

Conversely to coal, measures have been taken to enable it to fall back into its proper place. For example, in the regional supply forecasts for 2020, it is only placed in the first rank in three forecasts compared to 6 in 1978.

It is, thereby, hoped to make it return to the place it occupied in about 1960, in the same time as one would push coal to recover its past splendour, which also blossomed (at least after the War) in that period: 4 first places and 2 second places in 1960. The fundamental question which occurs behind the smoke-screen of forecasts is indeed this: are we witnessing a comeback towards equilibrium, with coal and oil recovering in 2020 their respective positions of 1960 after having interchanged in about 1975 (although, of course, recovering these positions at incomparably higher levels)?

Or is this not the more profound sign that the era of oil (and that of gas) was only a "historical parenthesis" in energy evolution of some 75 years, before one was to return to the heavy and stable tendencies which continue to give pride of place to resources with superabundant reserves such as coal (or later, nuclear power)?

Even if it will turn out to be so (and only the future will prove it), another observation seems to be necessary in the context of this study. Assuming

(a) that the consumption forecasts envisaged for 2020 do, indeed, become reality, and
(b) if we only retain Scenario II (where in total, we succeed in containing world oil consumption at near 1978 levels),

one then realises, with some anxiety, that in order to achieve this *tour de force,* we must *simultaneously,* between 1978 and 2020, achieve the following:

(i)   the increase of coal consumption from 1.7 GTOE to 4.4 GTOE (a 2.5 multiplication)
(ii)   that consumption of gas doubles from 1.2 GTOE to 2.4 GTOE
(iii) that hydro consumption increases from 0.4 GTOE to 1 GTOE (a 2.5 multiplication)
(iv)   that nuclear consumption increases from 0.15 GTOE to 1.6 GTOE (a 10 multiplication)
(v)   that consumption of new energies from almost nothing to 0.8 GTOE (a multiplication by 100)
(vi)   that wood consumption increases from 0.5 to 0.7 GTOE
(vii) that waste consumption increases from 0.25 to 0.4 GTOE

In other words, that there is a total increase of non-commercial energies, from 0.7 GTOE to 1.1 GTOE, of nearly 50%. It is under these conditions, and *only* these conditions, that oil demand could be reduced from 2.7 to 2.4 GTOE.

Of course, it is always possible to imagine that the partial failing of one source is compensated by another. But it is the need for the conjunction of all these successes which renders the global objective very risky: can one indeed reasonably count on all these successes together? One can imagine that the probability of that happening is largely asymmetrical: unfortunately, there are far greater probabilities of not being able to achieve each of these forecasts than being able to exceed them. However, only combined successes on all the alternative sources would allow us to relieve the pressure on oil.

At the global level, one is very far from the sterile rivalries between advocates of nuclear energy and those of renewable energies, since we will not have enough of all these energy options to reverse the tendency and set in motion the decline of the quantities of oil. It will also be noted, incidentally, that if these forecasts for oil were to be achieved, we would very quickly have to abandon the illusion that the price of oil could continue to fall for a long time. At the slightest economic upturn, it will again rise up since the pressure on reserves can only remain persistent. This may be ascertained by the fact that, by 2020, the cumulative total consumption (even in Scenario II which is the slowest) will be of the order of 100 GTOE for proven reserves calculated today at less than 90 GTOE. In the context of growth and if all other resources were to develop considerably, a world government with free access to all the oil rigs of the world, would have exhausted reserves in forty years.

What can one say both of Scenario I (where the corresponding quantities will turn around 120 GTOE) and, of course, of the political divisions which force one to extract oil even today from much more expensive deposits. Consequently, are we not today counting our (energy) chickens before they are hatched?

This is because the comparison of cumulative consumption figures with those of available reserves according to source (cf. A. 22 and A. 23) reveals notable difficulties, especially at very long-term.

This is not so much the case with coal, where, it seems, the room for manoeuvre is still large. Indeed, of the over 460 GTOE of proven reserves, 120–140 will be extracted by 2020, and an additional 170–220 GTOE by 2050. In this event there would still remain some 70 (I) to 200 GTOE (II) of proven reserves in 2050.

However, the situation for nuclear and natural gas appears more precarious. As far as nuclear power is concerned, proven reserves will be on the way to being almost exhausted by 2020 (35–45 GTOE consumed out of 45), the deficit in 2050 on the total proven reserves and additional reserves amounting to 60 GTOE in (I) (0 in II). Consequently, one conceives the necessity of developing the fast breeder option if one wishes nuclear power to take its proper very long-term oil substitution role.

From the point of view of quantities, fast breeders present the double advantage of producing about 50 times more energy per tonne of uranium and thus lessen the use of reserves proportionately. It also allows one access to, and the use of, much poorer deposits which would not be commercially operated by present-day reactors.

Gas is threatened even earlier: with 60 GTOE of proven reserves, the cumulative demand for up to 2020 is of the order of 70–90 GTOE. It is thus necessary to rely, even before the end of the forecasting period, on additional reserves. Up to 2050, 70 to 115 GTOE will be needed (out of 110–90 GTOE remaining). The 2050 shortfall will, therefore, be −25 GTOE in (I) as against +40 GTOE in (II).

However, the main difficulties concentrate on oil. They are mentioned above.

Up to 2020, the cumulative consumptions, as they appear in forecasts, represent 110–130 GTOE since 1978 against 85 GTOE of proven reserves. Of these, a shortfall of 25 to 45 GTOE will therefore exist in 2020. Even if one analyses the decline of oil beyond that date, some additional 50 to 90 GTOE will be required up to 2050, out of 50 GTOE of additional reserves. The shortfall will thus be 25 GTOE (in II) and 85 GTOE (in I) in 2050. Consequently, we will have to rely, and that very early, on dearer and dearer reserves (such as non-conventional oil, for instance, with its 80 GTOE).

Even as depressed a context as Scenario II does not allow one to avoid, in the best of cases, serious supply difficulties for world oil demand, even if it were entirely transparent and independent of all political influences. Does not this very consideration mean that world energy demand is too high and that it is necessary to drastically revise it downwards to be able to supply it, given the constraints on supply?

We are then forced back on questions about demand: if they have the courage or if they are compelled by circumstances, we know that the industrialised countries can save even more than is here forecasted (on the hypotheses of unchanged economic growth) on their consumption and can reduce their elasticities by a corresponding amount (but at the price of what effort?).

On the other hand, how can we be deaf to the voice of the hungry Third World, crushed by energy misery, dependent on non-commercial sources on the way to exhaustion? Is it decent to ask it to reduce its already minimal needs, beyond the elimination of blatant wastage? Any restructuring of the demand patterns must be done almost exclusively by the countries of the North: the equilibrium of the world, if not simple justice, needs it to be so.

(D) What can one conclude from all this? Let us leave to the decision-makers the task of drawing up the catalogue of the measures to be taken and of establishing a timetable of priorities. Let us content ourselves with observing that the question marks heavily outnumber positive statements: "grey", even "black", upon "pink". "The future is not what it used to be": it has become worrying, and the world, ruled by the law of the strongest, has become dangerous. But must one, for all that, resign oneself to catastrophe or to the apocalypse?

Energy is only one facet of a much graver and profound problem. It acts as a revelator of political (and spiritual) tensions which cross the planet. The future patterns which emerge at the end of this three year long research effort must, nevertheless, not drive us towards pessimism, still less plunge us into despair. They must instead incite us to take the necessary measures to preserve the future without delay. Our societies, rendered fragile by the present crisis, have already proved in the most dramatic of circumstances their capacity for adaptation and for responding to threatening events.

The key to the problem is, however, not just technological or political. It is also, whether one wishes it or not, spiritual: will reconciliation and sharing conquer individual and collective divisions and egoisms? With St. Paul we know that: "everything is ours: life or death, the present or the future". Neither neutral nor fatal, the future is what we will make it. It is for us to choose resolutely hope, justice and life.

Part II

METHODOLOGY

Chapter V

METHODOLOGY

In this part of the study, we will outline the methodology employed in detail and analyse the problems encountered during each of the steps of the forecasting process.

5.1 REGIONALISATION

Once the idea of a decentralised long-term global energy forecasting study had been launched, it was, therefore, necessary to organise proper regional groupings to give substance to the research undertaken and to create working teams. This problem is discussed in Annex 1: "Composition of the 10 regions".

At this point, we should simply like to note that there can be no perfect division and that any aggregation of countries into regions is a more or less happy compromise between global and technical comparison criteria; these criteria are always a little schematic since one has to work at some distance on countries which each have their own personality, and on delicate political sensibilities.

There is nothing original in the regional division which we have adopted: it is a conventional division which is close to that of the United Nations' and to that of other large-scale global studies. It is rather with regard to the nations on the boundaries of regions that questions arose with regard to the propriety of adding a country or whether to add it to the adjacent region.

We have thus solved several "frontier" problems by placing these countries in the following regions: Puerto Rico with North America; Yugoslavia with Western Europe; Albania with Eastern Europe; Israel with Western Europe; Sudan and Iran with North Africa/Middle East; Mauritania, Djibouti and Somalia with Sub-Saharan Africa; Burma, Taiwan and the Islands of the Pacific with South-East Asia. For diplomatic reasons, South Africa was given separate treatment; it was attached to Sub-Saharan Africa only for the purpose of final results.

Finally, the decentralised phase led to the revision of certain initial placements: thus Turkey and Cyprus were transferred from North Africa/Middle East to Western Europe.

It is hardly necessary to point out that the regions which have been created in this way only represent technological divisions; these divisions do not, in any way, prejudice the level of their political or economic integration.

5.2 PERIOD OF REFERENCE

The choice of the *past reference period* for the study was essentially influenced by *technological considerations*.

(A) It was necessary to adopt a period which would be sufficiently long to be coherent with the projected forecasting time-span (about 20 and 40 years, thus up to 2000 and 2020).

It was thus necessary to concentrate on a period of some 20 years. It was tempting to adopt the year *1973* as the turning point, in so far as that year marked a decisive break with the energy equilibrium of the past.

(B) But the launching of the study in 1981 enabled one at best to base oneself on the year *1978,* as a final year, in order to avail oneself of the complete range of the economic, demographic and energy data required by the study. Indeed, the data for 1979 (and even more so, for 1980) were only very incompletely available in 1981, despite the obvious desirability of basing oneself on the most recent year which could provide valuable additional information. Consequently, one could not realistically envisage basing oneself on the years 1973-8, a five year period which is much too short to be able to give an account of sufficient stability of current structural inflexions.

(C) It was necessary to find an earlier date, which might make it possible to base long-term arguments more soundly. *1960* was thus much more advantageous in that a large part of the task of collecting and transcribing statistical data had already been done for that year in the study presented in 1980 in Munich.[11]

In the end, therefore, the period *1960-1978* was adopted as a reference period, since it integrates over 5 years the discontinuities at the end of 1973, while at the same time weighing them up against the heavy growth tendencies which had marked the previous years and which continue to produce lasting effects on the development of the energy system.

5.3 THE FORECASTING HORIZON

The delineation of the forecasting horizon was dictated by the study's objective and by the history of the Commission's activities.

(A) The objective pursued (namely, to plan the long-term evolution of regional energy balances in context) implied a very clear focus on 2000 in a sector as heavy as energy, and also that one should explore evolutions beyond that date. All the time, it was borne in mind that, in matters of forecasting, every horizon that is chosen highlights a very different category of problems.

(B) Moreover, the previous studies of the Commission presented in Istanbul and Munich had used the years *2000* and *2020* as future forecasting landmark dates. It was thus natural for us to adopt these dates as "horizon years", in that they enabled us to ensure continuity and in that they facilitated the comparison of the results of previous studies with ours.

(C) It will be noted that the work of this study was made by means of a cross-section analysis: that is to say, the forecasts were made by reference to two years only (2000 and 2020), excluding, thereby, any detailed study based on trajectories over time. Likewise, as regards the past, only two years were examined: 1960 and 1978. The evolutions between the reference years of the past and the horizon years of the future are only expressed through progressions measured in annual average global growth rates without further precision.

This "photographic approach" to the future was justified by the size of the subject analysed and by our concern at achieving the crystallisation of the opinions of regional experts on a limited number of data.

5.4 FIELD OF STUDY

At what level of energy system (technological or geographical) did we choose to work?

(A) The objective aimed at in this study is centred on the problem of the global adjustment of energy supply and demand, and of the resulting possible disequilibrium of trade. Furthermore, a *sine qua non* condition for the success of the study was to go through the decentralised stage and to give the regional teams maximum chance to take the forecasting process to the very end.

From that fact, previous experience and the quasi-inexistence of budgetary means inclined us to opt for a *limitation* of the effort imposed on the teams, both in terms of research and in terms of time.

As a result, the decision was made at the start to keep to the first link in the energy chain, that of primary energy balances, without considering transformation sectors, secondary sectors (hence the total absence of electricity) and of final consumptions. The wisdom of this choice was amply confirmed during the study.

One will not find any sectoral consumption analysis (split up in industry, transport, housing, agriculture, etc.) and still less any analysis of useful energies. Only *partial balances* are provided, therefore, and they are limited to primary supply.

(B) Moreover, for technological and political reasons, only regional results appear and no explicit references are made to national results. This was made necessary to obtain flexibility with tables and overall command of the coherence of estimates. On the other hand, the presentation of results on a regional basis alone avoided all contraction, even all intemperate political blocking of the supply and publication of long-term data.

In freeing experts from national official constraints, this policy enabled the study to keep its "unofficial" or at least "non-governmental" character, which is to say a guarantee of its success, even if, in the actual work of the groups, much basic data for the past and in forecasting were naturally borrowed from studies that were, properly, national ones.

In the end, the uncertainties encountered in statistics relating to the past led us to refrain from presenting the base of reference country by country, as we had initially envisaged; instead, we simply published regional values, which are much less controversial and less liable to be treated without caution.

5.5 FORECASTING METHOD

This was thoroughly and schematically introduced in Part 1, at 1.3.2.

It is interesting to describe it in greater detail at this point, and to comment on the evolution which our method underwent during the course of our work. (A) We started off with an inevitably slightly theoretical and mechanical conception of the procedure to be envisaged, in so far as there was hardly any comparable methodologial experience available, and in that we were also anxious to take precautions against the risk of failure (for example, in abandoning the study because of the failure of a regional group).

We should like to underline in this regard that our determination *to succeed* (our means, in other words) did not mean that we anticipated the conclusions to be arrived at. This study was performed without prejudging any final result (and this was one of its most interesting, if not innovative, aspects). It was only at the time of the final adding up of the separate estimates of the working groups (in summer 1982 at the latest) that the outlines of the world energy panorama began to emerge, just as the picture on a jigsaw reconstitutes and reveals itself from its fragments.

No doubt, one can immediately object that such a procedure runs the risk of being incoherent: what guarantee has one of the homogeneity of the isolated estimates of each region? We took care in the procedure adopted to ensure that there would be an indispensable, overall harmony, while at the same time giving free rein to the freedom of thought and expression of all participants. This is the reason why it was decided to structure the working method of each of the RWT (Regional Working Teams) according to:

(i)   a homogeneous base of reference for the past (Annex 5);
(ii) a common system of units for definitions and equivalences (Annex 4);
(iii) two general forecasting scenarios giving a framework to the energy analysis (Annexes 6 and 7).

(B) This method was concretely reflected in a *regional dossier,* containing these different points of information, given to each member of the RWTs (Annex 8).
This regional dossier was divided into 4 parts:

(i)   *Part I* This outlined the general objectives of the study, a summary of the methodology employed and a draft timetable. It also contained a description of the composition of the relevant region, of that of the dossier itself as well as a description, and a booklet of instructions for the use of the central element of the dossier: the *"forecasting sheets".*
(ii) *Part II* This provided the relevant statistical data for each country of the region, figures for the *1960-1978 reference base* in terms of population and GNP; figures of commercial and non-commercial energy consumptions according to the different sources, for primary energy consumption, as well as the system of units and equivalences adopted.
(iii) *Part III* This contained *Scenario I* ("normative-cooperation") and contained a general description of the scenario; and the overall global hypotheses (in figures) for the industralised countries and the Third World, designed to give direction to the thoughts of the RWTs' members. Finally, it contained the forecasting sheet for Scenario I which brought

together in a synthesising table the 1978 data (on a regional basis only) linked to the tendencies for 1960–1978. In addition, it contained the provisional forecasts of the central team for 2000 and 2020 with regard to:

- Population
- GNP
- energy consumption according to 8 sources

There were no forecasts for production, imports and exports (even if these data were provided for 1978). Each RWT was given the responsibility of modifying consumption forecasts and to add its own forecasts for production and trade.

For some regions, we also provided the global results of a "high scenario" which was simply the result of the adding on of national forecasts (either official or published by experts, mostly limited to 2000) and which, up to a point, measured the resulting sum of regional optimisms. This obviously led to rather high resultant figures. The aim of this "high scenario" was to create a framework for the thoughts of the RWTs' members.

(iv) *Part IV* This repeated the elements of Scenario I but adapted to *Scenario II* (and without a "high scenario").

(C) For a study which was concerned with the primary energy balances, the simplest and best adapted forecasting tool remained a *global model*, despite all its imperfections. *A priori*, therefore, the following working method was adopted, and applied to each of the 10 RWTs:

1st Stage: Regional Dossier   2nd Stage

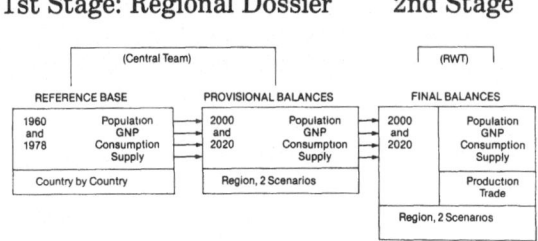

The general architecture of the global forecasting model may be summarised by the following relations, which were based on the methodology employed in ref. 11:

(1) Total primary energy consumption (PEC) = commercial consumption (CEC) + non-commercial consumption (NCEC).
(2) CEC = CEC/inh × Population (Pop)
(3) Δ CEC/inh = Δ GNP /inh × elasticity e.
(4) CEC = NCEC/inh × Population

The central team provided each region with the CEC, NCEC, and thus the PEC, of 1960 and 1978 of the countries of the region, expressed in terms of 8 energy sources (solid mineral fuels, oil, natural gas, hydropower, nuclear, new energies, fire-wood, animal and vegetable waste), as well as corresponding populations and GNP.

For 2000 and 2020, the Central Team estimated, as a first analytical method:
(1) population
(2) GNP according to the two Scenarios I and II
(3) e I and e II

thus

(4) total CECI and CECII
(5) total NCECI and NCECII (forecasted autonomously), therefore, PECI and PECII
(6) the first breakdown of PEC according to the 8 sources.

(D) The forecasts of the Central Team originated from different sources:

(I)   *Population:* forecasts of the U.N. Population Division (cf. Annex 6).
(II)   *GNP:* for the Third World, from the hypotheses of the Scenarios B and C published in the study, "Energy Horizons for the Third World", for 1978-2000 rates of growth, corrected by reference to the events of 1976-1978; as regards the years 2000-2020, the same tendencies used in the previous study were extrapolated further in time.
        For the Industrialised Countries, from values taken from the recent works of international organisations.
(III)   *CEC:* for the Third World, the methodology was borrowed from the "Horizons" study. For each of the 6 regions, one started from the per capita growth rates for 1976-2000 ascertained by the study (Scenarios B and C); these were then corrected by reference to the events of the years 1976-1978, thereby giving the rate for 1978-2000. It was then multiplied by the new forecast for population in 2000, to arrive at the proposed CEC for 2000. For 2020, the 2000-2020 per capita rate was kept and applied to the demographic projection for 2020.
        For the Industrialised Countries, basing oneself on the GNP growth rates, a forecast was made of total income elasticity which enables one to progress to the CEC for 2000 and 2020.
(IV)   NCEC: for the Third World, we proceeded on a per-capita basis (according to the "Horizons" study). Without modifying the NEC ratios, we limited ourselves to correcting the population forecasts for 2000 and 2020, for both fire-wood and for animal and vegetable waste.
(V)   *Supply patterns:* to the new CEC, we applied the patterns for 2000-2020 obtained from the "Horizons" study vigorously, for the Third World regions. For the Industrialised Countries, we proposed an initial breakdown, based on the most recent international studies.

(E) The general idea was that there should be no modification by the RWTs of the data of the reference base, or of the demographic and economic forecasts, but that these should be accepted as the fixed inputs of the study.

The RWTs' freedom of manoeuvre started at the forecasting of total income eleasticities (e) and thus of the formulation of future CECs, as well as the forms of the NCEC/inh. The RWT then had total freedom in the definition of the pattern of supply of the 8 sources, and in determining the levels of production and trade (which were not in fact discussed in the preparatory dossiers). One of the key points of the process was thus the forecast of elasticities (e), on which the energies of the RWTs were particularly directed.

It will be noted that one is here concerned with income elasticity. Indeed we did not introduce any elasticity explicity referable to energy prices in this global approach. However, we defined a general situation clearly characterised by a *progressive and continous tendency towards long-term increase* (after the current years of decrease) of the price of the leading energy source, oil (a tendency which was more marked in Scenario I than in Scenario II). Some regional teams integrated explicit hypotheses of long-term price increases in their models. Most of the teams limited themselves to the elaboration of their forecasts in this global context.

To be concerned with an income elasticity does not, for all that, indicate that any reference to price was ignored. This factor was indirectly taken into account by the growth of the GNP, partly affected as it is by increase in energy prices. Furthermore, the orientation towards diminished income elasticities equally translates the consumption savings effort induced by expensive energy.

(F) In the event, things did not follow the procedure defined *a priori*. The Central Team, indeed, provided each RWT with a regional dossier containing all the elements envisaged. But it is at the level of the RWTs work that the course of the project evolved in a very different manner than anticipated. Indeed:

(i) It transpired that the statistical basis for the past period was sometimes incompatible with the information obtained by the RWT itself. Since absolute priority was given to decentralised work, it was thus necessary to amend the data for the years 1960 and 1978 for certain regions, when this was justified by the inaccuracies of international statistics.

(ii) Moreover, while demographic forecasts were almost completely accepted by the RWTs, it was necessary to alter some of the economic growth forecasts which did not tally with the information obtained by the regions.

The inputs of the study underwent certain changes, therefore, in a manner contrary to the principle adopted at the start.

(iii) In addition, far from following the proposed procedure (which was to give pride of place to forecasts of total income elasticities), most of the RWTs adopted their own forecasting methodology which depended on the data available to them, on their specific approach to the problem, on the means of investigation which they possessed to participants in research.

In the end, therefore, there occurred a marked shift away from the somewhat rigid framework which had been determined at the start to shape the thoughts and to guide the work of the RWTs.

Certain RWTs (like the North American RWT) had recourse to complex simulaton models; others (like RWT 6) kept closer to the proposed procedure.

However, one of the successes of this study, it will be noted, is that there is a large gap between the results of the provisional balance elaborated by the Central Team, and the final results brought out by the RWTs. Each region made an in-depth study of past data and of forecasting research, which enabled it to emancipate itself from the suggested framework, while at the same time respecting the overall coherence of the study guaranteed by the active presence of the Central Team in all the meetings of the RWTs (cf.Annex 20).

At the end of the second phase, the main objective was to transfer forecasting responsibility from the Central Team to the RWT, so that *final forecasts* should be published as originating from the RWT except for RWT 4, where the Central Team kept official responsibility for the figures produced.

5.6. FORECASTING SCENARIOS

The definition and the contents of the development scenarios constituted one of the central elements which conditioned energy forecasts.

(A) In a similar manner as for the study "Energy Horizons for the Third World", we used as an initial starting point the excellent study, produced by OECD to elaborate consistent global scenarios for 2000, published in 1978 under the title "Interfutures". From the range of the scenarios explored by OECD, we initially selected two to provide guidelines for forecasting evolutions. One can summarise the essential elements of the Scenarios "B" and "C" as follows:

Scenario B
'• Collective management of interests and conflicts within the developed nations.
• Increased commercial liberalism.
• Increasingly strong participation of the Third World in international trade, but with internal differentiation.
• Moderate "traditional" growth of the Industrialised Countries caused more by the difficulties of structural adaptation than by conscious desire and by a unanimously accepted change of values.
• Convergence of the relative productivities of Industrialised Nations.
• No recovery of production losses due to the present recession.
• Retention of one part of current unemployment.
• Aid maintained at its present level.
• Increased internationalisation of investment by multinational companies.
• Slowdown of world industrial production, with increased differentiation between Developing Countries.
• Increase of the debt total and reduced solvency of Developing Countries.'

Scenario C
'• Decoupling strategy in most Developing Countries and development of a policy of collective autonomy.
• Break of numerous links with the Industrialised Countries.
• Efforts towards collective management among Industrialised Nations, resulting in greater trade liberalisation between them; slower growth without appreciable modification of values.
• Non-convergence of relative productivities.
• Among the Developing Countries, stress placed on equality and the poor.
• Reduction of aid, partially compensated by OPEC.
• Substantial loss of income by the Industrialised Nations and the Developing Countries.

- Indispensable independence as regards food by the Developing Countries, hence initial priority given to agriculture.
- Search for autocentred development models and for total control of accumulation process.
- Necessity for the Developing Countries to develop research on "adapted" technologies since technological transfers diminish.
- Substantial contraction of North–South trade.
- Increased differentiation between the regions of the Third World.'

In so far as the uncertainties of the future can hardly be reduced to two scenarios, one could, of course, have multiplied the scenarios. However, the need for manoeuvrability and efficiency necessitated the reduction of our research to two characteristic trajectories.

(B) Indeed, these Scenarios B and C were not applied in this study without modification: they merely helped as starting points. It should chiefly be borne in mind that they enabled us to characterise two rather different types of evolutions:

(i) Firstly, a scenario which could be described as "optimistic" or "desirable" which was called "normative-cooperation" in the study. This describes a long-term picture of a less stressful world where economic growth, after the present crisis, recovers and returns to a level compatible with the common aspirations of the Industralised Countries and of the Third World.

(ii) Secondly, another scenario which can be described as "pessimistic" but unfortunately "possible", which we have called "the Increasing Stresses Scenario", so to place it in relation with the first scenario. In a more difficult overall situation, uncertainties and rivalries increase, there is greater reliance on relative positions of strength, and new areas of conflict emerge, especially between the better protected industrialised world and the more threatened Third World.

The "Interfutures" scenarios were thus mainly used to define the "spirit" of each type of development — all the more so as the little amount of quantified data, associated with the "Interfutures" hypotheses, which was provided as "background" information to the RWTs soon turned out to be generally too high for the period up to 2000. In fact, these two scenarios were transposed by only retaining their central concepts and without giving further details of their numerical data. The "spirit" of the two scenarios was only translated in terms of the GNP for each region (but which could be corrected by each region). Two dangerous pitfalls had to be avoided:

(i) imprisoning the RWT in a too rigid set of hypotheses so that they would be forced to adopt certain energy consumption forecasts. To give them complete freedom of thought, we finally presented them with the scenarios under their most general "optimistic" or "pessimistic" aspects, leaving to each RWT the task of gauging their significance for its region, in terms of GNP development possibility, consumption or energy production.

(ii) developing extreme scenarios, either unwarrantedly optimistic or much too catastrophic the probability of which is, in the final analysis, minimal. Therefore, care was taken to moderate the context of each scenario, by stressing its pertinent and plausible character.

Overall, one could say that Scenarios I and II, as they appear in the final form in the study, represent two realistic versions of the future, one rosy and one grey. It is not difficult to guess which is the more desirable. However, the prolongation of the present recession tends to direct the future increasingly towards the grey scenario, which consequently no longer appears to be so apocalyptic. However, it is a grey scenario that is much closer to a "bad" rose scenario than to Scenario C in its original form, namely with a collegial reorganisation of interests in the North or in the South. The situation of the poor regions of the South in II, therefore, particularly risks being even more precarious than one could have supposed *a priori* in C.

5.7 UNITS AND EQUIVALENCES

It will be noted that all energy figures have been expressed in terms of tonnes of oil equivalent or TOE.

This TOE is defined by its heat content which in this study is conventionally set at 10500×10^{10} kcal.

All other forms of energy are expressed in TOE by means of the system adopted by the UN, as described in the *Yearbook of World Energy Statistics* (1979 edition). The past reference base for 1960–1978 with regard to national energy consumption was elaborated on the basis of this system of equivalences. An extract of this system is to be found in Annex 4.

However, two important modifications were made to the UN system, in order to bring it into line with the decisions taken by the Conservation Commission:

(A) the UN reference TOE is the equivalent of 10180×10^3 kcal. All UN figures expressed in TOE were thus modified by a coefficient of $10180/10500 \simeq 0.97$ to transpose them into TOE at 10500×10^3 kcal.

(B) Moreover, as equivalent for kWh, we chose the primary energy quantity used in the most modern thermal power station to produce 1 kWh of electricity, instead of the theoretical calorific equivalent recommended by the UN of the kWh used by the "Joule effect", which is equal to 860 kcal per kWh or 860×10^3 kcal for 1000 kWh).

In any case this equivalence is used simultaneously with calorific equivalence by OECD. It is adopted by UNIPEDE, the World Bank and the IAEA. The World Energy Conference recommended its use in Munich ("Guide to Substitutes between Energy Forms").

We have, therefore, retained an average equivalence of:

$$1000 \text{ kWh} = 0.222 \text{ TOE} = 2330 \times 10^3 \text{ kcal,}$$

whatever the means of electricity production in question (hydro, nuclear, geothermal, etc.).

When precise equivalences are not specified according to product (e.g. as was the case in forecasting researches), we applied the following general system:

1 tonne of crude oil (or petroleum products) = 10000×10^3 kcal = 0.952 TOE.
1 tonne of solid mineral fuels = 7000×10^3 kcal = $\frac{2}{3}$ TOE
1000 m³of natural gas = 0.85 TOE = 9000×10^3kcal.
1000 kWh of electricity (primary or secondary) = 0.222TOE = 2330×10^3 kcal.

Even if we did not produce figures in SI units, which are recommended by the UN but little used in the energy sector since they do not *adequately* represent the degrees of magnitude to which professionals are used, one can easily make the TOE to Joules transposition by means of the following table:

1 TOE = 44×10^9 J = 44 GJ
1GTOE = 44×10^{18} J = 44 EJ
1 EJ = 22.8 MTOE
1 TJ = 22.8 TOE

In BTU (British Thermal Units) one also has the following:

1 million BTU = 252 kcal
10^{15} BTU = 1000 billion BTU = 24 MTOE
1 MTOE = 4.18×10^{13} BTU

Part III

ANNEXES

COMPOSITION OF THE REGIONS

The problem of regional divisions occurs as soon as one has to analyse the evolution of the great geographic zones within the world system.

Different approaches can be envisaged on the aim and the task of the body responsible for determining the divisions. The UN places the emphasis on geographical proximities, continental frontiers, and the types of economic system, and it had introduced distinctions between developed and developing countries, according to the stage of development. The World Bank, another body with world standing, places more emphasis on levels of development and creates a hierarchy of countries by income categories, beyond their continental appurtenances and their political affinities.

The division adopted in this study is close to that of the UN. Indeed, in matters of energy, geographic proximity plays an important role, if only because of the climatic, sociological and cultural constraints common to a region. Likewise, the type of economic organisation (market or centrally planned economy) and the respective levels of development have an influence on technologies on the range of practices in use.

Between the unity of the whole and the infinite diversity of the parts, one must define a set of divisions which is representative of the problems and at the same time pertinent and operational, despite its unavoidable arbitrary elements.

As a result, we arrived at a relatively traditional reorganisation of the world in 10 regions which were deemed more or less "homogeneous" from the points of view of geography, economy, geopolitics and sociology:

Regions:			
	R1	North America	NA
	R2	Western Europe	WE
	R3	Industrialized Countriesof the Pacific	PIC
	R4	Eastern Europe	EE
	R5	North Africa/Middle East	NAME
	R6	Sub-Saharan Africa	SSA
	R7	South Asia	SA
	R8	South-East Asia	SEA
	R9	Centrally Planned Asian Countries	CPAC
	R10	Latin America	LA

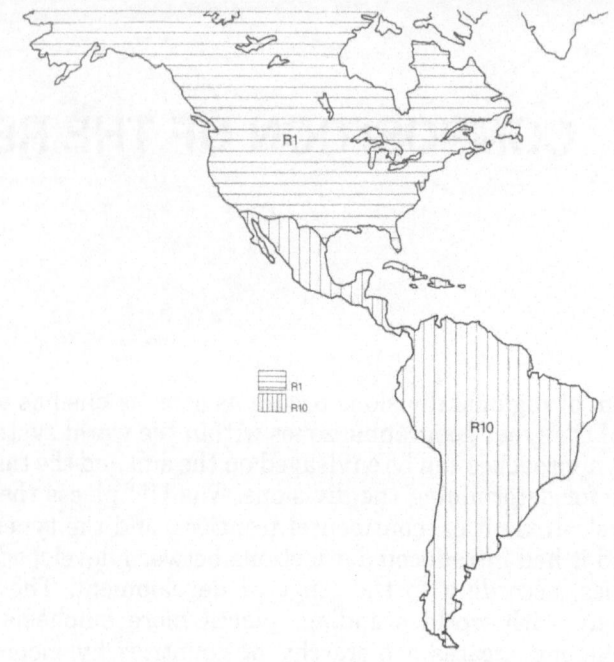

REGIONS

R1 NORTH AMERICA

CANADA
PUERTO RICO
U.S.A.

R2 WESTERN EUROPE

AUSTRIA
BELGIUM
CYPRUS
DENMARK
FINLAND
FRANCE
GERMANY (Fed. Rep.)
GREECE
ICELAND
IRELAND
ISLANDS: Channel Islands
            Isle of Man
            Greenland
            (Gibraltar)
ISRAEL
ITALY
LUXEMBOURG
MALTA
NETHERLANDS
NORWAY
PORTUGAL
SPAIN
SWEDEN
SWITZERLAND
TURKEY
UNITED KINGDOM
YUGOSLAVIA

R3 INDUSTRIALIZED COUNTRIES OF THE PACIFIC

AUSTRALIA
JAPAN
NEW ZEALAND

R4 EASTERN EUROPE

ALBANIA
BULGARIA
CZECHOSLOVAKIA
GERMANY (Dem. Rep.)
HUNGARY
POLAND
ROMANIA
USSR

R5 NORTH AFRICA/MIDDLE EAST

ALGERIA
BAHRAIN
EGYPT
IRAQ
IRAN
JORDAN
KUWAIT
LEBANON
LIBYA
MOROCCO
OMAN
QATAR
SAUDI ARABIA
SUDAN
SYRIAN ARAB REP.
TUNISIA
UNITED ARAB EMIRATES
YEMEN ARAB REP.
YEMEN DEM. REP.

R6 SUB-SAHARAN AFRICA

SOUTH AFRICA
ANGOLA
BENIN
BOTSWANA
BURUNDI
CAMEROON
CAPE VERDE
CENTRAL AFRICAN REP.
CHAD
COMOROS
CONGO
DJIBOUTI
EQUATORIAL GUINEA
ETHIOPIA
GABON
GAMBIA
GHANA
GUINEA
GUINEA-BISSAU
IVORY COAST
KENYA
LESOTHO
LIBERIA
MADAGASCAR
MALAWI
MALI
MAURITANIA
MAURITIUS
MOZAMBIQUE
NAMIBIA
NIGER

NIGERIA
REUNION
RWANDA
SAO TOME & PRINCIPE
SENEGAL
SEYCHELLES
SIERRA LEONE
SOMALIA
SWAZILAND
TANZANIA
TOGO
UGANDA
UPPER VOLTA
ZAIRE
ZAMBIA
ZIMBABWE

R7 SOUTH ASIA

AFGHANISTAN
BANGLADESH
BHUTAN
INDIA
MALDIVES
NEPAL
PAKISTAN
SRI LANKA

R8 SOUTH-EAST ASIA

BRUNEI
BURMA
FIJI
HONG KONG
INDONESIA
ISLANDS: Solomon
   Guam
   French Polynesia
   Tonga
   Western Samoa
   New Hebrides
   Vanatu
   New Caledonia
   Gilbert-Kiribati
   Pacific US Trust
   American Samoa
MACAO
MALAYSIA
PAPUA NEW GUINEA
PHILIPPINES
SINGAPORE
SOUTH KOREA
TAIWAN
THAILAND

R9 CENTRALLY PLANNED ASIA

CHINA
KAMPUCHEA
LAO
MONGOLIA
NORTH KOREA
VIETNAM

R10 LATIN AMERICA

ARGENTINA
BAHAMAS
BELIZE
BERMUDA
BOLIVIA
BRAZIL
CHILE
COLOMBIA
COSTA RICA
CUBA
DOMINICAN REP.
ECUADOR
EL SALVADOR
FRENCH GUYANA
GUATEMALA
GUYANA
HAITI
HONDURAS
ISLANDS: Guadeloupe
   St. Vincent
   Martinique
   Virgin Islands
   Netherlands Antilles
   Dominica
   Barbados
   Antigua
   Santa Lucia
   St. Kitts
   Grenada
JAMAICA
MEXICO
NICARAGUA
PANAMA
PARAGUAY
PERU
SURINAME
TRINIDAD/TOBAGO
URUGUAY
VENEZUELA

Moreover, when it becomes useful to argue on a more general plan for the sake of the analysis presented in the study, we used a distinction between four great categories of countries.

(A)   "Industrialised Countries with Market Economies": this category cor-responds in total to the regions R1, R2, and R3, with the addition of South Africa. In the tables, they will be abreviated to IWC ("Industrialised Western Countries").

(B)   "Total Industrialised Countries": this category includes, in addition to the previous category, R4 (Eastern Europe), and is abbreviated to IC.

(C)   "Total Third World": this category joins R5, R6, R7, R8, R9, and R10. It was thought interesting to make a special category within this vast category ("TW") of the Centrally Planned Asian Countries (R9), which have a particular character by virtue of their population and by their style of development.

Consequently a final category was introduced:

(D)   "Third World excluding R9": this only includes R5, R6,* R7, R8, and R10. This category is often called "Developing Countries", but it will be abbreviated here to "TW-R9".

(1)   In many world studies one finds the category "other nations with market economies", which contains a rather heterogeneous group of "western" countries which are difficult to integrate within their geographical regions. Most often, this means South Africa, Israel, Japan, Australia and New Zealand.

In this study, we have decided to place Israel within R2 "Western Europe"; South Africa within R6 "Sub-Saharan Africa" for regional results; on the other hand, we transferred South Africa to the industrialised countries with market economies for the drawing up of total figures for economic zones.

The "Third World" (TW) thus means R5 + R6 − South Africa + R7 + R8 + R9 + R10. And "WIC" means R1 + R2 + R3 + South Africa.

For well-known political reasons, the R6's RWT did not contain any representatives from South Africa and only was concerned with prospects for Black Africa excluding South Africa.

A separate consultation enabled us to gather complementary forecasts for South Africa, which were integrated with the forecasts for "Black Africa", to produce total figures for R6.*

It was also decided to leave the "Industrialized Countries of the Pacific" (Japan, Australia and New Zealand) grouped together in the same distinct region R3.

(2)   The regional division adopted at the beginning of the study, which was mainly inherited from ref. 11, underwent certain modifications as a result of the work during the decentralised stage.

In particular, this occurred with R5 "North Africa/Middle East", which initially contained Cyprus and Turkey. During discussions with the RWT it transpired that its members were much more familiar with, and that their forecasting studies were consequently better adapted to,

---

*Except South Africa.
*This regrouping is purely geographical; it in no way reflects the political position of authors.

the concept of the "Arab World"—this term incorporates 21 countries or all the countries of Region 5, except Cyprus and Turkey, Iran, Djibouti, Mauritania and Somalia, the last three belonging to R6.

In the comparison of the Arab World/R5 forecasts, we have neglected the difference created by these three last-mentioned countries allocated to R6. However, it was decided to transfer Cyprus and Turkey to Western Europe, a step which was quite justifiable in the case of Turkey, in view of its political membership of OECD. Iranian forecasts were treated separately. The RWT for R5 was thus responsible for the "Arab World" part, while the task fell to the Central Team to adjust estimates by introducing the Iranian element. Likewise in region 2, "Western Europe", the RWT expressed views for all the nations defined at the outset of this study for its region; the Central Team added a separate estimate of the group "Israel, Cyprus and Turkey" to this forecast.

COMPOSITION OF THE REGIONAL WORKING TEAMS (RWT)

Regions National Committees	RWT Members (Assistants)	Affiliation	City
R1 North America			
Canada	Ron Erdmann	Minister of Energy	Toronto
	Ram Sahi	Minister of Energy	
USA	Robert Finger	Exxon USA	Houston
USA	Wray Smith	DOE	Washington
	(C. Kilgore)	DOE	
	(B. Michelski)	DOE	
R2 Western Europe			
	Klaus Brendow	UNECE	Geneva
	(E. Mayorga-Alba)	UNECE	
	Jean-Paul Cleron	IEA	Paris
	Robert de Bauw	EEC	Brussels
	(C. Waeterloos)	EEC	
R3 Industrialized **Countries of the Pacific**			
Japan	Teruhiro Umezu	Inst. Elec. Power	Tokyo
Australia	Neale Taylor	ESSO	Sydney
	(I. Walker)	ESSO	
New Zealand	Basil Walker	Minister of Energy	Wellington
R4 Eastern Europe			
	Klaus Brendow	UNECE	Geneva
	(E. Mayorga-Alba)	UNECE	
Hungary	L. Horvarth	Minister of Industry	Budapest
Romania	Calin Mihaileanu	Inst. Ener. Research	Bucarest
USSR	S. N. Yatrov	Inst. Ener. Analysis	Moscow
R5 North Africa/ **Middle East**			
Algeria	Ali Aissaoui	Minister of Energy	Algiers
	Nour-Eddine Berrah	SONELGAZ	Algiers
	Ali Alwattari	OAPEC	Kuwait
	(Ibrahim B. Ibrahim)	OAPEC	
	Nouhad Baroudi	UN-ECWA	Beyrouth
Jordan	H. Khatib	Jordan Elec. Authority	Amman
R6 Sub-Saharan Africa			
	Murtada Diallo	UNAEC	Addis-Abeba
	Eric Guillon	Club de Dakar	Paris
Ivory Coast	Marcel Kroko Diby	EECI	Abidjan
	Kana Mutombo	UPDEA	Abidjan

Regions National Committees	RWT Members (Assistants)	Affiliation	City
R7 South Asia			
Bangladesh	S. M. Al-Husainy	Minist. FC, WR & Power	Dacca
Sri Lanka	Mohan Munasinghe	World Bank/Min. Ener.	Wash./Colombo
India	T. L. Sankar	Public Entr. Institute	Hyderabad
R8 South-East Asia			
Malaysia	M. Jalaluddin	Lembaya Lek. Negara	Kuala Lumpur
	(Dr. Rozali)	Lembaya Lek. Negara	
Malaysia	Bebe Chooi	PETRONAS	Kuala Lumpur
	Stuart R. MacGill	ESSO	Kuala Lumpur
Singapore	Michel Khor	PUB	Singapore
	(Ng Heng Liat)	PUB	
Singapore	Hans Dumoulin	SHELL	Singapore
South Korea	Dr. Rhee Seung-Yoon	Inst. Energy Res.	Seoul
Thailand	Pravit Ruyabhorn	NEA	Bangkok
	(Dr. I. Bijayendrayodhin)	NEA	
	H. M. C. Quick	SHELL International	London
R9 Centrally Planned Asian Countires			
	Vaclav Smil	Univ. of Manitoba	Winnipeg
	Kim Woodard	China Energy Ventures	Washington
R10 Latin America			
Mexico	Gerardo Bazan	Petroleos Mexicanos	Mexico
Brazil	Pietro Erber	ELECTROBRAS	Rio
	Alfredo del Valle	University of Science	Santiago
	Gabriel Sanchez	OLADE/World Bank	Quito/Washing.
	Alvaro Umana	OLADE/Univ. Costa Rica	Quito/San Jose

Special Advisers: A. R. Dykes (Paris)
Balint Balkay (Budapest)

ANALYSIS OF THE COMPOSITION OF THE REGIONAL WORKING GROUPS

20 National Committees took part in the study: 8 from Industrialised Nations and 12 from Developing Countries.

50 Persons directly contributed to the study: 40 official members, 8 of their assistants and 2 special advisers.

Of these 50 persons, 41 attended the meetings, 3 corresponded and 6 were associated with research.

According to types of activity, the 50 persons were involved in the following manner:

Energy administration:	11
Oil:	9
Electricity:	9
Regional and International organisations:	14
Universities and Research Institutes:	7

The regions were represented in the following way:

North America:	8
Latin America:	6

Western Europe: 7
Eastern Europe   4
Africa: 6
Asia: 16
Oceania: 3

MEETINGS OF THE REGIONAL WORKING TEAMS

CHRONOLOGY

06 / 81	KUALA LUMPUR	R7 – R8	03 / 82	DOUALA	R6
07 / 81	SYDNEY	R3	04 / 82	BUDAPEST	R4
07 / 81	WELLINGTON	R3	04 / 82	WASHINGTON	R1
09 / 81	CANCUN	R10	05 / 82	GENEVA	R4
01 / 82	WASHINGTON	R1	05 / 82	KUWAIT	R5
01 / 82	WASHINGTON	R9	06 / 82	HYDERABAD	R7
02 / 82	QUITO	R10	06 / 82	SINGAPORE	R8
03 / 82	PARIS	R2	06 / 82	PARIS	R5
03 / 82	PARIS	R4			

In total, 17 decentralised meetings were organised in one year (06/81 – 06/82):

3 in North America	1 in Africa
2 in Latin America	4 in Asia
5 in Europe	2 in Oceania

REGIONAL CALENDAR

R1	01 / 82	WASHINGTON	R6	03 / 82	DOUALA
	04 / 82	WASHINGTON	R7	06 / 81	KUALA LUMPUR
R2	03 / 82	PARIS		06 / 82	HYDERABAD
R3	07 / 81	SYDNEY	R8	06 / 81	KUALA LUMPUR
	07 / 81	WELLINGTON		06 / 82	SINGAPORE
R4	03 / 82	PARIS	R9	01 / 82	WASHINGTON
	04 / 82	BUDAPEST	R10	09 / 81	CANCUN
	05 / 82	GENEVA		02 / 82	QUITO
R5	05 / 82	KUWAIT			
	06 / 82	PARIS			

UNITS — EQUIVALENCES

The reader will find below the tables of conversion factors used by the UN to express its world energy statistics[9]. These factors were used to elaborate the energy reference base of the 1960–1978 study with regard to consumption of fossil fuels, petroleum products, natural gas and hydro-power. However, (and cf. Chapter V: Methodology) we have altered the heating value of the TOE of reference (10500×10^3 kcal), as well as the equivalence for electricity by establishing it at 1000 kWh = 0.222 TOE = 2330×10^3 kcal.

For the forecasts, where the units were not specified, the following system was adopted:

1 tonne of crude oil or petroleum products = 10000×10^3 kcal = 0.952 TOE
1 tonne of solid mineral fuel = 7000×10^3 kcal = $\frac{2}{3}$ TOE
$1000 m^3$ of natural gas = 9000×10^3 kcal = 0.85 TOE
1000 kWh = 0.222 TOE = 2330×10^3 kcal

with:

1 TOE = 44×10^9 J
10^{12} J = 1 TJ = 22.8 TOE

Extracted from: Yearbook of World Energy Statistics, UN, 1979.

Table A4.1

Conversion factor tables

Tableaux des facteurs de conversion

Summary of coefficients used to convert fuels into coal equivalent

Sommaire des coefficients utilisés pour convertir les différents combustibles
en equivalent charbon

Primary solids — solides primaires

HARD COAL (PRODUCTION, EXPORTS, BUNKERS AND CHANGES IN
STOCKS) HOUILLE (PRODUCTION, EXPORTATIONS, SOUTAGES ET
VARIATIONS DES STOCKS)

Standard factor/Facteur standard	1.000
Argentina/Argentine	0.843
Australia/Australie	0.943
Belgium/Belgique	0.914
Brazil/Brésil	0.714–0.823
Canada/Canada	0.900
Chile/Chili	0.986
China/Chine	0.857
Czechoslovakia/Tchécoslovaquie	0.851–0.862
German Democratic Rep./Rép. Allemande Démocratique	0.600–0.608
Hungary/Hongrie	0.529–0.600
India/Inde	0.714
Japan/Japon	0.901–0.984
Korea Rep. of/République de Corée	0.757
Mexico/Mexique	0.714
New Zealand/Nouvelle-Zélande	0.833–0.841
Norway/Norvège	0.957
Pakistan/Pakistan	0.700
Philippines/Philippines	0.675
South Africa/Afrique du Sud	0.798–0.913
Spain/Espagne	0.800–0.929
Sweden/Suède	0.929
USSR/URSS	0.808–0.839
UK/Royaume-Uni	0.879–0.904
USA/Etats-Unis d'Amérique	0.884–0.975
Yugoslavia/Yougoslavie	0.943
Zambia/Zambie	0.843

HARD COAL IMPORTS
IMPORTATIONS DE HOUILLE

Standard factor/Facteur standard	1.000
Argentina/Argentine	1.029
Austria/Autriche	0.900
Bangladesh/Bangladesh	0.714
Belgium/Belgique	0.971
Brazil/Brésil	1.014
Bulgaria/Bulgarie	0.843
Canada/Canada	1.024
Chile/Chili	0.986
Czechoslavakia/Tchécoslovaquie	0.914

Denmark/Danemark	0.971
Egypt/Egypte	0.871
Finland/Finlande	0.914
France/France	0.943
German Democratic Rep./Rép Démocratique Allemande	0.929
German Federal Rep./Rép Fédérale d'Allemagne	0.957
Hungary/Hongrie	0.927–0.976
Italy/Italie	0.943
Japan/Japon	0.929
Korea Rep. of/République de Corée	0.929
Netherlands/Pays-Bas	0.929
Poland/Pologne	0.759–0.791
Romania/Roumanie	0.914
Spain/Espagne	0.929
Sweden/Suède	0.900
UK/Royaume-Uni	0.917–1.036
USA/Etats-Unis d'Amérique	0.914
Yugoslavia/Yougoslavie	0.943

RECOVERED SLURRIES (PRODUCTION)
SCHLAMMS RECUPERES (PRODUCTION)

Belgium/Belgique	0.600
Brazil/Brésil	1.000
Czechoslovakia/Tchécoslovaquie	0.600
France/France	0.700
Hungary/Hongrie	0.515
Spain/Espagne	0.700
Turkey/Turquie	0.500
UK/Royaume-Uni	0.744

LIGNITE AND BROWN COAL (PRODUCTION, EXPORTS AND CHANGES IN
STOCKS)
LIGNITE (PRODUCTION, EXPORTATIONS ET VARIATIONS DES STOCKS)

Standard factor/Facteur standard	0.385
Albania/Albanie	0.500
Australia/Australie	0.330
Austria/Autriche	0.500
Bulgaria/Bulgarie	0.500
Canada/Canada	0.524
Chile/Chili	0.586
Czechoslovakia/Tchécoslovaquie	0.429–0.461
Denmark/Danemark	0.300–0.330
France/France	0.600
German Democratic Rep./République Démocratique Allemande	0.303–304
Germany Federal Rep./République Fédérale d'Allemagne	0.300
Greece/Grèce	0.190–0.197
Hungary/Hongrie	0.371–0.467
India/Inde	0.330
Italy/Italie	0.360
Japan/Japon	0.330
Korea D.R./République Démocratique de Corée	0.600
Mongolia/Mongolie	0.330
New Zealand/Nouvelle-Zélande	0.500–0.573
Romania/Roumanie	0.330
Spain/Espagne	0.529–0.600
Thailand/Thailande	0.330
Turkey/Turquie	0.330
USSR/URSS	0.500
USA/Etats-Unis d'Amérique	0.470–0.571
Yugoslavia/Yougoslavie	0.407–0.487

LIGNITE/BROWN COAL IMPORTS
IMPORTATIONS DE LIGNITE

Standard factor/Facteur standard	0.385
Austria/Autriche	0.500
Czechoslovakia/Tchécoslovaquie	0.304–0.450
Denmark/Danemark	0.300–0.330
France/France	0.300
German Democratic Rep./Rép. Démocratique Allemande	0.300
German Federal Rep./Rép. Fédérale d'Allemagne	0.330
Hungary/Hongrie	0.300
Italy/Italie	0.500
Japan/Japon	0.330
Norway/Norvège	0.503
Papua New Guinea/Papouasie Nouvelle-Guinée	0.330
Singapore/Singapour	0.330
Spain/Espagne	0.600
USSR/URSS	0.500
USA/Etats-Unis d'Amérique	0.524
Yugoslavia/Yougoslavie	0.300

SECONDARY SOLIDS — SOLIDES SECONDAIRES

HARD COAL BRIQUETTES (IMPORTS, EXPORTS AND CHANGES IN STOCKS)
BRIQUETTES DE HOUILLE (IMPORTATIONS, EXPORTATIONS ET VARIATIONS DES STOCKS)

Standard factor/Facteur standard	1.000
New Zealand/Nouvelle-Zélande	0.992

COKE-OVEN COKE (EXPORTS AND CHANGES IN STOCKS)
COKE DE FOUR (EXPORTATIONS ET VARIATIONS DES STOCKS)

Standard factor/Facteur standard	0.900
Argentina/Argentine	0.971
Australia/Australie	0.942–0.944
Austria/Autriche	0.957–0.971
Canada/Canada	0.971
Czechoslovakia/Tchécoslovaquie	0.935–0.950
German Democratic Rep./République Démocratique Allemande	0.869–0.941
Hungary/Hongrie	0.924–0.980
Mexico/Mexique	0.953
New Zealand/Nouvelle-Zélande	0.957
South Africa/Afrique du Sud	0.952
Sweden/Suède	0.957
USA/Etats-Unis d'Amérique	0.957
Yugoslavia/Yougoslavie	1.029

COKE-OVEN COKE IMPORTS
IMPORTATIONS DE COKE DE FOUR

Standard factor/Facteur standard	0.900
Argentina/Argentine	0.971
Canada/Canada	0.971
Sweden/Suède	0.957

GAS COKE (IMPORTS, EXPORTS AND CHANGES IN STOCKS)
COKE DE GAZ (IMPORTATIONS, EXPORTATIONS ET VARIATIONS DES STOCKS)

Standard factor/Facteur standard	0.900
Czechoslovakia/Tchécoslovaquie	0.769
German Democratic Rep./Rép. Démocratique Allemande	0.871–0.910
Hungary/Hongrie	0.844–0.866
South Africa/Afrique du Sud	0.873

LIGNITE/BROWN COAL BRIQUETTES (EXPORTS AND CHANGES IN STOCKS)
BRIQUETTES DE LIGNITE (EXPORTATIONS ET VARIATIONS DES STOCKS)

Standard factor/Facteur standard	0.670
Australia/Australie	0.747–0.754
Czechoslovakia/Tchécoslovaquie	0.714–0.776
Germany Democratic Rep./Rép. Démocratique Allemande	0.666–0.744
Hungary/Hongrie	0.726–0.744

BROWN COAL BRIQUETTES (IMPORTS)
BRIQUETTES DE CHARBON BRUN (IMPORTATIONS)

Standard factor/Facteur standard	0.670
Hungary/Hongrie	0.653–0.729

COKES OF BROWN COAL OR LIGNITE (IMPORTS, EXPORTS AND CHANGES IN STOCKS)
COKES DE CHARBON BRUN ET DE LIGNITE (IMPORTATIONS, EXPORTATIONS ET VARIATIONS DES STOCKS)

Standard factor/Facteur standard	0.670
German Democratic Rep./République Démocratique Allemande	0.786

PEAT BRIQUETTES (IMPORTS, EXPORTS AND CHANGES IN STOCKS)
BRIQUETTES DE TOURBE (IMPORTATIONS, EXPORTATIONS ET VARIATIONS DES STOCKS)

Standard factor/Facteur standard	0.500

OTHER — AUTRES

Crude petroleum/Pétrole brut	1.454
Natural gas liquids (weighted average)/Les condensats provenant du gaz (moyenne pondérée)	1.542
Liquified petroleum gases/Gaz de pétrole liquéfiés	1.554
Natural gasolene/Gazoline naturelle	1.532
Condensate and other/Condensats et autres	1.512
Gasolenes/Gazolines	1.5
Kerosene and jet fuels/Kérosène et carburéacteurs	1.474
Gas-Diesel oils/Gas oils et Fuel oils fluides	1.450
Residual fuel oil/Fuel oils résiduels	1.416
Natural gas (Terajoules)a//Gas naturel (Térajoules)	34.131
Manufactured gases (Terajoules) a//Gaz d'usine (Térajoules)	34.131
Hydro, nuclear and geothermal electricity (1000 kWh)/ Electricité hydraulique, nucléaire et géothermique	0.123

a/ The average calorific value of gas is measured in kcal/m^3 (st), i.e., in kcal per cubic metre of gas under standard conditions of 15°C, 1,013.25 mbar, dry.

a/ La valeur calorifique moyenne du gaz en kcal/m^3 (st), c'est-à-dire en kcal par mètre cube de gaz dans les conditions normales 15°C, 1,013.25 mbar/sec.

Table A4.2

Summary of specific gravities of crude petroleum
Sommaire des poids spécifiques du petrole brut

Country / Pays	Specific Gravity / Poids spécifique	Country / Pays	Specific Gravity / Poids spécifique
Albania/Albanie	0.94	Japan/Japon	0.86
Algeria/Algérie	0.802	Kuwait/Koweit	0.863
Angola/Angola	0.851	Libyan A.J./J.A. Libyenne	0.829
Argentina/Argentine	0.885	Malaysia/Malaisie	0.82
Australia/Australie	0.816	Mexico/Mexique	0.886
Austria/Autriche	0.90	Mongolia/Mongolie	0.86
Bahrain/Bahreïn	0.858	Morocco/Maroc	0.83
Barbados/Barbade	0.86	Netherlands/Pays-Bas	0.92
Bolivia/Bolivie	0.80	New Zealand/Nouvelle-Zélande	0.781
Brazil/Brésil	0.837	Nigeria/Nigéria	0.851
Brunei/Brunéi	0.84	Norway/Norvège	0.845
Bulgaria/Bulgarie	0.86	Oman/Oman	0.839
Burma/Birmanie	0.89	Pakistan/Pakistan	0.86
Canada/Canada	0.846	Peru/Pérou	0.85
Chile/Chili	0.84	Poland/Pologne	0.85
China/Chine	0.86	Qatar/Qatar	0.834
Colombia/Colombie	0.89	Romania/Roumanie	0.84
Congo/Congo	0.84	Saudi Arabia/Arabie Saoudite	0.86
Cuba/Cuba	0.95	Spain/Espagne	0.84
Czechoslovakia/Tchécoslovaquie	0.93	Sweden/Suède	0.97
Denmark/Danemark	0.82	Syrian A.R./R.A. Syrienne	0.906
Ecuador/Equateur	0.873	Thailand/Thaïlande	0.86
Egypt/Egypte	0.869	Trinidad and Tobago/ Trinité et Tobago	0.89
France/France	0.86	Tunisia/Tunisie	0.82
Gabon/Gabon	0.873	Turkey/Turquie	0.88
German D.R./R.D. Allemande	0.86	USSR/URSS	0.856
Germany, F.R./Allemagne, R.F.	0.87	UAE/EAU	0.846
Guatemala/Guatemala	0.86	United Kingdom/Royaume-Uni	0.86
Hungary/Hongrie	0.94	USA/EUA	0.848
India/Inde	0.83	Venezuela/Venezuela	0.901
Indonesia/Indonésie	0.848	Yugoslavia/Yougoslavie	0.85
Iran/Iran	0.864	Zaire/Zaïre	0.86
Iraq/Irak	0.845	Unspecified origin/ De source non spécifiée	0.86
Israel/Israël	0.84		
Italy/Italie	0.92		

Table A4.3

Selected conversion factors for petroleum products

Quelques facteurs de conversion pour les produits pétroliers

(To convert from metric tons into the following units,
multiply by the factor in the appropriate column)

(Pour convertir à partir des tonnes métriques vers les unités
suivantes, multiplier par le facteur dans la colonne appropriée)

Commodity Produit	Litre Litre	U.S. Gallon Gallon E.-U.	Imperial Gallon Gallon Brit.	Barrel Baril	Barrels/ Day Barils/ jour	Cubic Metre Métre cubé
Liquefied petroleum gas/Gaz de pétrole liquéfiés	1852	489	407	11.65	0.03192	1.852
Natural gasolene Gazoline naturelle	1590	420	350	10.00	0.02740	1.590
Plant condensate Condensat d'usine	1429	378	314	8.99	0.02463	1.429
Naphthas/Naphtas (undifferentiated) (non différenciés)	1389	367	306	8.74	0.02395	1.389
Aviation gasolene Essence aviation	1370	362	301	8.62	0.02362	1.370
Gasolenes/Gazolines (undifferentiated) (non différenciés)	1351	357	297	8.50	0.02329	1.351
Motor gasolene Essence auto	1351	357	297	8.50	0.02329	1.351
Paraffin wax Paraffine	1250	330	275	7.86	0.02153	1.250
Jet fuel Carburéacteurs	1235	326	272	7.77	0.02129	1.235
Kerosene Kérosène	1235	326	272	7.77	0.02129	1.235
White spirit White-spirit	1235	326	272	7.77	0.02129	1.235
Gas-diesel oil Gas oils et fuel oils fluides	1149	304	253	7.23	0.01981	1.149
Lubricants Lubrifiants	1111	294	244	6.99	0.01915	1.111
Fuels oils (undifferentiated) (non différenciés)	1099	290	242	6.91	0.01893	1.099

Commodity	Litre	U.S. Gallon	Imperial Gallon	Barrel	Barrels/ Day	Cubic Metre
Produit	Litre	Gallon E.-U.	Gallon Brit.	Baril	Barils/ jour	Métre cubé
Residual fuel oil Fuels oils résiduels	1053	278	232	6.62	0.01814	1.053
Bitumen (asphalt) Bitume (bral)	962	254	212	6.05	0.01658	0.962
Petroleum coke Coke de pétrole	877	232	193	5.52	0.0151	0.877

REFERENCE BASE

1960–1978

Any forecasting has to begin by examining previous patterns of evolution. The future can only be explored in terms of the shape of the past, and the present study was no exception to the rule. To facilitate the formulation of forecasts, a firm base had to be set up by highlighting the structural tendencies of explanatory variables (population and GNP) and of energy consumptions.

Such was the aim of the work which was done at the start of the study using the years 1960–1978 as a reference base. It is described in some detail in the paragraphs which follow.

It should be remembered that the data were examined country by country for the 160 nations making up the world community; however, the data are published here on a regional basis.

1 DEMOGRAPHY AND ECONOMY

Demographic and economic estimates for 1960 and 1978 are to be found in the specialised Annexes 6 and 7 and not in this annex; in Annexes 6 and 7 they are compared with population and economic growth forecasts.

It will be noted that demographic data was extracted from the World Bank's 1980 Atlas which indicates the population level of each country in mid-1978, along with the rate of growth over the period 1970–1978.

The 1960 population was determined on the basis of the 1960–1976 growth rate published in the 1978 Edition of the Atlas.

Compared to the figures produced by the UN for the same years, these estimates have proved themselves to be robust and without any appreciable discrepancy.

The corresponding per capita and total GNP level at market price is also to be found in the 1980 Atlas, as well as an estimate of its 1970–1978 evolution in real terms. (This GNP is measured in 1978 at the rates of exchange of that year.)

In the same way, we went back to 1960 by using the 1960–1976 progressions produced in the 1978 Edition of the Atlas.

2 COMMERCIAL ENERGY CONSUMPTIONS

The estimates were derived from UN statistics. For 1978, they were extracted from the 1979 *Yearbook of World Energy Statistics* (table 9 in TOE), and for 1960, the 1976 edition of *World Energy Supplies 1950-1974* (table 2 in TOE).

These statistics were harmonised and transposed in the unit system used in this study (cf. Annex 4).

We have collected under the heading "Commercial Energies" all the energy consumptions of solid mineral fuels, of petroleum products, of natural gas, of primary electricity (hydropower and nuclear) and of new energies (solar, geothermal, wind, tidal, wave, biomass, alcohol, etc.)

The consumption of petroleum products represents what are here described as purely energy consumptions. This excludes consumptions of non-energy products (naphthas, bitumen, lubricants, etc.) where oil is used as a primary chemical raw material, as well as bunkers and stock variations.

With regard to primary electricity, measurements were made including losses at the ex-power station stage (possible foreign trade being excluded).

3 NON-COMMERCIAL ENERGY CONSUMPTIONS

It is much more difficult to ascertain the quantities involved here, as there is a dearth of statistics in the area (since, by definition, a great deal of consumption does not go through the habitual trade channels). Two types of fuels were collected under this category: fuel wood and its derivants (charcoal) and all the types of vegetable and animal waste. Human and animal traction are thus excluded from our analysis.

What method did we use to produce a plausible and coherent (if inexact) estimate?

The author's publication of June 1980 *The Evolution of Energy Consumption in the World: a 1960-1976 Retrospective* (published by Electricité de France) may be consulted for a detailed explanation. We will confine ourselves to reiterating the principles adopted to estimate the approximate amount of consumption.

3.1 Fuel wood

The basis of the estimates for wood were provided by FAO's *Yearbook of Forest Products* (1978 edition, table on p.87 and the 1971 edition for 1960).

In fact, the *Yearbook* provides an order of magnitude for fuel wood production, country by country and expressed in m³. This estimate contains the equivalent in round wood of charcoal (coefficient 6 to calculate the tons of charcoal expressed from the m³ volume units). International sales of fuel wood or of charcoal which represent only 3% of production were ignored, and consumption and production were thus assimilated.

One goes from an m³ measure to the TOE measure by a general equation of the following type:

Consumption of fuel wood (in TOE) = production of fuel wood (in m³)
$$\times\ C_1\ (m^3 \rightarrow tonne)$$
$$\times\ C_2\ (tonne \rightarrow TOE)$$

In fact, the correspondence C_1"volume → weight" varies according to the type of wood.

The average convention recommended by FAO was adopted: 725 kg by m^3 (for charcoal, the equivalence is 167 kg by m^3).

As for the equivalence C_2: "weight → heating content", several studies allow one to use it. It is known that it varies according to the level of humidity in the wood used. For a tonne of dry wood, it may be fixed at 0.45 TOE.

We should like to reiterate that we are here concerned with a potential calorific value at the input stage of this final use, which in no way prejudges the real efficiency in useful energy resulting from the combustion process. It is known that, in practice, efficiency is of the order of 5-10% (as in cooking, for example). The same method is also used when one imparts to road transport a consumption calculated in litres of oil while, in reality, the return in "useful" energy of the engines is of barely 10-20%. Depending on the region, a variable C_2 coefficient was retained here in ranges of 0.45 to 0.35 TOE to the tonne, drawing inspiration from the in-depth studies produced by the OLADE for all Latin America, as well as those of Tillman ("Wood as an energy resource"). One progresses, for example, from 0.36 TOE/T for R10 to 0.38 TOE/T for R6 and 0.45 TOE/T for R7.

The TOE consumption of heating wood can thus be derived from the m^3 production provided by FAO, country by country, by an average coefficient of $C_1 \times C_2$ or $0.725C_2$.

This calculation was applied to the 1978 production data (published in the *1978 Yearbook*). For 1960, we used the 1960 values gathered in the June 1980 Electricité de France Study, brought back to the regions own $0.725C_2$ coefficient. We modified them proportionately to the changes in absolute values introduced for 1978 in certain regions which had specific estimates of their fuel wood consumption (e.g. R9).

3.2 Vegetable waste

The reference base of estimate was published in the June 1980 Electricité de France Study.The 1960 evaluation was retained. For 1978, we simply extrapolated the 1960-1976 tendency.

In fact, the complexity of the calculation justified a reactualisation by a simple procedure, since the patterns of waste consumption are stable and only vary very little over time.

We shall only reiterate the principle of calculation.

For each non-commercial source studied, the energy consumption from commercial production was derived by means of the following equation:

$CE_i = P_i \times C_{WP} \times C_{EV} \times HC$ (cf, the publications of J. K. Parikh[12]).

where CE_i expresses the energy quantity extracted from the vegetable product "i" (in TOE);

P_i = the commercial production of the product "i" (in tonnes);

C_{WP} = the corresponding production of waste (compared to the commercial P_i tonnage);

C_{EV} = the proportion of this waste employed for energy uses;

HC = the heating content of the waste (TOE/tonne), not in useful energy but in potential energy.

The series of P_i were provided by the FAO Yearbook of Production.

The C_{WP} and C_{EV} volumes were particularly inspired by the SEMA's work on Black Africa ("Evaluation of New Energies for the Development of African States", 1977).

Finally, the HC came from the cross-checking of SEMA's work and from various sources (Tillman (13), B. A. Stout for FAO (15), etc.). For the entire range of products, they are of the order of 4000 kcal per tonne of waste (except for cereals: 0.33 TOE/T for rice; 0.35 TOE/T for the other cereals; 0.22 TOE/T for the bagass of humid sugar cane).

For each product, we attributed the average P_i harvest for 1975, 1976 and 1977 to the year 1976, so as to corrrect the disturbing effect of climatic variations. The same was done for 1960: the average harvest for 1961–65 was attributed to 1963, then after a calculation of the 1963–76 tendency, we extrapolated it retrospectively up to 1960, and then forward to 1978.

Wood waste for fuel in the wood industry was also integrated within vegetable waste (a detailed calculation being provided in the Electricité de France Study cited above). Eventually, the use of the waste (excluding cereal waste) occurs mainly at the industrial level as fuel recovered in the agro-alimentary production process.

The entire range of vegetable products taken into account and whose energy production was modified according to variable coefficients is the following:

• oil plants (peanuts, castor oil, sunflower, colza, sesame, linseed, safflower, cotton seeds)
• coconut
• palm
• coffee
• cocoa
• sugar cane

The different equations resulting from the combination of the three coefficients $C_{WP} \times C_{EV} \times HC$ according to the different species are the following: (P being the agricultural producton of the product "i" and CE the energy production-consumption from "i"):

CE = 0.048 P	rice in industrialized countries
CE = 0.383 P	rice in Third World nations
CE = 0.021 P	other cereal (wheat, barley, maize, rye, oats, millet, sorghum) in the industrialised countries.
CE = 0.168 P	other cereals in the countries of the Third World
CE = 0.07 P	oil plants
CE = 0.09 P	coconut
CE = 0.45 P	palm
CE = 0.13 P	coffee
CE = 0.04 P	cocoa
CE = 0.04 P	sugar cane

The non-commercial recuperation and use of the vegetable biomass is in issue here. The production of alcohol as an oil substitute is classified in the "new energy" category.

3.3 Animal waste

The use of animal waste for energy occurs mainly in Asian countries. This use is linked to objective factors:

(i) the importance and density of livestock,
(ii) religious factors (Hindu or Buddhist tradition).

The combination of these factors explains the development of this practice.

The density of livestock lease plays a preponderant role. In zones with intensive breeding, one can envisage the collection of dung more easily. In zones with extensive breeding (e.g. South America or Black Africa), this is hardly feasible. This "concentration" will be measured by a coefficient comparing the (permanent) acreages of pasture in hectares with the number of heads (bovines and buffalo). As a threshold of animal dung use, the countries with value lower than 0.20 ha/head will be used. Thus, one finds 0.05 ha/head in India; 0.20 ha/head in Pakistan; 1 ha/head in Niger; 2.5 in Argentina.

China does not fit this very global criteria. However, the RWT estimated its use of dung from various publications. Likewise, the Indian representative made some correction to initial theoretical estimates.

According to the information provided by R. K. Pachauri[43], by Tillman[23] confirmed by R. Carillon of the CNEEMA, an Indian bovine will produce the equivalent of 85 KEO per annum (being a combination of dry dung production with its energy use coefficient 1/3).

For the other Asian countries a lower base value was used: 60 KEO/per annum for 1976 and 1978. For 1960, it was estimated that this practice was less developed: 65 KEO for India and 40 KEO for the other countries. This "traditional" production, entering the category of non-commercial sources, does not prejudge possible ulterior developments aimed at the improvement of the efficiency of the process by means of investments (e.g. units of biogas production).

In a general way, one should not be deceived by the apparent precision of the figures provided. In fact, these are only very approximate orders of magnitude but which are better formulated than ignored, if one wishes to obtain a plausible picture of real energy situations, particularly in the Third World.

In the following tables, one will find the division in absolute values (MTOE) and in % of primary energy consumption (PEC) according to the 8 sources studied, themselves regrouped in CEO (Consumption of commercial energy = BC + DVA). The tables of annual average growth for 1960–1978 for each post were added.

4 ENERGY PRODUCTION

The 1978 energy production was essentially extracted from the UN *Yearbook of Energy Statistics* (1979 Edition). Again with regard to the energy consumption, the values recorded were placed at the level: TOE at $10500 \, 10^3$ kcal and electricity calculated at 0.222 TOE for 1000 kWh.

A few alterations (e.g. in R5 or R9) were nevertheless made to the UN figures to reconcile them to the more precise statistics published by certain countries.

Non-commercial sources are reckoned to be used on the spot and not to be traded internationally. As a result, we have postulated an equality of consumption and production.

The same was done with primary electricity which is thought to be traded only within each region.

One therefore, arrives at an estimate of the total production of primary energy (PEP), according to each of the 8 sources studied, and of CEP (Production of commercial energy = CMS + GN + HY +NU +EN), and NCEP (Production of non-commercial energy = BC + DVA).

5 INTERNATIONAL ENERGY TRADE

For 1978, trade involves all fossil fuels: solid mineral fuels, oil and natural gas. Indeed, it was assumed that non-commercial energies are not traded interregionally (cf. above). Moreover, electricity trading occurs at the frontiers of adjacent nations. Trade for non-commercial energies and electricity between the great regions was thus ignored.

For oil, it will be noted that only trade in crude oil, is in issue. Trade in refined products appears in the balance factor (cf. Annex 14). We should like to stress that the following tables only express trade between regions. Indeed, trade between countries within a region compensates itself (imports cancel exports).

The columns E (exports) and I (imports) only take into account the region's trade with other regions, to the exclusion of intraregional fluctuations. We also placed the balance B per region which is equal to E—I (positive if E > I; negative if E < I).

The division of inter-and intraregional fluctuations for 1978 was the object of a specific study, of which the total results were as follows: in 1978, the total interregional trade equalled 53% of total global coal trade, 88% of that trade for crude oil, 33% for that of natural gas.

Table   : 1
Appendix : 5

ENERGY CONSUMPTIONS (Mtoe)

- 1960 -

(Mtoe)	SMF	O	NG	HY	NU	NS	FW	VAW	CEC	NCEC	PEC
R1	249	491	335	57	-	-	35	5	1132	40	1172
R2	359	183	10	61	-	-	31	7	613	38	651
R3	60	38	1	15	-	-	2	1	114	3	117
R4	341	125	53	13	-	-	37	9	532	46	578
R5	1	18	2	1	-	-	4	7	22	11	33
R6	26	9	-	1	-	-	56	8	36	64	100
R7	27	9	1	2	-	-	50	30	39	80	119
R8	6	15	3	2	-	-	43	20	26	63	89
R9	128	9	1	4	-	-	87	61	142	148	290
R10	6	66	12	8	-	-	60	7	92	67	159
WIC	690	715	346	133	-	-	70	13	1884	83	1967
IC	1031	840	399	146	-	-	107	22	2416	129	2545
TW	172	123	19	18	-	-	298	133	332	431	763
TW-R9	44	114	18	14	-	-	211	72	190	283	473
WORLD	1203	963	418	164	-	-	405	155	2748	560	3308

Table   : 2
Appendix : 5

ENERGY CONSUMPTIONS (Mtoe)

- 1978 -

(Mtoe)	SMF	O	NG	HY	NU	NS	FW	VAW	CEC	NCEC	PEC
R1	358	914	528	129	77	2	30	5	2008	35	2043
R2	260	627	173	97	40	-	16	14	1197	30	1227
R3	83	247	24	27	14	-	1	1	395	2	397
R4	575	404	334	42	12	-	30	16	1367	46	1413
R5	2	79	29	6	-	-	8	10	116	18	134
R6	43	27	1	7	-	-	86	13	78	99	177
R7	48	29	5	13	1	-	71	47	96	118	214
R8	14	92	7	5	-	-	63	38	118	101	219
R9	306	89	10	14	-	-	126	93	419	219	638
R10	12	168	50	51	2	6	55	12	289	67	356
WIC	740	1799	725	254	131	2	49	21	3651	70	3721
IC	1315	2203	1059	296	143	2	79	37	5018	116	5134
TW	386	473	102	95	3	6	407	212	1065	619	1684
TW-R9	80	384	92	81	3	6	281	119	646	400	1046
WORLD	1701	2676	1161	391	146	8	486	249	6083	735	6818

ENERGY CONSUMPTIONS

REGIONAL STRUCTURE %
- 1960 -

Table : 3
Appendix : 5

(%)	SMF	O	NG	HY	NU	NS	FW	VAW	CEC	NCEC	PEC
R1	21	42	29	5	-	-	3	ϵ	97	3	100
R2	55	28	2	9	-	-	5	1	94	6	100
R3	51	32	1	13	-	-	2	1	97	3	100
R4	59	22	9	2	-	-	6	2	92	8	100
R5	3	55	6	3	-	-	12	21	67	33	100
R6	26	9	-	1	-	-	56	8	36	64	100
R7	23	7	1	2	-	-	42	25	33	67	100
R8	7	17	3	2	-	-	48	23	29	71	100
R9	44	3	ϵ	2	-	-	30	21	49	51	100
R10	4	42	7	5	-	-	38	4	58	42	100
WIC	35	36	17	7	-	-	4	1	96	4	100
IC	40	33	16	6	-	-	4	1	95	5	100
TW	23	16	3	2	-	-	39	17	44	56	100
TW-R9	9	24	4	3	-	-	45	15	40	60	100
WORLD	36	29	13	5	-	-	12	5	83	17	100

ENERGY CONSUMPTIONS

REGIONAL STRUCTURE %
- 1978 -

Table : 4
Appendix : 5

(%)	SMF	O	NG	HY	NU	NS	FW	VAW	CEC	NCEC	PEC
R1	18	45	26	6	4	ϵ	1	ϵ	99	1	100
R2	21	51	14	8	4	-	1	1	98	2	100
R3	21	62	6	7	4	-	ϵ	ϵ	100	ϵ	100
R4	41	28	24	3	1	-	2	1	97	3	100
R5	2	59	22	4	-	-	6	7	87	13	100
R6	24	15	1	4	-	-	49	7	44	56	100
R7	23	14	2	6	ϵ	-	33	22	45	55	100
R8	7	42	3	2	-	-	29	17	54	46	100
R9	48	14	2	2	-	-	20	14	66	34	100
R10	3	47	14	14	1	2	16	3	81	19	100
WIC	20	48	19	7	4	ϵ	1	1	98	2	100
IC	25	43	21	6	3	ϵ	1	1	98	2	100
TW	23	28	6	6	ϵ	ϵ	24	13	63	37	100
TW-R9	8	37	9	8	ϵ	ϵ	27	11	62	38	100
WORLD	25	39	17	6	2	ϵ	7	4	89	11	100

ENERGY CONSUMPTIONS

FUEL STRUCTURE %

- 1960 -

Table : 5
Appendix : 5

(%)	SMF	O	NG	HY	NU	NS	FW	VAW	CEC	NCEC	PEC
R1	21	51	80	35	-	-	9	3	41	7	35
R2	30	19	2	37	-	-	8	4	23	7	20
R3	5	4	ε	9	-	-	ε	1	4	1	4
R4	28	13	13	8	-	-	9	6	19	8	17
R5	ε	2	1	1	-	-	1	5	1	2	1
R6	2	1	-	1	-	-	14	5	1	11	3
R7	2	1	ε	1	-	-	12	19	2	14	3
R8	1	1	1	1	-	-	11	13	1	11	3
R9	10	1	ε	2	-	-	21	39	5	27	9
R10	1	7	3	5	-	-	15	5	3	12	5
WIC	58	74	82	81	-	-	17	8	69	15	60
IC	86	87	95	89	-	-	26	14	88	23	77
TW	14	13	5	11	-	-	74	86	12	77	23
TW-R9	4	12	5	9	-	-	53	47	7	50	14
WORLD	100	100	100	100	-	-	100	100	100	100	100

ENERGY CONSUMPTIONS

FUEL STRUCTURE %

- 1978 -

Table : 6
Appendix : 5

(%)	SMF	O	NG	HY	NU	NS	FW	VAW	CEC	NCEC	PEC
R1	21	34	45	33	53	25	6	2	33	5	30
R2	15	24	15	25	27	-	3	6	20	4	18
R3	5	9	2	7	10	-	ε	ε	6	ε	6
R4	34	15	29	10	8	-	6	6	22	6	21
R5	ε	3	2	2	-	-	2	4	2	2	2
R6	3	1	ε	2	-	-	18	5	1	14	3
R7	3	1	1	3	1	-	15	19	2	16	3
R8	1	4	1	1	-	-	13	15	2	14	3
R9	18	3	1	4	-	-	26	38	7	30	9
R10	1	6	4	13	1	75	11	5	5	9	5
WIC	43	67	62	65	90	25	10	9	60	10	55
IC	77	82	91	75	98	25	16	15	82	16	76
TW	23	18	9	25	2	75	84	85	18	84	24
TW-R9	5	15	8	21	2	75	58	47	11	54	15
WORLD	100	100	100	100	100	100	100	100	100	100	100

ENERGY CONSUMPTIONS

- AVERAGE ANNUAL RATES OF GROWTH (%) -

- 1960 - 1978 -

Table : 7
Appendix : 5

(%)	SMF	O	NG	HY	NU	NS	FW	VAW	CEC	NECE	PEC
R1	2,0	3,5	2,6	4,6	-	-	-0,8	0	3,2	-0,7	3,1
R2	-1,8	7,1	17,2	2,6	-	-	-3,6	3,9	3,8	-1,3	3,6
R3	1,8	11,0	19,3	3,3	-	-	-3,8	0	7,1	-2,2	7,0
R4	2,9	6,7	10,8	6,7	-	-	-1,2	3,2	5,4	0	5,1
R5	4,0	8,6	16,0	10,5	-	-	3,9	2,0	9,7	2,8	8,1
R6	2,8	6,3	-	11,4	-	-	2,4	2,7	4,4	2,5	3,2
R7	3,2	6,7	9,4	11,0	-	-	2,0	2,5	5,1	2,2	3,3
R8	4,8	10,6	4,8	5,2	-	-	2,1	3,6	8,8	2,7	5,1
R9	5,0	13,6	13,6	7,2	-	-	2,1	2,4	6,2	2,2	4,5
R10	4,0	5,3	8,3	10,8	-	-	-0,5	3,0	6,6	0	4,6
WIC	0,4	5,3	4,2	3,7	-	-	-2,0	2,7	3,7	-0,9	3,6
IC	1,4	5,5	5,6	4,0	-	-	-1,7	2,9	4,1	-0,6	4,0
TW	4,6	7,8	9,8	9,7	-	-	1,7	2,6	6,7	2,0	4,5
TW - R9	3,4	7,0	9,5	10,2	-	-	1,6	2,8	7,0	1,9	4,5
WORLD	1,9	5,8	5,8	4,9	-	-	1,0	2,7	4,5	1,5	4,1

ENERGY PRODUCTION (Mtoe)

- 1978 -

Table : 8
Appendix : 5

(Mtoe)	SMF	O	NG	HY	NU	NS	FW	VAW	CEP	NCEP	PEP
R1	380	574	527	129	77	2	30	5	1689	35	1724
R2	220	92	151	97	40	-	16	14	600	30	630
R3	71	25	10	27	14	-	1	1	147	2	149
R4	599	575	344	42	12	-	30	16	1572	46	1618
R5	1	1182	59	6	-	-	8	10	1248	18	1266
R6	53	113	1	7	-	-	86	13	174	99	273
R7	49	12	8	13	1	-	71	47	83	118	201
R8	12	101	18	5	-	-	63	38	136	101	237
R9	307	104	12	14	-	-	126	93	437	219	656
R10	8	240	52	51	2	6	55	13	359	68	427
WIC	721	691	688	254	131	2	49	21	2487	70	2557
IC	1320	1266	1032	296	143	2	79	37	4059	116	4175
TW	380	1752	150	95	3	6	407	213	2386	620	3006
TW-R9	73	1648	138	81	3	6	281	120	1949	401	2350
WORLD	1700	3018	1182	391	146	8	486	250	6445	736	7181

ENERGY PRODUCTION

REGIONAL STRUCTURE %

- 1978 -

(%)	SMF	O	NG	HY	NU	NS	FW	VAW	CEP	NCEP	PEP
R1	22	33	31	8	4	ε	2	ε	98	2	100
R2	35	15	24	15	6	-	3	2	95	5	100
R3	47	17	7	18	9	-	1	1	98	2	100
R4	37	35	21	3	1	-	2	1	97	3	100
R5	ε	93	4	1	-	-	1	1	98	2	100
R6	19	41	ε	3	-	-	32	5	63	37	100
R7	24	6	4	7	ε	-	35	24	41	59	100
R8	5	43	7	2	-	-	27	16	57	43	100
R9	47	16	2	2	-	-	19	14	67	33	100
R10	2	56	12	12	1	1	13	3	84	16	100
WIC	28	27	27	10	5	ε	2	1	97	3	100
IC	32	30	25	7	3	ε	2	1	97	3	100
TW	13	58	5	3	ε	ε	14	7	79	21	100
TW-R9	3	70	6	4	ε	ε	12	5	83	17	100
WORLD	24	42	17	5	2	ε	7	3	90	10	100

ENERGY PRODUCTION

FUEL STRUCTURE %

- 1978 -

(%)	SMF	O	NG	HY	NU	NS	FW	VAW	CEP	NCEP	PEP
R1	22	19	44	33	53	25	6	2	26	5	24
R2	13	3	13	25	27	-	3	6	9	4	9
R3	4	1	1	7	10	-	ε	ε	2	ε	2
R4	35	19	29	11	8	-	6	6	24	6	22
R5	ε	39	5	2	-	-	2	4	19	2	18
R6	3	4	ε	2	-	-	18	5	3	14	4
R7	3	1	1	3	1	-	15	19	1	16	3
R8	1	3	2	1	-	-	13	15	2	14	3
R9	18	3	1	3	-	-	26	38	7	30	9
R10	1	8	4	13	1	75	11	5	6	9	6
WIC	42	23	58	65	90	25	10	9	39	10	36
IC	77	42	87	76	98	25	16	15	63	16	58
TW	23	58	13	24	2	75	84	85	37	84	42
TW-R9	5	55	12	21	2	75	58	47	30	54	33
WORLD	100	100	100	100	100	100	100	100	100	100	100

Table : 11
Appendix : 5

INTER REGIONAL ENERGY EXCHANGES (Mtoe)

- 1978 -

(Mtoe)	SMF			CO			NG			TOTAL		
	E	I	B	E	I	B	E	I	B	E	I	B
R1	27	5	22	2	343	-341	1	2	-1	30	350	-320
R2	1	41	-40	54	656	-602	-	22	-22	55	719	-664
R3	7	19	-12	-	252	-252	-	14	-14	7	285	-278
R4	26	1	25	58	26	32	17	8	9	101	35	66
R5	-	1	-1	1002	-	1002	27	-	27	1029	1	1028
R6	9	-	9	97	22	75	-	-	-	106	22	84
R7	-	-	-	-	22	-22	3	-	3	3	22	-19
R8	-	2	-2	73	85	-12	11	-	11	84	87	-3
R9	1	-	1	13	1	12	-	-	-	14	1	13
R10	-	5	-5	75	66	9	-	-	-	75	71	4
WORLD	71	74	-3	1374	1473	-99	59	46	13	1504	1593	-89

Table : 12
Appendix : 5

INTER REGIONAL ENERGY EXCHANGES

REGIONAL STRUCTURE %
- 1978 -

(%)	SMF			CO			NG			TOTAL		
	E	I	B	E	I	B	E	I	B	E	I	B
R1	90	1		7	98		3	1		100	100	
R2	2	6		98	91		-	3		100	100	
R3	100	7		-	88		-	5		100	100	
R4	26	3		57	74		17	23		100	100	
R5	-	100		97	-		3	-		100	100	
R6	8	-		92	100		-	-		100	100	
R7	-	-		-	100		100	-		100	100	
R8	-	2		87	98		13	-		100	100	
R9	7	-		93	100		-	-		100	100	
R10	-	7		100	93		-	-		100	100	
WORLD	5	5		91	92		4	3		100	100	

INTER REGIONAL ENERGY EXCHANGES

FUEL STRUCTURE %

- 1978 -

(%)	SMF			CO			NG			TOTAL		
	E	I	B	E	I	B	E	I	B	E	I	B
R1	38	7		Ɛ	23		2	4		2	22	
R2	1	55		4	44		-	48		3	45	
R3	10	26		-	17		-	31		1	18	
R4	37	1		4	2		29	17		7	2	
R5	-	1		73	-		46	-		68	Ɛ	
R6	13	-		7	2		-	-		7	1	
R7	-	-		-	2		5	-		Ɛ	1	
R8	-	3		5	6		18	-		6	6	
R9	1	-		1	-		-	-		1	Ɛ	
R10	-	7		6	4		-	-		5	5	
WORLD	100	100		100	100		100	100		100	100	

DEMOGRAPHIC FORECASTS

The demographic forecasts used in this study (and which are one of the determining factors of the evolution of energy consumption) were extracted from a study of the Population Division of the UN,[10] using the average variant.

This study provided, on a five year by five year basis, all the population projections up to 2020, country by country. We made no attempt to differentiate the demographic forecasts according to the two scenarios, which consequently present one single set of hypotheses.

Nevertheless, the 1960 - 1978 economic demographic base was established from World Bank data (cf. Annex 4). Consequently, we were faced with the problem of reconciling the demographic statistics of the UN and the World Bank, but it remained circumscribed in a few particular cases which were easily resolved.

Moreover, all the RWTs confirmed the accuracy of the estimates contained in the reference base. Only the R1 and R3 projections were slightly altered after the regional stage, the corrections nevertheless remaining of the order of ±2% in 2020 for both regions. Overall, the corpus of the demographic forecasts formulated by the UN was thus almost entirely adopted. In the following table, all the regional estimates for 1960 - 2020 will be found, expressed in levels (millions of inhabitants) and in rates of growth for sub-periods. Compared to one study,[11] one will note a substantial slowdown in growth, especially beyond 2000, when the difference of barely −60 Minh for 2000 for the world population will increase to −580 Minh in 2020.

This slowdown is essentially caused by the Third World (−40 Minh in 2000 and −550 Minh in 2020 compared to the "Munich" hypotheses). The Third World growth rate correspondingly decreases from 2.35% between 1960 and 1978 to 1.9% in 2000 and 1.4% beyond (as opposed to 2% and 1.8%, respectively, in the previous study), thus marking the appearance of a clear tendency towards relief of global long-term demographic perspectives.

Table :   1
Appendix :   6

DEMOGRAPHY FORECASTS

	LEVELS (Minhb)				RATES OF GROWTH (% /year)		
	1960	1978	2000	2020	1960-1978	1978-2000	2000-2020
R1	200,9	245,0	296	349	1,11	0,9	0,8
R2	357,1	417,2	463	489	0,87	0,5	0,3
R3	107,5	132,3	149	152	1,16	0,6	0,1
R4	313,9	372,4	420	462	0,95	0,6	0,5
R5	111,4	183,8	320	465	2,82	2,5	1,9
R6	219,5	347,8	681	1 152	2,59	3,1	2,7
R7	560,4	848,6	1 284	1 644	2,33	1,9	1,2
R8	220,6	337,6	503	631	3,03	1,8	1,1
R9	714,5	1 032,0	1 359	1 576	2,06	1,3	0,7
R10	209,6	344,5	564	802	2,80	2,2	1,8
WIC	681,8	822,2	959	1 071	1,05	0,7	0,6
IC	995,7	1 194,6	1 379	1 533	1,02	0,65	0,5
TW	2 019,7	3 066,6	4 660	6 189	2,35	1,9	1,4
TW-R9	1 305,2	2 034,6	3 301	4 613	2,50	2,2	1,7
WORLD	3 015,4	4 261,6	6 039	7 722	1,94	1,6	1,2

ECONOMIC FORECASTS

Economic forecasts linked to forecasts of energy consumption are presented in the two following tables (GNP total and per capita).

The 1960 and 1978 values were extracted from the World Bank's Atlas (cf. Annex 5 reference base). The two forecasting scenarios were the fruit of the successive work of the Central Team and of the RWTs.

(A) Their first version provided in the regional dossiers (cf. Annex 9) came from 2 sources, according to the zones:

(i) For the 6 regions of the Third World, it was the result of a revision of the economic hypotheses contained in the study: "Energy Horizons of the Third World". Starting from the per-capita results for 1976 – 2000 (Scenarios B and C of the study, themselves largely based on the Scenarios B and C in OECD's study "Interfutures"), we integrated the 1976 – 1978 realisations, and we extracted the corresponding 1978 – 2000 rates, retaining the rates for 2000 – 2020. By multiplying the 2000 and 2020 per-capita values by the new demographic projections, we thus obtained the total 2000 and 2020 GNP.

(ii) For the four industrialised regions, we referred to the most recent information on economic development hypotheses, while at the same time retaining an overall coherence with "Interfutures" estimates relating to these zones and to Developing Countries. However, we in fact only provided forecasts to R1 where they were extracted from an official document.

In this centralised phase of the study, energy consumption forecasts were derived from these various hypotheses through the medium of coefficients of income elasticity, which were variable according to region and period.

(B) Each according to its own method, the regional teams analysed these forecasts to encapsulate them or, more often, to modify them in the context of the scenarios proposed to them.

Depending on the forecasting procedure adopted by the teams, some kept to the initial evolution of economic forecasts towards energy forecasts, others emancipated themselves from it by starting off by determining energy forecasts and then, a *posteriori*, returning to the economic forecasts which appeared to them to be compatible with the energy results (although more as an illustration than as a definite conclusion in the latter case).

All the regions thus came to formulate their own economic forecasts at some stage of their work. The only exception is Region R4 where the growth scenario which occurs in its case for the sake of coherence was introduced by the Central Team and not by the RWT.

TOTAL ECONOMIC GROWTH

- LEVELS AND RATES OF GROWTH -

			LEVELS (G $ 78)				RATES OF GROWTH (% /year)				
			2000		2020			1978-2000		2000-2020	
	1960	1978	I	II	I	II	1960-1978	I	II	I	II
R1	1 162	2 109	4 048	3 633	6 073	4 941	4,0	3,0	2,5	2,0	1,5
R2	1 294	2 596	5 031	4 059	7 528	5 505	4,0	3,0	2,0	2,0	1,5
R3	224	1 017	2 359	1 949	3 865	2 625	8,8	3,9	3	2,5	1,5
R4 (*)	471	1 366	(3 237)	(2 617)	(5 846)	(3 888)	6,1	(4,0)	(3)	(3,0)	(2,0)
R5	61	292	1 048	779	2 824	1 577	9,1	6,0	4,6	5,1	3,6
R6	78	166	437	394	870	712	4,3	4,5	4,0	3,5	3,0
R7	77	150	365	294	722	506	3,8	4,1	3,1	3,5	2,5
R8	63	220	824	580	1 848	1 048	7,2	6,2	4,5	4,2	3,0
R9	97	247	558	332	1 217	528	5,3	3,8	1,35	4,0	2,35
R10	180	483	1 685	1 222	4 082	2 435	5,6	5,8	4,3	4,5	3,5
WIC	2 698	5 766	11 553	9 745	17 695	13 259	4,3	3,2	2,4	2,15	1,55
IC	3 169	7 132	14 790	12 362	23 541	17 147	4,6	3,4	2,5	2,35	1,65
TW	538	1 314	4 802	3 497	11 334	6 618	5,1	6,1	4,5	4,4	3,2
TW-R9	441	1 067	4 244	3 165	10 117	6 090	5,0	6,5	5,1	4,4	3,3
WORLD	3 707	8 446	19 592	15 859	34 875	23 765	4,7	3,9	2,9	2,9	2,0

(*) R4 : economic assumptions ex-post related to the energy forecasts by the central team

ECONOMIC GROWTH PER CAPITA

- LEVELS AND RATES OF GROWTH -

			LEVELS ($ 78/p.c.)				RATES OF GROWTH (% /year)				
			2000		2020			1978-2000		2000-2020	
	1960	1978	I	II	I	II	1960-1978	I	II	I	II
R1	5 786	8 608	13 661	12 274	17 391	14 158	2,8	2,1	1,6	1,2	0,7
R2	3 623	6 223	10 866	8 767	15 395	11 258	3,1	2,6	1,6	1,8	1,2
R3	2 086	7 680	15 780	13 035	25 480	17 300	7,5	3,3	2,4	2,4	1,4
R4 (*)	1 500	3 668	(7 707)	(6 230)	(12 654)	(8 415)	5,1	(3,4)	(2,4)	(2,5)	(1,5)
R5	544	1 588	3 275	2 434	6 073	3 391	6,1	3,3	2,0	3,1	1,7
R6	356	478	642	578	755	618	1,7	1,3	0,9	0,8	0,3
R7	137	177	284	229	439	308	1,4	2,2	1,2	2,2	1,2
R8	287	652	1 639	1 153	2 960	1 661	4,7	4,3	2,6	3,0	1,8
R9	136	239	411	244	772	335	3,2	2,5	0,1	3,2	1,6
R10	857	1 402	2 988	2 167	5 091	3 036	2,8	3,5	2,0	2,7	1,7
WIC	3 957	7 013	12 047	10 162	16 522	12 380	3,2	2,5	1,7	1,6	1,0
IC	3 183	5 970	10 725	8 964	15 356	11 185	3,6	2,7	1,9	1,8	1,1
TW	266	428	1 030	750	1 831	1 069	2,7	4,1	2,6	2,9	1,8
TW-R9	338	524	1 286	959	2 193	1 320	2,5	4,2	2,8	2,7	1,6
WORLD	1 229	1 982	3 244	2 626	4 516	3 078	2,7	2,3	1,3	1,7	0,8

(*) R4 : economic assumptions ex-post related to the energy forecasts by the central team

REGIONAL FILE*

1 Summary of the study
2 Composition of the region
3 Content of the regional file
4 Description and how to use the forecast tables
5 Forecast table model

I. REFERENCE BASIS 1960–1978

6 Population and economic growth 60–78
7 Commercial energy consumption (CEC)
8 Non commercial energy consumptions (CENC)
9 Primary energy consumptions 1978 — total and per capita (PEC)
10 Energy equivalences
11 Major forecast references

II. SCENARIO I "NORMATIVE — COOPERATION"

12 Description of the scenario
13 Main global assumptions
14 Forecast table
15 ("High Scenario")
16 Forecast form (to be filled in by the regional working team)

III. SCENARIO II "STRENGTHENING CONSTRAINTS"

17 Description of the scenario
18 Main global assumptions
19 Forecast table
20 Forecast form (to be filled in by the regional working team)

*The regional file presented in Annex 8 is incomplete: the data (chapters 6-7-8-19-14-15-19) and chapters 2 and 11, specific to each region, are omitted.

One sole forecasting format sheet (16 and 20) is given and chapter 5 has been separated, according to industrialised or Third World countries.

WORLD SURVEY OF LONG-TERM REGIONAL ENERGY BALANCES: SYNOPSIS

1 Objectives

- Following the Munich Conference (September 1980), the Conservation Commission decided to present to the forthcoming New Delhi Conference (September 1983) a long-term world Energy Survey based on regional balances.
- For each of the *ten major regions* making up the world totality, it is intended to establish, on a 1960–1978 reference basis, simplified 2000–2020 energy balances consisting of:
  - on the *demand* side, a forecast of the total primary energy consumption (without going down to the level of end-use sectors), from a forecast of economic growth on the one hand (within the framework of 2 development scenarios), and of the evolution of the global elasticity between energy growth and economic growth on the other,
  - on the *supply* side, and for each primary energy source (commercial and non-commercial), a forecast of corresponding available supply, taking into account both local future production and complementary import and export fluctuations.

The whole originality of the survey, when compared to similar attempts already made (MIT, IIASA, OECD, EXXON, etc . . .) rests on the priority given to the decentralised forecast procedure.

2 Methodology

This will be carried out in *two major stages:*
- A first stage in which a centralized team will establish, for each region, energy balance projections (demand, supply, exchanges*) within the framework of the two contemplated scenarios.
- A second stage in which those projections will be submitted to the scrutiny of regional working groups, made up essentially from regional participants in the work of the World Energy Conference for their particular region, who will determine, in consultation with the central team, the final anticipated balances, under their own responsibility. The task of putting together all regional findings into a coherent pattern will be that of the central team.

The real input of these working groups is of vital importance to the success of the project and will give it all its value.

3 Timing

1981 • setting up the reference base 1960–1978 (central team)
  • provisional energy balances (central team): 10 regions
  • building up the regional working teams
1982 • final assessment of the regional balances (regional teams) with central team
  • coherence work (central team)
1983 • completion of the final report (central team)
  • September: presentation in New Delhi

*At least for the base year 1978.

THE REGIONAL DOSSIER

The regional dossier contains three sub-files pertaining specifically to the region, as well as a general description of the study and its methodology, and a list of the countries comprising the region. The sub-files are:

1 Reference data base 1960–1978

For two base years, 1960 and 1978, the following data are given for each country in the region:

(1) the *population* (millions of inhabitants) and the *GNP:* total and per capita (US \$ 1978). The source of this data is the World Bank Atlas 1980.

(2) the *commercial energy consumption* by fuel type: solid fuels, liquid fuels, natural gas and primary electricity (nuclear + hydroelectric + geothermal). These are given in millions of tons of oil equivalent and are published in the United Nations 1979 Yearbook of World Energy Statistics.

(3) the *non-commercial energy consumption,* specifying fuelwood (and wood waste), and vegetable and animal waste products (crop residues, cow dung, etc). These figures are also given in TOE. The methodology adopted for estimating non-commercial energy consumption is described in the report (published in French): *Evolution des consommations d'énergie dans le monde: une retrospective 1960–1976* (Electricité de France, juin 1980). The 1978 consumption figures are extrapolated from the 1976 figures based upon the 1976 and 1978 crop production figures published by the FAO.

(4) the total *primary energy consumption* (sum of (2) and (3)), and the primary energy consumption per capita.

(5) the *energy conversion factors* utilized in this study.

(6) finally, a list of the principal references used for the demographic, economic and energy consumption projections.

2 Forecast Scenario I "normative-cooperative"

This sub-file contains:

(1) firstly, a *general description* of the international environment assumed in this scenario.

(2) a set of *world macro-economic assumptions* for the years 1978, 2000 and 2020, and for the periods 1960–78, 1978–2000, and 2000–2020. Different assumptions are given for industrialised and developing *countries* (including China).

• population (millions of inhabitants, UN projected growth rates),

• economic growth rates (OECD projections — "Interfutures"),

• an initial projection of primary energy consumption (estimates previously presented by the Conservation Commission at the Munich Conference),

• as well as the corresponding energy production by principal fuel types,

• and, for developing countries, a summary of the forecasts included in "Third World Energy Horizons", presented at the World Energy Conference in Munich, 1980, Editions Techniques et Economiques, Paris.

(3) the *forecasting format sheets* (see description and instructions following).

(4) whenever sufficient information was available, a regional energy consumption projection *(high scenario)* was calculated, for the purpose of

comparison, by aggregating *national* projections taken from long-term energy forecasting reports originating from the countries themselves.

3 Forecast Scenario II "strengthening constraints"

This sub-file contains the same elements as the ones described above (with the exception of the regional projection based upon different national projections — point (4)):
(1) a *general description* of the international environment.
(2) *world macro-economic assumptions* (1978, 2000, 2020),
(3) *forecasting format sheets* (see description and instructions following).

DESCRIPTION AND INSTRUCTIONS FOR THE FORECASTING FORMAT SHEETS

The forecasting format sheets for Scenarios I and II contain all the necessary elements for the work of the Regional Working Teams. It is essential that the projections be made only at the regional level, and that no national projections should appear in the figures.

Each forecasting format sheet is composed of two tables:

(A) The first table contains demographic, economic and primary energy consumption projections (base year 1978; 2000, 2020, and the periods 1960-1978; 1978-2000 and 2000-2020).
(1) demographic projections: population (millions of inhabitants — Source: Population Division of the UN). These projections are input data for the study.
(2) economic growth projections:
(3) total (billion US $78 — constant)
(4) per capita (US $78 — constant)
These projections are derived from the OECD "Interfutures" assumptions for its Scenario B and C, corresponding to Scenarios I and II, respectively, in this study; the OECD projections were readjusted to reflect recent economic trends, and extrapolated beyond 2000, taking into account a general trend towards slower economic growth. For each of the 6 developing regions, the assumptions for 2000-2020 were drawn from the report "Third World Energy Horizons 2000-2020". These projections should be considered as data inputs to the study. They should not be modified without the permission of the Central Team.
(5) primary energy consumption (millions of tons of oil equivalent: MTOE).
(6) elasticities
(7) elasticity of commercial energy demand per capita

$$eC/hb = \frac{\text{growth rate of commercial energy consumption/capita}}{\text{growth rate of GNP per capita}}$$

For developing regions, this assumption is crucial since it conditions, together with (4), the evolution of commercial energy consumption growth per capita (see ref. 13): it is calculated from the results of the study "Horizons".
(8) elasticity of primary energy demand

$$eT = \frac{\text{growth rate of total primary energy consumption}}{\text{growth rate of total GNP}}$$

For industrialized regions, this principal hypothesis will determine, on the basis of projected global economic growth, the future level of energy consumption (since non-commercial energy consumption is very small in these regions).

(9) total primary energy consumption (MTOE)

It is the sum of (10) + (11).

(10) total commercial energy consumption (CEG, MTOE)

CEC is calculated by: (13) × (1).

(11) total non-commercial energy consumption (CENC, MTOE)

It is calculated in two ways: globally for industrialized regions; and by multiplying (14) × (1) for developing regions.

(12) energy consumption per capita

It is the sum of (13) + (14); or (9) divided by (1).

(13) commercial energy consumption per capita (CEC/capita, TOE)

For developing countries, this is a key element and is calculated by multiplying (4) × (7).

(14) non-commercial energy consumption per capita (CENC/capita, TOE)

The estimation is directly originated from the assumptions of the study: "Energy Horizons".

(B) The second table contains: firstly, the energy supply required to meet the projected energy consumption by fuel type; and secondly, the corresponding projection of regional production and *inter-regional* trade.

Eight primary energy sources are distinguished:

SMF   = solid mineral fuels (coal + lignite + peat)
PP   = petroleum products
NG   = natural gas
HY   = hydropower
NU   = nuclear
NS   = "new" primary energy sources (solar, geothermal, wind, biomass, tide and wave energies, etc . . .)
FW   = fuelwood (plus charcoal and wood residues)
VAW   = vegetable and animal waste products

They are combined as follows:

CEC   = commercial energy consumption (SMF + PP + NG + HY + NU + NS)
NCEC = non-commercial energy consumption (FW + VAW)
PEC   = primary energy consumption (CEC + NCEC)

It is assumed that only commercial fossil fuels (SMF, PP and NG) are traded between regions. For these three fossil fuels, estimations (1978, 2000 and 2020) are given not only for regional consumption (C), but also for regional production (P), interregional imports (I) and interrregional exports (E), such that:

• for SMF and NG:
  C = P + I − E

- for petroleum products:
  $$C + BF = P + I - E$$

where BF = balance factor, composed of the following elements:

Production of non-energy products	P_{NE}
+ imports of refined products	I_{RP}
− exports of refined products	E_{RP}
+ refinery losses	L_R
+ stock variation	Δs
+ bunkers	B

In fact, (C) is the consumption of refined petroleum products while (P), (I), and (E) measure the production and trade of crude oil.

It is assumed that trade of refined products occurs only within a region; therefore $I_{PR} - E_{PR}$ is neglected (similarly, trade of non-energy products); in the future, $\Delta s = 0$. Therefore, for 2000 and 2020:

$$BF = P_{NE} + L_R + B$$

P_{NE} = production of non-energy products

L_R = refinery losses uniformly estimated at 7% of $(P + I - E)$

B = bunkers

Depending on the production level of the region, BF should be integrated into either the projection of P or I. For the purpose of unit consistency, P_{NE} should be recorded in MTOE. The total of BF should be specified at the bottom of the table.

For other primary energy sources (HY, NU, NS, FW, VAW), it is assumed that any trade occurs only within the region; therefore, I and E are neglected.

The provisional figures in the energy supply table are drawn from:

- for developing countries, the supply structure (in %) calculated in the study "Horizons" for Scenarios B and C, which were applied here, without modification, to the new projections of primary energy consumption,
- for industrialized countries, the forecasts of various national and international organizations were used.

When the information is insufficient, the Regional Working Team should determine the respective levels of production, imports and exports (this is the case for developing countries).

Recall that a supplementary table is provided for the majority of developing regions in order to facilitate the exercise: this is the high scenario described in point 4 of Scenario I.

In summary, the work required from the Regional Working Team (RWT) should concern only the figures included in the following table:

/ : historical data,
X : projections given by the Central Team (theoretically non-modifiable),
1 : projections of the RWT 1st step: energy consumption

2 : projections of the RWT 2nd step: supply structure,
3 : projections of the RWT 3rd step: production, interregional trade of
      commercial fossil fuels,
§§ : figures calculated from above.

REFERENCE BASIS 1960–1978

ENGLISH GLOSSARY I: TO USE TABLES OF THE REFERENCE BASIS

1 POPULATION ET CROISSANCE ECONOMIQUE = POPULATION AND ECONOMIC GROWTH

Mhab = million inhabitants
Taux 60–78% /an = average yearly growth rate 1960–1978
PNB monnaie constante = GNP constant money
Par hb = per capita
Taux/hb 60–78% /an = average yearly growth per capita 1960–1978

2 CONSOMMATIONS D'ENERGIES COMMERCIALES = COMMERCIAL ENERGY CONSUMPTIONS (CEC)

(MTEP) = million tons of oil equivalent (MTOE)
Combustibles minéraux solides = solid mineral fuels
Produits pétroliers = oil products
Gaz naturel = natural gas
Electricité primaire = primary electricity
Consommation énergie commerciale (CEC) = commercial energy consumption (CEC)

3 CONSOMMATIONS D'ENERGIE NON-COMMERCIALES = NON COMMERCIAL ENERGY CONSUMPTIONS (CENC)

(MTEP) = million tons of oil equivalent (MTOE)
Bois de chauffage = firewood
Déchets végétaux et animaux = vegetable and animal wastes

4 CONSOMMATIONS D'ENERGIE PRIMAIRE = PRIMARY ENERGY CONSUMPTIONS (CEP)

CEC = commercial energy consumptions
CENC = non-commercial energy consumptions
CEP = primary energy consumptions
TEP = ton of oil equivalent (TOE)
par habitant = per capita

// FORECAST TABLES //

SCENARIO :

REGION : INDUSTRIALIZED COUNTRIES (R1 - R2 - R3 - R4)

	1960-1978 (%)	1978	1978-2000 (%)	2000	2000-2020 (%)	2020
(1) POPULATION (M.inhb)						
(2) ECONOMIC GROWTH						
(3) - total (G $ 78)						
(4) - per cap. ($ 78)						
(5) ENERGY CONSUMPTION						
(6) Elasticities						
(7) . commercial /p.c.						
(8) . total						
(9) Total (Mtoe)						
(10) . commercial						
(11) . non-commercial						
(12) Per capita (toe)						
(13) . commercial						
(14) . non-commercial						

(Mtoe)	SMF	PP	NG	HY	NU	NS	FW	VAW	CEC	NCEC	PEC
1978											
(15) Consumption						-					
(16) Production						-					
(17) + Imports											
(18) - Exports											
2000											
(19) Consumption											
(20) Production											
(21) + Imports											
(22) - Exports											
2020											
(23) Consumption											
(24) Production											
(25) + Imports											
(26) - Exports											

(*) of which balance factor BF : 1978 = ⬜ Mtoe ; 2000 = ⬜ Mtoe ; 2020 = ⬜ Mtoe.

-oOo-

// FORECAST TABLES //

SCENARIO :

REGION : THIRD WORLD (R5-R6-R7-R8-R9-R10)

	1960-1978 (%)	1978	1978-2000 (%)	2000	2000-2020 (%)	2020
(1) POPULATION (M.inhb)						
(2) ECONOMIC GROWTH						
(3) - total (G $ 78)						
(4) - per cap. ($ 78)						
(5) ENERGY CONSUMPTION						
(6) .Elasticities						
(7) . commercial /p.c.						
(8) . total						
(9) Total (Mtoe)						
(10) . commercial						
(11) . non-commercial						
(12) Per capita (toe)						
(13) . commercial						
(14) . non-commercial						

(Mtoe)	SMF	PP	NG	HY	NU	NS	FW	VAW	CEC	NCEC	PEC
1978											
(15) Consumption						-					
(16) Production						-					
(17) + Imports											
(18) - Exports											
2000											
(19) Consumption											
(20) Production											
(21) + Imports											
(22) - Exports											
2020											
(23) Consumption											
(24) Production											
(25) + Imports											
(26) - Exports											

(*) of which balance factor BF : 1978 = ⬜ Mtoe ; 2000 = ⬜ Mtoe ; 2020 = ⬜ Mtoe.

-oOo-

ENERGY EQUIVALENCES

The system used is the one recommended by the Conservation Commission.
The basic unit is the "ton of oil equivalent" (TOE):

$$1 \text{ TOE} = 10500 \text{ Mcal}$$

with reference to the ton of coal equivalent (TCE).

$$1 \text{ TCE} = 7000 \text{ MCal}$$
$$\rightarrow 1 \text{TOE} = 1.5 \text{ TCE}$$

All the other primary energy sources are calculated with the same unit (for fossil fuels, from the UNO conversion system as defined in the Energy Statistical Yearbook 1981; for non-commercial fuels as defined in the report: *L'évoluton de consommations d'énergie dans le monde: une rétrospective 1960-1976,* Electricité de France, juin 1980).

As far as electricity is concerned, the equivalence is the practical one in the modern thermal plants, recommended by the WEC — UNIPEDE Guide,

$$1000 \text{ kWh} = 0.222 \text{ TOE} \approx 2330 \text{ Mcal}$$

(and not the thermal equivalence by effect Joule: 1000 Kkh = 860 Mcal).

The results can be easily transposed in SI units:

$$1 \text{ TOE} = 44 \text{ GJ}$$
$$1 \text{ EJ} = 10^{18} \text{ J} = 22.8 \text{ MTOE}$$

ENGLISH GLOSSARY II: TO USE TABLES OF THE SCENARIOS

1 Hypotheses d'encadrement = general assumptions

Population = population
Monde = world
Pays industrialisés = industrialised countries
Pays du Tiers Monde = Third World countries
Croissance économique = economic growth; % /an = % per year
Consommation d'énergie = energy consumption
  Totales (GTEP) = Total (GTOE)
  Par tête (TEP) = Per capita (TOE)
Production d'énergie = energy production
  (GTEP) = GTOE
  Monde = world
Approvisionnements = supply
  (GTEP) = GTOE
  Tiers Monde = Third World
  CMS = solid mineral fuels
  PP = petroleum products
  GN = natural gas
  HY = hydraulic energy
  NU = nuclear energy
  EN = "new sources"
  ENC = non commercial energy consumption
  PEP = primary energy production

2 Feuille de prevision = forecast table

Titles of the lines:
  see no. 5: "FORECAST TABLE MODEL"

Titles of the columns
  SMF = solid mineral fuels
  PP = petroleum products
  NG = natural gas
  HY = hydraulic energy
  NU = nuclear energy
  NS = "new sources"
  FW = firewood
  VAW = vegetable and animal wastes
  CEC = commercial energy consumption (SMF + PP + NG + HY + NU + NS)
  NCEC = non-commercial energy consumption (FW + VAW)
  PEC = primary energy consumption (CEC + NCEC)

3 Scenario haut = "high" scenario

"NORMATIVE-COOPERATION"
SCENARIO I
INTERNATIONAL ENVIRONMENT

1. Description

The "normative-cooperation" scenario is, in a first approach, derived from the Scenario B studied in the report: "Third World Energy Horizons 2000–2020". This Scenario B was very close to one of the OECD scenarios in "Facing the futures"; its main characteristics were:

"Cooperative management of interests and conflicts within the industrialised countries.
Intensified free trade.
The Third World takes one more part in the world trade but this development is inequally distributed among the countries.
Moderate and 'traditional' growth of the industrialised countries due to a difficult structural change more than to a wish and/or a shift in the traditional value system.
The relative productivities in the development countries tend to get closer.
The low GNP growth due to the present economic crisis is not compensated in the future.
A certain structural unemployment remains unreduced.
The aid to LDCs remains at its present level.
The investment of the transnational companies intensify throughout the world.
The world industrial production slows down, this move affecting inequally the different LDCs.
The total debt of the LDCs is increasing and their capability of reimbursement degradating."

Scénario "NORMATIF-COOPERATION"

2. HYPOTHESES D'ENCADREMENT (∗∗)   3 01

A 8

	1960-1976 (%)	1976	1976-2000 (%)	2000	2000-2020 (%)	2020
/POPULATION/						
(Mhb)						
Monde	1,9	4070	1,7	6100	1,5	8300
Pays Industrialisés	1,0	1130	0,7	1330	0,5	1460
Pays du Tiers Monde	2,3	2940	2,0	4770	1,8	6840
/CROISSANCE ECONOMIQUE/						
(% / an)						
Monde	4,9		4,5		3,0	
Pays Industrialisés	4,6		4,1		2,2	
Pays du Tiers Monde	6,5		6,0		4,6	
/CONSOMMATION D'ENERGIE/						
TOTALES (Gtep)						
Monde		6,8		12,9		20,0
Pays Industrialisés		5,1		7,6		9,6
Pays du Tiers Monde		1,7		5,3		10,4
PAR TETE (tep)						
Monde		1,7		2,1		2,4
Pays Industrialisés		4,5		5,7		6,6
Pays du Tiers Monde		0,6		1,1		1,5

		CMS	PP	GN	HY	NU	EN	ENC	PEP	(∗)
/PRODUCTION D'ENERGIE/										
(Gtep)										
Monde	1976	1,9	3,0	1,3	0,4	0,1	-	0,7	7,4	
	2000	3,6	3,5	2,8	0,8	1,2	0,2	0,9	13,0	
	2020	5,6	3,7	2,8	1,3	4,5	1,0	1,1	20,0	
/APPROVISIONNEMENTS/										
(Gtep)										
Tiers Monde	1976	0,5	0,4	0,1	0,1	-	-	0,6	1,7	
	2000	1,4	1,8	0,5	0,3	0,2	0,1	0,8	5,3	
	2020	2,6	3,0	1,1	0,7	0,9	0,6	1,1	10,0	

(∗)   CMS = charbon   EN = énergies nouvelles
        PP = pétrole   ENC = énergies non-commerciales
        GN = gaz naturel   PEP = production d'énergie primaire
        HY = hydraulique
        NU = nucléaire

(∗∗) Estimations extraites directement du Scénario B de l'étude "Horizons Energétiques du Tiers Monde".

-o0o-

// FORECAST TABLES //

A 8

SCENARIO :

REGION :

	1960-1978 (%)	1978	1978-2000 (%)	2000	2000-2020 (%)	2020
(1) POPULATION (M.inhb)						
(2) ECONOMIC GROWTH						
(3) - total (G $ 78)						
(4) - per cap. ($ 78)						
(5) ENERGY CONSUMPTION						
(6) Elasticities						
(7) . commercial /p.c.						
(8) . total						
(9) Total (Mtoe)						
(10) . commercial						
(11) . non-commercial						
(12) Per capita (toe)						
(13) . commercial						
(14) . non-commercial						

(Mtoe)	SMF	PP	NG	HY	NU	NS	FW	VAW	CEC	NCEC	PEC
/1978/											
(15) Consumption											
(16) Production											
(17) + Imports											
(18) - Exports											
/2000/											
(19) Consumption											
(20) Production											
(21) + Imports											
(22) - Exports											
/2020/											
(23) Consumption											
(24) Production											
(25) + Imports											
(26) - Exports											

(∗) of which balance factor BF : 1978 =   Mtoe ; 2000 =   Mtoe ; 2020 =   Mtoe.

"STRENGTHENING CONSTRAINTS"
SCENARIO II
INTERNATIONAL ENVIRONMENT

1. Description

The "strengthening constraints" scenario is, in a first approach, derived from the Scenario C studied in the report: "Third World Energy Horizons 2000-2020". This Scenario C was very close to one of the OECD scenarios in "Facing the futures"; its main characteristics were:

"The strategy of the main LDCs tends to a more autonomous growth and to the implementation of a common self reliant policy.

Interdependence with industrialised countries is reduced.

The industrialised countries tend to a more cooperative management and increase their internal trade: limited growth without noticeable shift in the value system.

More and more diverging relative productivities.

More concern for quality and poverty in the LDCs.

The aid from industrialised countries is reduced and partially compensated by OPEC.

The revenue's growth level is drastically reduced both for the industrialised countries and the LDCs.

The LDCs have to be self-reliant for food; thus, the agriculture gets the top priority.

New types of self reliant development styles are sought and the accumulation process tends to be more under control.

Need for the LDCs to develop their R and D on appropriate technologies due to a decrease in the technology transfers.

Drastic decrease in North–South trade.

Accentuated divergence in the development stages of Third World countries."

A 8

Scénario "TENSIONS ACCRUES"

2. HYPOTHESES D'ENCADREMENT (**)

	1960-1976 (%)	1976	1976-2000 (%)	2000	2000-2020 (%)	2020
POPULATION						
(Mhb)						
Monde	1,9	4070	1,7	6100	1,5	8300
Pays Industrialisés	1,0	1130	0,7	1330	0,5	1460
Pays du Tiers Monde	2,3	2940	2,0	4770	1,8	6840
CROISSANCE ECONOMIQUE						
(% / an)						
Monde	4,9		3,65		2,6	
Pays Industrialisés	4,6		3,2		1,8	
Pays du Tiers Monde	6,5		5,3		4,25	
CONSOMMATION D'ENERGIE						
TOTALES (Gtep)						
Monde		6,8		11,8		17,8
Pays Industrialisés		5,1		7,0		8,8
Pays du Tiers Monde		1,7		4,8		9,0
PAR TETE (tep)						
Monde		1,7		2,0		2,2
Pays Industrialisés		4,5		5,3		6,0
Pays du Tiers Monde		0,6		1,0		1,3

		CMS	PP	GN	HY	NU	EN	ENC	PEP	(*)
PRODUCTION D'ENERGIE										
(Gtep)										
Monde	1976	1,9	3,0	1,3	0,4	0,1	-	0,7	7,4	
	2000	3,2	3,3	2,5	0,7	0,9	0,3	1,1	12,0	
	2020	5,0	3,0	2,5	1,4	3,3	1,3	1,5	18,0	
APPROVISIONNEMENTS										
(Gtep)										
Tiers Monde	1976	0,5	0,4	0,1	0,1	-	-	0,6	1,7	
	2000	1,4	1,3	0,4	0,3	0,2	0,2	1,0	4,8	
	2020	2,5	2,0	0,9	0,8	0,5	0,9	1,4	9,0	

(*) CMS = charbon   EN = énergies nouvelles
     PP   = pétrole   ENC = énergies non-commerciales
     GN   = gaz naturel   PEP = production d'énergie primaire
     HY   = hydraulique
     NU   = nucléaire

(**) Estimations extraites directement du Scénario C de l'étude "Horizons Energé-
     tiques du Tiers Monde".

-oOo-

FIRST STAGE OF ENERGY FORECASTS (CENTRALISED STAGE)

This comprises of all the tables, prepared by the Central Team, which occurred in the forecasting files sent to each region.

One will therefore find 10 tables for Scenario (I) and 10 others for Scenario (II).

// FORECAST TABLES //

Table : 1
Appendix : 9

SCENARIO : I "NORMATIVE-COOPERATION"

REGION : /NORTH AMERICA/ (R1)

	1960-1978 (%)	1978	1978-2000 (%)	2000	2000-2020 (%)	2020
(1) POPULATION (M.inhb)	1,12	245,4	0,98	304,0	0,60	342,6
(2) ECONOMIC GROWTH						
(3) - total (G $ 78)	4,0	2347,9	2,55	4081,8		
(4) - per cap. ($ 78)	2,83	9570	1,55	13427		
(5) ENERGY CONSUMPTION						
(6) Elasticities						
(7) . commercial /p.c.	0,69		0,18			
(8) . total	0,75		0,51			
(9) Total (Mtoe)	3,0	1981,4	1,3	2625	1,4	3475
(10) . commercial	3,1	1955,0	1,25	2573	1,4	3417
(11) . non-commercial	-0,3	26,4	3,1	52	0,5	58
(12) Per capita (toe)	1,9	8,07	0,3	8,63	0,8	10,14
(13) . commercial	1,94	7,96	0,28	8,46	0,82	9,97
(14) . non-commercial	-1,4	0,11	2,0	0,17	0	0,17

(Mtoe)	SNF	PP	NG	HY	NU	NS	FW	VAW	CEC	NCEC	PEC
1978/											
(15) Consumption	389,9	831,9	552,2	181,0		-	6,2	20,2	1955,0	26,4	1981,4
(16) Production	370,5	539,9(*)	544,5								
(17) + Imports	1,8	325,7	1,9								
(18) - Exports	23,8	2,1	1,2								
2000/											
(19) Consumption	927	604	475	170	304	93	13	39	2573	52	2625
(20) Production	1021	624 (*)	475	170	304	93	13	39	2687	52	2739
(21) + Imports	-	126							126		126
(22) - Exports	94	-							94		94
2020/											
(23) Consumption	1743	251	419	200	504	300	14	44	3417	58	3475
(24) Production	1841	410 (*)	419	200	504	300	14	44	3674	58	3732
(25) + Imports	-	28							28		28
(26) - Exports	98	-							98		98

(*) of which balance factor BF : 1978 = 31,6 Mtoe ; 2000 = 146 Mtoe ; 2020 = 187 Mtoe.

// FORECAST TABLES //

SCENARIO : I "NORMATIVE-COOPERATION"   Table   : 2
                                                                         Appendix : 9
REGION : /WESTERN EUROPE/ (R2)

	1960-1978 (%)	1978	1978-2000 (%)	2000	2000-2020 (%)	2020
(1) POPULATION (M.inhb)	0,71	373,4	0,28	392,3	0,03	394,5
(2) ECONOMIC GROWTH						
(3) - total (G $ 78)	3,90	2540,4				
(4) - per cap. ($ 78)	3,17	6800				
(5) ENERGY CONSUMPTION						
(6) Elasticities						
(7) . commercial /p.c.	0,94					
(8) . total	0,90					
(9) Total (Mtoe)	3,50	1124,5	2	1736	1,5	2334
(10) . commercial	3,71	1107,2	2	1712	1,5	2306
(11) . non-commercial	-3,25	17,3	1,5	24	0,8	28
(12) Per capita (toe)	2,79	3,02	1,7	4,42	1,5	5,92
(13) . commercial	2,98	2,97	1,7	4,36	1,5	5,85
(14) . non-commercial	-3,32	0,05	0,8	0,06	0,8	0,07

(Mtoe)	SMF	PP	NG	HY	NU	NS	FW	VAW	CEC	NCEC	PEC
/1978/											
(15) Consumption	237,8	558,6	176,8	96,1	36,8	-	8,3	9,1	1107,2	17,3	1124,5
(16) Production	205,1	86,3	154,1								
(17) + Imports	39,6	576,8(*)	24,6								
(18) - Exports	0,9	13,5	-								
/2000/											
(19) Consumption	428	599	308	111	223	43	11	13	1712	24	1736
(20) Production											
(21) + Imports											
(22) - Exports											
/2020/											
(23) Consumption	690	460	346	180	530	160	13	15	2306	28	2334
(24) Production											
(25) + Imports											
(26) - Exports											

(*) of which balance factor BF : 1978 =   91   Mtoe ; 2000 =   Mtoe ; 2020 =   Mtoe.

// FORECAST TABLES //

SCENARIO : I "NORMATIVE-COOPERATION"   Table   : 3
                                                                         Appendix : 9
REGION : /PACIFIC INDUSTRIALIZED COUNTRIES/ (R3)

	1960-1978 (%)	1978	1978-2000 (%)	2000	2000-2020 (%)	2020
(1) POPULATION (M.inhb)	1,2	132,4	0,6	151,1	0,15	156,1
(2) ECONOMIC GROWTH						
(3) - total (G $ 78)	8,8					
(4) - per cap. ($ 78)	7,5					
(5) ENERGY CONSUMPTION						
(6) Elasticities						
(7) . commercial /p.c.	0,67					
(8) . total	0,67					
(9) Total (Mtoe)	5,9	348,9	3,1	680	1,1	850
(10) . commercial	6,3	343,2	3,1	675	1,1	845
(11) . non-commercial	-3,2	5,7	-0,6	5	0	5
(12) Per capita (toe)	4,7	2,63	2,5	4,50	1,0	5,44
(13) . commercial	5,0	2,59	2,5	4,47	1,0	5,41
(14) . non-commercial	4,4	0,04	-1,3	0,03	0	0,03

(Mtoe)	SMF	PP	NG	HY	NU	NS	FW	VAW	CEC	NCEC	PEC
/1978/											
(15) Consumption	71,8	210,9	23,5	37,0		-	0,6	5,1	343,2	5,7	348,9
(16) Production	67,8	22,5	10,2								
(17) + Imports	18,1	237,4(*)	13,3								
(18) - Exports	7,1										
/2000/											
(19) Consumption	177	292	81	29	78	18	1	4	675	5	680
(20) Production	276	34	23	29	78	18	1	4	458	5	463
(21) + Imports	26		58								
(22) - Exports	125								125		125
/2020/											
(23) Consumption	284	108	60	33	266	94	1	4	845	5	850
(24) Production	412	26	24	33	266	94	1	4	855	5	860
(25) + Imports	62		36								
(26) - Exports	190								190		190

(*) of which balance factor BF : 1978 = 49,0   Mtoe ; 2000 =   Mtoe ; 2020 =   Mtoe.

// FORECAST TABLES //

SCENARIO :   I   "NORMATIVE-COOPERATION"

REGION :   /EASTERN EUROPE/ (R4)

Table   : 4
Appendix : 9

	1960-1978 (%)	1978	1978-2000 (%)	2000	2000-2020 (%)	2020
(1) POPULATION (M.inhb)	0,95	371,9	0,55	419,5	0,5	462,3
(2) ECONOMIC GROWTH						
(3) - total (G $ 78)	6,1					
(4) - per cap. ($ 78)	5,1					
(5) ENERGY CONSUMPTION						
(6) Elasticities						
(7) . commercial /p.c.	0,86					
(8) . total	0,84					
(9) Total (Mtoe)	5,1	1413,8	3,3	2902	1,85	4192
(10) . commercial	5,4	1367,7	3,4	2864	1,9	4160
(11) . non-commercial	0	46,1	-0,9	38	-0,9	32
(12) Per capita (toe)	4,1	3,80	2,8	6,92	1,4	9,07
(13) . commercial	4,4	3,68	2,85	6,83	1,4	9,00
(14) . non-commercial	-1,3	0,12	-1,3	0,09	-1,3	0,07

(Mtoe)	SMF	PP	NG	HY	NU	NS	FW	VAW	CEC	NCEC	PEC
/1978/											
(15) Consumption	574,7	403,9	334,5	54,6	-		30,3	15,8	1367,7	46,1	1413,8
(16) Production	613,0	574,6	350,4								
(17) + Imports	1,3	26,0	8,1								
(18) - Exports	25,9	57,9	17,3								
/2000/											
(19) Consumption	1003	621	785	77	307	71	25	13	2864	38	2902
(20) Production	1190	731 (*)	973	77	307	71	25	13	3349	38	3387
(21) + Imports	-	136	-						136		136
(22) - Exports	187	60	188						435		435
/2020/											
(23) Consumption	1248	624	1248	125	707	208	21	11	4160	32	4192
(24) Production											
(25) + Imports											
(26) - Exports											

(*) of which balance factor BF : 1978 = 138,8 Mtoe ; 2000 = 186 Mtoe ; 2020 = 156 Mtoe.

// FORECAST TABLES //

SCENARIO : I "NORMATIVE-COOPERATION"

REGION : /NORTH AFRICA / MIDDLE EAST/ (R5)

Table   : 5
Appendix : 9

	1960-1978 (%)	1978	1978-2000 (%)	2000	2000-2020 (%)	2020
(1) POPULATION (M.inhb)	2,75	227,6	2,75	413,6	1,9	606,0
(2) ECONOMIC GROWTH						
(3) - total (G $ 78)	8,6	347,3	6,3	1345,4	4,5	3230
(4) - per cap. ($ 78)	5,71	1526	3,5	3253	2,5	5330
(5) ENERGY CONSUMPTION						
(6) Elasticities						
(7) . commercial /p.c.	1,04		0,90		0,75	
(8) . total	1,03		0,86		0,80	
(9) Total (Mtoe)	7,34	153,6	5,4	492	3,6	1000
(10) . commercial	8,85	125,4	6,0	451	3,8	958
(11) . non-commercial	3,33	28,2	1,7	41	0,1	42
(12) Per capita (toe)	4,47	0,675	2,6	1,19	1,65	1,65
(13) . commercial	5,94	0,551	3,15	1,090	1,875	1,581
(14) . non-commercial	0,57	0,124	-1,0	0,10	-1,8	0,07

(Mtoe)	SMF	PP	NG	HY	NU	NS	FW	VAW	CEC	NCEC	PEC
/1978/											
(15) Consumption	7,2	87,3	24,5	6,4	-	-	15,9	12,3	125,4	28,2	153,6
(16) Production	6,2	1215,3	41,9								
(17) + Imports	1,0	1,1									
(18) - Exports		1103,1	17,4								
/2000/											
(19) Consumption	10	247	130	16	22	26	24	17	451	41	492
(20) Production											
(21) + Imports											
(22) - Exports											
/2020/											
(23) Consumption	13	467	293	27	82	21	21	21	958	42	1000
(24) Production											
(25) + Imports											
(26) - Exports											

(*) of which balance factor BF : 1978 = 26,0 Mtoe ; 2000 = Mtoe ; 2020 = Mtoe.

// FORECAST TABLES //

SCENARIO : I "NORMATIVE-COOPERATION"

REGION : /AFRICA SOUTH OF THE SAHARA/ (R6)
(excluding South Africa)

Table : 6
Appendix : 9

	1960-1978 (%)	1978	1978-2000 (%)	2000	2000-2020 (%)	2020
(1) POPULATION (M.inhb)	2,6	320,1	3,1	630,1	2,7	1071,1
(2) ECONOMIC GROWTH						
(3) - total (G $ 78)	4,0	122,3	5,2	372,4	4,2	852,6
(4) - per cap. ($ 78)	1,43	382	2	591	1,5	796
(5) ENERGY CONSUMPTION						
(6) Elasticities						
(7) . commercial /p.c.	2,01		1,5		1,5	
(8) . total	0,75		0,62		0,64	
(9) Total (Mtoe)	3,0	138,5	3,2	278	2,7	475
(10) . commercial	5,5	27,0	6,2	102	5,0	271
(11) . non-commercial	2,5	111,5	2,1	176	0,7	204
(12) Per capita (toe)	0,4	0,432	0,1	0,44	0	0,44
(13) . commercial	2,87	0,084	2	0,162	2,25	0,253
(14) . non-commercial	0	0,348	-1,0	0,28	-1,9	0,19

(Mtoe)	SMF	PP	NG	HY	NU	NS	FW	VAW	CEC	NCEC	PEC
/1978/											
(15) Consumption	3,2	16,8	0,6	6,4	-	-	100,1	11,4	27,0	111,5	138,5
(16) Production											
(17) + Imports											
(18) - Exports											
/2000/											
(19) Consumption	8	54	6	29	1	4	158	18	102	176	278
(20) Production											
(21) + Imports											
(22) - Exports											
/2020/											
(23) Consumption	17	116	23	83	13	19	178	26	271	204	475
(24) Production											
(25) + Imports											
(26) - Exports											

(*) of which balance factor BF : 1978 =   11,3   Mtoe ; 2000 =   Mtoe ; 2020 =   Mtoe.

// FORECAST TABLES //

SCENARIO : I "NORMATIVE-COOPERATION"

REGION : /SOUTH ASIA/ (R7)

Table : 7
Appendix : 9

	1960-1978 (%)	1978	1978-2000 (%)	2000	2000-2020 (%)	2020
(1) POPULATION (M.inhb)	2,3	848,6	1,9	1284,1	1,25	1643,8
(2) ECONOMIC GROWTH						
(3) - total (G $ 78)	3,8	149,9	4,1	374,7	2,75	627,9
(4) - per cap. ($ 78)	1,4	177	2,2	284	1,5	382
(5) ENERGY CONSUMPTION						
(6) Elasticities						
(7) . commercial /p.c.	1,91		1,23		0,81	
(8) . total	1,03		0,78		0,69	
(9) Total (Mtoe)	3,9	239,2	3,2	475	1,9	698
(10) . commercial	5,1	95,3	4,6	257	2,5	419
(11) . non-commercial	3,2	143,9	1,9	218	1,25	279
(12) Per capita (toe)	1,5	0,282	1,25	0,370	0,7	0,425
(13) . commercial	2,7	0,112	2,7	0,200	1,2	0,255
(14) . non-commercial	0,9	0,170	0	0,17	0	0,17

(Mtoe)	SMF	PP	NG	HY	NU	NS	FW	VAW	CEC	NCEC	PEC
/1978/											
(15) Consumption	49,2	27,7	6,5	11,9		-	54,0	89,9	95,3	143,9	239,2
(16) Production	49,2	11,4	8,6								
(17) + Imports		20,7(*)									
(18) - Exports			2,1								
/2000/											
(19) Consumption	96	81	16	31	22	11	78	140	257	218	475
(20) Production											
(21) + Imports											
(22) - Exports											
/2020/											
(23) Consumption	165	102	15	53	55	29	94	185	419	279	698
(24) Production											
(25) + Imports											
(26) - Exports											

(*) of which balance factor BF : 1978 =   4,4   Mtoe ; 2000 =   Mtoe ; 2020 =   Mtoe.

// FORECAST TABLES //

SCENARIO : I "NORMATIVE-COOPERATION"

REGION : /SOUTH-EAST ASIA/ (R8)

	1960-1978 (%)	1978	1978-2000 (%)	2000	2000-2020 (%)	2020
(1) POPULATION (M.inhb)	2,4	337,6	1,8	502,7	1,15	631,3
(2) ECONOMIC GROWTH						
(3) - total (G $ 78)	7,2	220,2	6,2	823,9	4,2	1848,6
(4) - per cap. ($ 78)	4,7	652	4,3	1639	3,0	2960
(5) ENERGY CONSUMPTION						
(6) Elasticities						
(7) . commercial /p.c.	1,37		0,915		0,67	
(8) . total	0,72		0,71		0,62	
(9) Total (Mtoe)	5,2	218,7	4,4	561	2,6	937
(10) . commercial	8,9	118,1	5,8	410	3,2	767
(11) . non-commercial	2,6	100,6	1,9	151	0,6	170
(12) Per capita (toe)	2,7	0,648	2,5	1,116	1,4	1,485
(13) . commercial	6,4	0,350	3,92	0,816	2,0	1,215
(14) . non-commercial	0,2	0,298	ε	0,30	-0,5	0,27

(Mtoe)	SMF	PP	NG	HY	NU	NS	FW	VAW	CEC	NCEC	PEC
/1978/											
(15) Consumption	13,9	92,4	7,2	4,6	-	-	62,5	38,1	118,1	100,6	218,7
(16) Production	11,8	101,1(*)	18,0								
(17) + Imports	2,6	85,2									
(18) - Exports	0,5	73,5	10,8								
/2000/											
(19) Consumption	39	255	38	19	47	12	83	68	410	151	561
(20) Production											
(21) + Imports											
(22) - Exports											
/2020/											
(23) Consumption	59	370	93	69	130	46	79	91	767	170	937
(24) Production											
(25) + Imports											
(26) - Exports											

(*) of which balance factor BF : 1978 = 20,4   Mtoe ; 2000 =   Mtoe ; 2020 =   Mtoe.

// FORECAST TABLES //

SCENARIO : I "NORMATIVE-COOPERATION"

REGION : /CENTRALLY PLANNED ASIAN COUNTRIES/ (R9)

	1960-1978 (%)	1978	1978-2000 (%)	2000	2000-2020 (%)	2020
(1) POPULATION (M.inhb)	1,86	1034,5	1,25	1359,3	0,74	1576,3
(2) ECONOMIC GROWTH						
(3) - total (G $ 78)	5,1	247,1	4,8	691,9	3,3	1314,6
(4) - per cap. ($ 78)	3,18	239	3,5	509	2,5	834
(5) ENERGY CONSUMPTION						
(6) Elasticities						
(7) . commercial /p.c.	1,98		1,3		0,9	
(8) . total	1,27		1,08		0,85	
(9) Total (Mtoe)	6,47	638,0	5,2	1937	2,8	3374
(10) . commercial	8,26	495,9	4,55	1733	3,0	3137
(11) . non-commercial	2,72	142,1	1,65	204	0,75	237
(12) Per capita (toe)	4,52	0,616	3,9	1,425	2,05	2,14
(13) . commercial	6,29	0,479	4,55	1,275	2,25	1,990
(14) . non-commercial	0,86	0,137	0,4	0,15	0	0,15

(Mtoe)	SMF	PP	NG	HY	NU	NS	FW	VAW	CEC	NCEC	PEC
/1978/											
(15) Consumption	385,1	79,2	12,2	19,4	-	-	57,3	84,8	495,9	142,1	638,0
(16) Production	385,1	100,9(*)	12,2								
(17) + Imports											
(18) - Exports											
/2000/											
(19) Consumption	1115	444	53	33	53	35	74	130	1733	204	1937
(20) Production											
(21) + Imports											
(22) - Exports											
/2020/											
(23) Consumption	1828	679	136	118	219	157	78	159	3137	237	3374
(24) Production											
(25) + Imports											
(26) - Exports											

(*) of which balance factor BF : 1978 = 21,7   Mtoe ; 2000 =   Mtoe ; 2020 =   Mtoe.

// FORECAST TABLES //

SCENARIO : I "NORMATIVE-COOPERATION"

Table : 10
Appendix : 9

REGION : /LATIN AMERICA/ (R10)

	1960-1978 (%)	1978	1978-2000 (%)	2000	2000-2020 (%)	2020
(1) POPULATION (M.inhb)	2,8	344,5	2,3	564,1	1,4	802,4
(2) ECONOMIC GROWTH						
(3) - total (G $ 78)	5,6	482,8	5,8	1685,4	4,3	3928,6
(4) - per cap. ($ 78)	2,8	1402	3,5	2988	2,5	4896
(5) ENERGY CONSUMPTION						
(6) Elasticities						
(7) . commercial /p.c.	1,18		0,95		0,64	
(8) . total	0,77		0,79		0,73	
(9) Total (Mtoe)	4,3	370,0	4,6	1000	3,15	1858
(10) . commercial	6,1	270,4	5,7	910	3,4	1778
(11) . non-commercial	1,2	99,6	-0,5	90	-0,6	80
(12) Per capita (toe)	1,5	1,074	2,3	1,773	1,35	2,316
(13) . commercial	3,25	0,785	3,33	1,613	1,60	2,216
(14) . non-commercial	-1,5	0,289	-2,7	90	-2,3	0,10

(Mtoe)	SMF	PP	NG	HY	NU	NS	FW	VAW	CEC	NCEC	PEC
/1978/											
(15) Consumption	14,3	173,9	41,5	40,7		-	66,0	33,6	270,4	99,6	370,0
(16) Production	9,9	246,5(*)	41,5								
(17) + Imports	3,4	110,1									
(18) - Exports		59,2									
/2000/											
(19) Consumption	26	518	140	103	75	48	49	41	910	90	1000
(20) Production											
(21) + Imports											
(22) - Exports											
/2020/											
(23) Consumption	45	846	283	208	234	162	30	50	1778	80	1858
(24) Production											
(25) + Imports											
(26) - Exports											

(*) of which balance factor BF : 1978 = 123,5 Mtoe ; 2000 =   Mtoe ; 2020 =   Mtoe.

// FORECAST TABLES //

SCENARIO : II "INCREASED TENSIONS"

Table : 11
Appendix : 9

REGION : /NORTH AMERICA/ (R1)

	1960-1978 (%)	1978	1978-2000 (%)	2000	2000-2020 (%)	2020
(1) POPULATION (M.inhb)	1,12	245,4	0,98	304,0	0,60	342,6
(2) ECONOMIC GROWTH						
(3) - total (G $ 78)	4,0	2347,9	2,4	3963,7		
(4) - per cap. ($ 78)	2,83	9570	1,4	13038		
(5) ENERGY CONSUMPTION						
(6) Elasticities						
(7) . commercial /p.c.	0,69		-			
(8) . total	0,75		0,42			
(9) Total (Mtoe)	3,0	1981,4	1,0	2470	1,25	3178
(10) . commercial	3,1	1955,0	1,0	2421	1,3	3123
(11) . non-commercial	-0,3	26,4	2,8	49	0,6	55
(12) Per capita (toe)	1,9	8,07	0,3	8,12	0,7	9,28
(13) . commercial	1,94	7,96	0	7,96	0,7	9,12
(14) . non-commercial	-1,4	0,11	1,7	0,16	0	0,16

(Mtoe)	SMF	PP	NG	HY	NU	NS	FW	VAW	CEC	NCEC	PEC
/1978/											
(15) Consumption	389,9	831,9	552,2	181,0		-	6,2	20,2	1955,0	26,4	1981,4
(16) Production	370,5	539,9(*)	544,5								
(17) + Imports	1,8	325,7	1,9								
(18) - Exports	23,8	2,1	1,2								
/2000/											
(19) Consumption	892	531	448	156	300	94	12	37	2421	49	2470
(20) Production	977	587 (*)	448	156	300	94	12	37	2562	49	2611
(21) + Imports	-	85	-						85		85
(22) - Exports	85	-	-						85		85
/2020/											
(23) Consumption	1568	227	387	180	474	287	13	42	3123	55	3178
(24) Production	1656	384 (*)	387	180	474	287	13	42	3368	55	3423
(25) + Imports	-	17	-						17		17
(26) - Exports	88	-	-						88		88

(*) of which balance factor BF : 1978 = 31,6 Mtoe ; 2000 = 141 Mtoe ; 2020 = 174 Mtoe.

Table   : 12
Appendix :   9

SCENARIO :   II "INCREASED TENSIONS"

REGION :   /WESTERN EUROPE/ (R2)

	1960-1978 (%)	1978	1978-2000 (%)	2000	2000-2020 (%)	2020
(1) POPULATION (M.inhb)	0,71	373,4	0,28	392,3	0,03	394,5
(2) ECONOMIC GROWTH						
(3) - total (G $ 78)	3,90	2540,4				
(4) - per cap. ($ 78)	3,17	6800				
(5) ENERGY CONSUMPTION						
(6) Elasticities						
(7) . commercial /p.c.	0,94					
(8) . total	0,90					
(9) Total (Mtoe)	3,50	1124,5	1,5	1563	1	1906
(10) . commercial	3,71	1107,2	1,5	1536	1	1874
(11) . non-commercial	-3,25	17,3	2	27	0,8	32
(12) Per capita (toe)	2,79	3,02	1,25	3,98	1,0	4,83
(13) . commercial	2,98	2,97	1,25	3,91	1,0	4,75
(14) . non-commercial	-3,32	0,05	1,5	0,07	0,7	0,08

(Mtoe)	SMF	PP	NG	HY	NU	NS	FW	VAW	CEC	NCEC	PEC
/1978/											
(15) Consumption	237,8	558,6	176,8	96,1	36,8	-	8,3	9,1	1107,2	17,3	1124,5
(16) Production	205,1	86,3	154,1								
(17) + Imports	39,6	576,8(*)	24,6								
(18) - Exports	0,9	13,5									
/2000/											
(19) Consumption	384	522	276	107	200	47	13	14	1536	27	1563
(20) Production											
(21) + Imports											
(22) - Exports											
/2020/											
(23) Consumption	544	360	240	130	450	150	15	17	1874	32	1906
(24) Production											
(25) + Imports											
(26) - Exports											

(*) of which balance factor BF : 1978 =   91   Mtoe ; 2000 =   Mtoe ; 2020 =   Mtoe.

Table   : 13
Appendix :   9

SCENARIO :   II "INCREASED TENSIONS"

REGION :   /PACIFIC INDUSTRIALIZED COUNTRIES/ (R3)

	1960-1978 (%)	1978	1978-2000 (%)	2000	2000-2020 (%)	2020
(1) POPULATION (M.inhb)	1,2	132,4	0,6	151,1	0,15	156,1
(2) ECONOMIC GROWTH						
(3) - total (G $ 78)	8,8					
(4) - per cap. ($ 78)	7,5					
(5) ENERGY CONSUMPTION						
(6) Elasticities						
(7) . commercial /p.c.	0,67					
(8) . total	0,67					
(9) Total (Mtoe)	5,9	348,9	2,1	551	0,6	624
(10) . commercial	6,3	343,2	2,1	545	0,6	617
(11) . non-commercial	-3,2	5,7	0,2	6	0,8	7
(12) Per capita (toe)	4,7	2,63	1,5	3,65	0,5	4,00
(13) . commercial	5,0	2,59	1,5	3,61	0,5	3,95
(14) . non-commercial	-4,4	0,04	0	0,04	1,1	0,05

(Mtoe)	SMF	PP	NG	HY	NU	NS	FW	VAW	CEC	NCEC	PEC
/1978/											
(15) Consumption	71,8	210,9	23,5	37,0		-	0,6	5,1	343,2	5,7	348,9
(16) Production	67,8	22,5	10,2								
(17) + Imports	18,1	237,4(*)	13,3								
(18) - Exports	7,1										
/2000/											
(19) Consumption	141	235	64	28	63	14	1	5	545	6	551
(20) Production	186	30	17	28	63	14	1	5	338	6	344
(21) + Imports	30		47								
(22) - Exports	75								75		75
/2020/											
(23) Consumption	203	79	49	29	187	70	1	6	617	7	624
(24) Production	259	23	22	29	187	70	1	6	590	7	597
(25) + Imports	59		27								
(26) - Exports	115								115		115

(*) of which balance factor BF : 1978 =   49,0   Mtoe ; 2000 =   Mtoe ; 2020 =   Mtoe.

// FORECAST TABLES //

Table   : 14
Appendix : 9

SCENARIO :   II "INCREASED TENSIONS"

REGION :   /EASTERN EUROPE/ (R4)

	1960-1978 (%)	1978	1978-2000 (%)	2000	2000-2020 (%)	2020
(1) POPULATION (M.inhb)	0,95	371,9	0,55	419,5	0,5	462,3
(2) ECONOMIC GROWTH						
(3) - total (G $ 78)	6,1					
(4) - per cap. ($ 78)	5,1					
(5) ENERGY CONSUMPTION						
(6) Elasticities						
(7) . commercial /p.c.	0,86					
(8) . total	0,84					
(9) Total (Mtoe)	5,1	1413,8	2,65	2509	1,35	3282
(10) . commercial	5,4	1367,7	2,7	2467	1,4	3245
(11) . non-commercial	0	46,1	-0,4	42	-0,6	37
(12) Per capita (toe)	4,1	3,80	2,1	5,98	0,9	7,10
(13) . commercial	4,4	3,68	2,15	5,88	0,9	7,02
(14) . non-commercial	-1,3	0,12	-0,8	0,10	-1,1	0,08

(Mtoe)	SMF	PP	NG	HY	NU	NS	FW	VAW	CEC	NCEC	PEC
/1978/											
(15) Consumption	574,7	403,9	334,5	54,6	-		30,3	15,8	1367,7	46,1	1413,8
(16) Production	613,0	574,6	350,4								
(17) + Imports	1,3	26,0	8,1								
(18) - Exports	25,9	57,9	17,3								
/2000/											
(19) Consumption	721	589	751	66	263	77	28	14	2467	42	2509
(20) Production	801	622 (*)	843	66	263	77	28	14	2672	42	2714
(21) + Imports	-	184	-						184		184
(22) - Exports	80	40	92						212		212
/2020/											
(23) Consumption	812	649	1038	97	487	162	25	12	3245	37	3282
(24) Production											
(25) + Imports											
(26) - Exports											

(*) of which balance factor BF : 1978 =   138,8 Mtoe ; 2000 = 177   Mtoe ; 2020 =   152   Mtoe.

// FORECAST TABLES //

Table   : 15
Appendix : 9

SCENARIO :   II "INCREASED TENSIONS"

REGION :   /NORTH AFRICA / MIDDLE EAST/ (R5)

	1960-1978 (%)	1978	1978-2000 (%)	2000	2000-2020 (%)	2020
(1) POPULATION (M.inhb)	2,75	227,6	2,75	413,6	1,9	606,0
(2) ECONOMIC GROWTH						
(3) - total (G $ 78)	8,6	347,3	5,8	1209,4	4,0	2633,0
(4) - per cap. ($ 78)	5,71	1526	5	2924	2	4345
(5) ENERGY CONSUMPTION						
(6) Elasticities						
(7) . commercial /p.c.	1,04		0,90		0,80	
(8) . total	1,03		0,88		0,82	
(9) Total (Mtoe)	7,34	153,6	5,1	459	3,3	885
(10) . commercial	8,85	125,4	5,5	410	3,6	824
(11) . non-commercial	3,33	28,2	2,5	49	1,1	61
(12) Per capita (toe)	4,47	0,675	2,3	1,11	1,4	1,46
(13) . commercial	5,94	0,551	2,7	0,990	1,6	1,360
(14) . non-commercial	0,57	0,124	-0,15	0,12	-0,9	0,10

(Mtoe)	SMF	PP	NG	HY	NU	NS	FW	VAW	CEC	NCEC	PEC
/1978/											
(15) Consumption	7,2	87,3	24,5	6,4	-	-	15,9	12,3	125,4	28,2	153,6
(16) Production	6,2	1215,3	41,9								
(17) + Imports	1,0	1,1									
(18) - Exports		1103,1	17,4								
/2000/											
(19) Consumption	12	210	122	19	16	31	30	19	410	49	459
(20) Production											
(21) + Imports											
(22) - Exports											
/2020/											
(23) Consumption	20	373	254	28	46	103	34	27	824	61	885
(24) Production											
(25) + Imports											
(26) - Exports											

(*) of which balance factor BF : 1978 =   26,0   Mtoe ; 2000 =   Mtoe ; 2020 =   Mtoe.

// FORECAST TABLES //

Table : 16
Appendix : 9

SCENARIO : II "INCREASED TENSIONS"

REGION : AFRICA SOUTH OF THE SAHARA (R6)
(excluding South Africa)

	1960-1978 (%)	1978	1978-2000 (%)	2000	2000-2020 (%)	2020
(1) POPULATION (M.inhb)	2,6	320,1	3,1	630,1	2,7	1071,1
(2) ECONOMIC GROWTH						
(3) - total (G $ 78)	4,0	122,3	4,6	327,0	3,9	705,8
(4) - per cap. ($ 78)	1,43	382	1,4	519	1,2	659
(5) ENERGY CONSUMPTION						
(6) Elasticities						
(7) . commercial /p.c.	2,01		1,5		1,2	
(8) . total	0,75		0,67		0,67	
(9) Total (Mtoe)	3,0	138,5	3,1	273	2,6	458
(10) . commercial	5,5	27,0	5,3	84	4,2	190
(11) . non-commercial	2,5	111,5	2,4	189	1,8	268
(12) Per capita (toe)	0,4	0,432	0	0,43	-0,1	0,43
(13) . commercial	2,87	0,084	2,1	0,133	1,45	0,177
(14) . non-commercial	0	0,348	-0,7	0,30	-0,9	0,25

(Mtoe)	SNF	PP	NG	HY	NU	NS	FW	VAW	CEC	NCEC	PEC
1978											
(15) Consumption	3,2	16,8	0,6	6,4	-	-	100,1	11,4	27,0	111,5	138,5
(16) Production											
(17) + Imports											
(18) - Exports											
2000											
(19) Consumption	10	30	5	33	1	5	171	18	84	189	273
(20) Production											
(21) + Imports											
(22) - Exports											
2020											
(23) Consumption	16	63	17	69	5	20	239	29	190	268	458
(24) Production											
(25) + Imports											
(26) - Exports											

(*) of which balance factor BF : 1978 = 11,3 Mtoe ; 2000 =   Mtoe ; 2020 =   Mtoe.

// FORECAST TABLES //

Table : 17
Appendix : 9

SCENARIO : II "INCREASED TENSIONS"

REGION : SOUTH ASIA (R7)

	1960-1978 (%)	1978	1978-2000 (%)	2000	2000-2020 (%)	2020
(1) POPULATION (M.inhb)	2,3	848,6	1,9	1284,1	1,25	1643,8
(2) ECONOMIC GROWTH						
(3) - total (G $ 78)	3,8	149,9	3,1	294,1	2,75	506,3
(4) - per cap. ($ 78)	1,4	177	1,2	229	1,5	308
(5) ENERGY CONSUMPTION						
(6) Elasticities						
(7) . commercial /p.c.	1,91		1,26		0,91	
(8) . total	1,03		0,97		0,76	
(9) Total (Mtoe)	3,9	239,2	3,0	455	2,1	694
(10) . commercial	5,1	95,3	3,4	198	2,6	332
(11) . non-commercial	3,2	143,9	2,7	257	1,7	362
(12) Per capita (toe)	1,5	0,282	1,0	0,354	0,9	0,422
(13) . commercial	2,7	0,112	1,47	0,154	1,37	0,202
(14) . non-commercial	0,9	0,170	0,7	0,20	0,5	0,22

(Mtoe)	SNF	PP	NG	HY	NU	NS	FW	VAW	CEC	NCEC	PEC
1978											
(15) Consumption	49,2	27,7	6,5	11,9		-	54,0	89,9	95,3	143,9	239,2
(16) Production	49,2	11,4	8,6								
(17) + Imports		20,7(*)									
(18) - Exports			2,1								
2000											
(19) Consumption	89	38	9	35	15	12	92	165	198	257	455
(20) Production											
(21) + Imports											
(22) - Exports											
2020											
(23) Consumption	166	28	5	59	33	41	128	234	332	362	694
(24) Production											
(25) + Imports											
(26) - Exports											

(*) of which balance factor BF : 1978 = 4,4 Mtoe ; 2000 =   Mtoe ; 2020 =   Mtoe.

// FORECAST TABLES //

SCENARIO :   II "INCREASED TENSIONS"

REGION :   /SOUTH-EAST ASIA/ (R8)

	1960-1978 (%)	1978	1978-2000 (%)	2000	2000-2020 (%)	2020
(1) POPULATION (M.inhb)	2,4	337,6	1,8	502,7	1,15	631,3
(2) ECONOMIC GROWTH						
(3) - total (G $ 78)	7,2	220,2	5,1	662,6	1,7	1363,6
(4) - per cap. ($ 78)	4,7	652	3,25	1318	2,5	2160
(5) ENERGY CONSUMPTION						
(6) Elasticities						
(7) . commercial /p.c.	1,37		0,85		0,61	
(8) . total	0,72		0,72		0,62	
(9) Total (Mtoe)	5,2	218,7	3,7	491	2,3	772
(10) . commercial	8,9	118,1	4,6	320	2,7	545
(11) . non-commercial	2,6	100,6	2,4	171	1,4	227
(12) Per capita (toe)	2,7	0,648	1,9	0,977	1,1	1,223
(13) . commercial	6,4	0,350	2,76	0,637	1,53	0,863
(14) . non-commercial	0,2	0,298	0,6	0,34	0,3	0,36

(Mtoe)	SMF	PP	NG	HY	NU	NS	FW	VAW	CEC	NCEC	PEC
/1978/											
(15) Consumption	13,9	92,4	7,2	4,6	-	-	62,5	38,1	118,1	100,6	218,7
(16) Production	11,8	101,1(*)	18,0								
(17) + Imports	2,6	85,2									
(18) - Exports	0,5	73,5	10,8								
/2000/											
(19) Consumption	47	168	34	21	33	17	95	76	320	171	491
(20) Production											
(21) + Imports											
(22) - Exports											
/2020/											
(23) Consumption	71	201	61	77	80	55	108	119	545	227	772
(24) Production											
(25) + Imports											
(26) - Exports											

(*) of which balance factor BF : 1978 = 20,4 Mtoe ; 2000 =   Mtoe ; 2020 =   Mtoe.

// FORECAST TABLES //

SCENARIO :   II "INCREASED TENSIONS"

REGION :   /CENTRALLY PLANNED ASIAN COUNTRIES/ (R9)

	1960-1978 (%)	1978	1978-2000 (%)	2000	2000-2020 (%)	2020
(1) POPULATION (M.inhb)	1,86	1034,5	1,25	1359,3	0,74	1576,3
(2) ECONOMIC GROWTH						
(3) - total (G $ 78)	5,1	247,1	3,8	558,7	2,25	873,3
(4) - per cap. ($ 78)	3,18	239	2,5	411	1,5	554
(5) ENERGY CONSUMPTION						
(6) Elasticities						
(7) . commercial /p.c.	1,98		1,2		1,0	
(8) . total	1,27		1,03		0,93	
(9) Total (Mtoe)	6,47	638,0	3,9	1479	2,1	2263
(10) . commercial	8,26	495,9	4,3	1248	2,3	1948
(11) . non-commercial	2,72	142,1	2,2	231	1,6	315
(12) Per capita (toe)	4,52	0,616	2,6	1,09	1,4	1,44
(13) . commercial	6,29	0,479	3,0	0,918	1,5	1,236
(14) . non-commercial	0,86	0,137	1,0	0,17	0,8	0,20

(Mtoe)	SMF	PP	NG	HY	NU	NS	FW	VAW	CEC	NCEC	PEC
/1978/											
(15) Consumption	385,1	79,2	12,2	19,4	-	-	57,3	84,8	495,9	142,1	638,0
(16) Production	385,1	100,9(*)	12,2								
(17) + Imports											
(18) - Exports											
/2000/											
(19) Consumption	874	226	34	29	35	50	83	148	1248	231	1479
(20) Production											
(21) + Imports											
(22) - Exports											
/2020/											
(23) Consumption	1266	252	63	95	97	175	110	205	1948	315	2263
(24) Production											
(25) + Imports											
(26) - Exports											

(*) of which balance factor BF : 1978 = 21,7 Mtoe ; 2000 =   Mtoe ; 2020 =   Mtoe.

// FORECAST TABLES //

SCENARIO : II "INCREASED TENSIONS"

REGION : /LATIN AMERICA/ (R10)

	1960-1978 (%)	1978	1978-2000 (%)	2000	2000-2020 (%)	2020
(1) POPULATION (M.inhb)	2,8	344,5	2,3	564,1	1,4	802,4
(2) ECONOMIC GROWTH						
(3) - total (G $ 78)	5,6	482,8	4,8	1361,7	3,6	2767,5
(4) - per cap. ($ 78)	2,8	1402	2,5	2414	1,8	3449
(5) ENERGY CONSUMPTION						
(6) Elasticities						
(7) . commercial /p.c.	1,18		0,92		0,62	
(8) . total	0,77		0,79		0,72	
(9) Total (Mtoe)	4,3	370,0	3,8	837	2,6	1410
(10) . commercial	6,1	270,4	4,6	730	2,9	1298
(11) . non-commercial	1,2	99,6	0,3	107	0,2	112
(12) Per capita (toe)	1,5	1,074	1,5	1,484	0,8	1,757
(13) . commercial	3,25	0,785	2,30	1,294	1,12	1,617
(14) . non-commercial	-1,5	0,289	-1,9	0,19	-1,5	0,14

(Mtoe)	SMF	PP	NG	IIY	NU	NS	FW	VAW	CEC	NCEC	PEC
/1978/											
(15) Consumption	14,3	173,9	41,5	40,7		-	66,0	33,6	270,4	99,6	370,0
(16) Production	9,9	246,5(*)	41,5								
(17) + Imports	3,4	110,1									
(18) - Exports		59,2									
/2000/											
(19) Consumption	39	356	117	106	49	63	58	49	730	107	837
(20) Production											
(21) + Imports											
(22) - Exports											
/2020/											
(23) Consumption	54	512	205	203	126	198	47	65	1298	112	1410
(24) Production											
(25) + Imports											
(26) - Exports											

(*) of which balance factor BF : 1978 = 123,5 Mtoe ; 2000 =   Mtoe ; 2020 =   Mtoe.

FINAL STAGE OF ENERGY FORECASTS (DECENTRALISED STAGE)

This annex contains the finalised regional forecasts such as they appear as a result of the discussions of the RWTs. These tables provided the basis for all the following annexes.

One will therefore find 11 tables for Scenario (I) and 11 tables for Scenario (II).

// FORECAST TABLES //

Table : 1
Appendix : 10

SCENARIO : I "NORMATIVE-COOPERATION"

REGION : /NORTH AMERICA/ (R1)

05/82 (2)*

	1960-1978 (%)	1978	1978-2000 (%)	2000	2000-2020 (%)	2020
(1) POPULATION (M.inhb)	1,15	245	0,86	296	0,82	349
(2) ECONOMIC GROWTH						
(3) - total (G $ 78)	4,0	2348	2,5	4048	2,05	6073
(4) - per cap. ($ 78)	2,83	9584	1,6	13661	1,21	17391
(5) ENERGY CONSUMPTION						
(6) Elasticities						
(7) . commercial /p.c.						
(8) . total	0,78		0,40		0,74	
(9) Total (Mtoe)	3,1	2043	1,19	2648	1,51	3575
(10) . commercial	3,2	2008	1,12	2554	1,54	3469
(11) . non-commercial	-0,7	35	4,59	94	0,60	106
(12) Per capita (toe)	2,0	8,3	0,84	8,9	0,68	10,2
(13) . commercial	2,1	8,2	0,75	8,6	0,71	9,9
(14) . non-commercial	2,0	0,1	5,12	0,3	0	0,3

(Mtoe)	SNF	PP	NG	HY	NU	NS	FW	VAW	CEC	NCEC	PEC
/1978/											
(15) Consumption	358	914	528	129	77	2	30	5	2008	35	2043
(16) Production	380	574 (*)	527	129	77	2	30	5	1689	35	1724
(17) + Imports	5	343	2						350		350
(18) - Exports	27	2	1						30		30
/2000/											
(19) Consumption	898	552	526	179	288	111	84	10	2554	94	2648
(20) Production	1067	570 (*)	502	179	288	111	84	10	2717	94	2811
(21) + Imports	-	162	24						186		186
(22) - Exports	169	2	-						171		171
/2020/											
(23) Consumption	1774	234	460	199	484	318	91	15	3469	106	3575
(24) Production	1958	484 (*)	460	199	484	318	91	15	3903	106	4009
(25) + Imports	-	51	-						51		51
(26) - Exports	184	2	-						186		186

(*) of which balance factor BF : 1978 =   1   Mtoe ; 2000 =   178   Mtoe ; 2020 =   299   Mtoe.

190

// FORECAST TABLES //

Table   :   2
Appendix :   10

SCENARIO :   I "NORMATIVE-COOPERATION"

REGION :   /WESTERN EUROPE/ (R2*)

07/82 (4)

	1960-1978 (%)	1978	1978-2000 (%)	2000	2000-2020 (%)	2020
(1) POPULATION (M.inhb)	0,87	417,2	0,5	463	0,3	489
(2) ECONOMIC GROWTH						
(3) - total (G $ 78)	3,94	2596,0	3,05	5031	2	7528
(4) - per cap. ($ 78)	3,05	6222	2,6	10866	1,75	15395
(5) ENERGY CONSUMPTION						
(6) Elasticities						
(7) . commercial /p.c.	0,95		0,56		0,63	
(8) . total	0,91		0,62		0,68	
(9) Total (Mtoe)	3,6	1226,3	1,9	1871	1,35	2446
(10) . commercial	3,8	1196,5	1,9	1819	1,35	2380
(11) . non-commercial	-1,3	29,8	2,6	52	1,2	66
(12) Per capita (toe)	2,7	2,94	1,45	4,04	1,1	5,00
(13) . commercial	2,9	2,87	1,45	3,93	1,1	4,87
(14) . non-commercial	-2,2	0,07	2,1	0,11	0,8	0,13

(Mtoe)	SMF	PP	NG	HY	NU	NS	FW	VAW	CEC	NCEC	PEC
/1978/											
(15) Consumption	260,3	626,8	173,1	96,6	39,7	-	15,6	14,2	1196,5	29,8	1226,3
(16) Production	220,2	91,6	151,6	96,6	39,7	-	15,6	14,2	599,7	29,8	629,5
(17) + Imports	41,0	656,0(*)	21,5						718,5		718,5
(18) - Exports	0,9	54,0	-						54,9		54,9
/2000/											
(19) Consumption	477	629	281	134	243	55	23	29	1819	52	1871
(20) Production	305	158	121	134	243	55	23	29	1016	52	1068
(21) + Imports	172	565 (*)	160						897		897
(22) - Exports	-	40	-						40		40
/2020/											
(23) Consumption	730	446	303	171	615	115	24	42	2380	66	2446
(24) Production	360	115	103	171	615	115	24	42	1479	66	1545
(25) + Imports	370	399 (*)	200						969		969
(26) - Exports	-	20	-						20		20

(*) of which balance factor BF : 1978 =   66,8 Mtoe ; 2000 =   54   Mtoe ; 2020 =   48   Mtoe.

// FORECAST TABLES //

Table   :   3
Appendix :   10

SCENARIO :   I "NORMATIVE-COOPERATION"

REGION :   /PACIFIC INDUSTRIALIZED COUNTRIES/ (R3)

06/82 (3)

	1960-1978 (%)	1978	1978-2000 (%)	2000	2000-2020 (%)	2020
(1) POPULATION (M.inhb)	1,2	132,3	0,55	149,5	0,07	151,7
(2) ECONOMIC GROWTH						
(3) - total (G $ 78)	8,8	1017	3,9	2359	2,5	3865
(4) - per cap. ($ 78)	7,5	7680	3,3	15780	2,4	25480
(5) ENERGY CONSUMPTION						
(6) Elasticities						
(7) . commercial /p.c.	0,84		0,58		0,58	
(8) . total	0,67		0,64		0,60	
(9) Total (Mtoe)	5,9	397	2,5	683	1,5	920
(10) . commercial	6,3	395	2,5	680	1,5	916
(11) . non-commercial	-3,2	2	1,9	3	1,4	4
(12) Per capita (toe)	4,7	3,00	1,9	4,57	1,4	6,06
(13) . commercial	5,0	2,98	1,9	4,55	1,4	6,04
(14) . non-commercial	4,4	0,02	0	0,02	0	0,02

(Mtoe)	SMF	PP	NG	HY	NU	NS	FW	VAW	CEC	NCEC	PEC
/1978/											
(15) Consumption	82,8	247,3	23,6	26,8	14,5	-	0,5	1,5	395,0	2,0	397,0
(16) Production	71,1	24,6	9,9	26,8	14,5	-	0,5	1,5	146,9	2,0	148,9
(17) + Imports	18,7	251,8(*)	13,7						284,2		284,2
(18) - Exports	7,0	-	-						7,0		7,0
/2000/											
(19) Consumption	200	225	96	35	109	15	1	2	680	3	683
(20) Production	180	35	30	35	109	15	1	2	404	3	407
(21) + Imports	95	218 (*)	66						379		379
(22) - Exports	75	-	-						75		75
/2020/											
(23) Consumption	290	185	165	40	200	36	1	3	916	4	920
(24) Production	290	50	60	40	200	36	1	3	676	4	680
(25) + Imports	120	162 (*)	105						387		387
(26) - Exports	120	-	-						120		120

(*) of which balance factor BF : 1978 =   29,1 Mtoe ; 2000 =   28   Mtoe ; 2020 =   27   Mtoe.

// FORECAST TABLES //

Table : 4
Appendix : 10

SCENARIO :   I   "NORMATIVE-COOPERATION"

REGION :   /EASTERN EUROPE/ (R4)

05/1982 (3)

	1960-1978 (%)	1978	1978-2000 (%)	2000	2000-2020 (%)	2020
(1) POPULATION (M.inhb)	0,95	372,4	0,55	420	0,5	462
(2) ECONOMIC GROWTH						
(3) - total (G $ 78)	6,1					
(4) - per cap. ($ 78)	5,1					
(5) ENERGY CONSUMPTION						
(6) Elasticities						
(7) . commercial /p.c.						
(8) . total						
(9) Total (Mtoe)	4,4	1413	2,5	2440	1,7	3400
(10) . commercial	4,6	1367	2,6	2405	1,7	3370
(11) . non-commercial	0	46	-1,2	35	-0,8	30
(12) Per capita (toe)	3,4	3,79	2,0	5,81	1,2	7,35
(13) . commercial	3,6	3,67	2,0	5,73	1,2	7,29
(14) . non-commercial	-1,3	0,12	-1,8	0,08	-1,4	0,06

(Mtoe)	SMF	PP	NG	HY	NU	NS	FW	VAW	CEC	NCEC	PEC
/1978/											
(15) Consumption	574,6	403,9	334,4	42,2	11,7	-	30,2	15,8	1366,8	46	1412,8
(16) Production	599,2	574,6(*)	343,7	42,2	11,7	-	30,2	15,8	1571,4	46	1617,4
(17) + Imports	1,3	26,0	8,1						35,4		35,4
(18) - Exports	25,9	57,9	17,3						101,1		101,1
/2000/											
(19) Consumption	715	585	770	70	240	25	25	10	2405	35	2440
(20) Production	790	595 (*)	835	70	240	25	25	10	2555	35	2590
(21) + Imports	-	70	15						85		85
(22) - Exports	75	-	80						155		155
/2020/											
(23) Consumption	930	530	1100	130	600	80	20	10	3370	30	3400
(24) Production	1000	520 (*)	1200	130	600	80	20	10	3530	30	3560
(25) + Imports	-	60	20						80		80
(26) - Exports	70	-	120						190		190

(*) of which balance factor BF : 1978 =   138,9   Mtoe   ; 2000 =   80   Mtoe   ; 2020 =   50   Mtoe.

// FORECAST TABLES //

Table : 5
Appendix : 10

SCENARIO :   I   "NORMATIVE-COOPERATION"

REGION :   /NORTH AFRICA / MIDDLE EAST/ (R5*)

06/1982 (2)

	1960-1978 (%)	1978	1978-2000 (%)	2000	2000-2020 (%)	2020
(1) POPULATION (M.inhb)	2,8	183,8	2,55	320	1,9	465
(2) ECONOMIC GROWTH						
(3) - total (G $ 78)	9,1	291,9	6,0	1048	5,1	2824
(4) - per cap. ($ 78)	6,1	1588	3,3	3275	3,1	6073
(5) ENERGY CONSUMPTION						
(6) Elasticities						
(7) . commercial /p.c.	1,10		1,36		0,84	
(8) . total	0,89		1,11		0,86	
(9) Total (Mtoe)	8,1	134,5	6,65	555	4,4	1309
(10) . commercial	9,7	116,5	7,15	533	4,5	1286
(11) . non-commercial	2,8	18,0	0,9	22	0,2	23
(12) Per capita (toe)	5,2	0,732	4,0	1,734	2,5	2,815
(13) . commercial	6,7	0,634	4,5	1,665	2,6	2,766
(14) . non-commercial	0	0,098	-1,6	0,069	-1,7	0,049

(Mtoe)	SMF	PP	NG	HY	NU	NS	FW	VAW	CEC	NCEC	PEC
/1978/											
(15) Consumption	2,3	79,2	29,2	5,8	-	-	7,7	10,3	116,5	18,0	134,5
(16) Production	1,1	1182,4(*)	58,6	5,8	-	-	7,7	10,3	1247,9	18,0	1265,9
(17) + Imports	1,2	-	-						1,2		1,2
(18) - Exports	-	1002,1	27,3						1029,4		1029,4
/2000/											
(19) Consumption	10	335	155	17	9	7	9	13	533	22	555
(20) Production	5	1340 (*)	261 (*)	17	9	7	9	13	1639	22	1661
(21) + Imports	5	-	-						5		5
(22) - Exports	-	930	95						1025		1025
/2020/											
(23) Consumption	66	595	449	26	95	55	9	14	1286	23	1309
(24) Production	8	1310 (*)	515 (*)	26	95	55	9	14	2009	23	2032
(25) + Imports	58	-							58		58
(26) - Exports	-	665	41						706		706

(*) of which balance factor BF : 1978 =   103,2 Mtoe   ; 2000 =   86   Mtoe   ; 2020 =   75   Mtoe.

// FORECAST TABLES //

Table   : 6
Appendix : 10

SCENARIO : I "NORMATIVE-COOPERATION"

REGION : /AFRICA SOUTH OF THE SAHARA/ (R6)
(excluding South Africa)   03/1982 (2)

	1960-1978 (%)	1978	1978-2000 (%)	2000	2000-2020 (%)	2020
(1) POPULATION (M.inhb)	2,6	320,1	3,1	630	2,7	1071
(2) ECONOMIC GROWTH						
(3) - total (G $ 78)	4,0	122,3	4,5	322	3,5	641
(4) - per cap. ($ 78)	1,4	382	1,3	511	0,8	598
(5) ENERGY CONSUMPTION						
(6) Elasticities						
(7) . commercial /p.c.	2,01		2,38		2,75	
(8) . total	0,75		0,78		0,83	
(9) Total (Mtoe)	3,0	122,4	3,5	259	2,9	463
(10) . commercial	5,5	27,0	6,3	103	4,9	270
(11) . non-commercial	2,5	95,4	2,3	156	1,1	193
(12) Per capita (toe)	0,4	0,38	0,3	0,41	0,2	0,43
(13) . commercial	2,9	0,08	3,1	0,16	2,2	0,25
(14) . non-commercial	0	0,30	-0,8	0,25	-1,6	0,18

(Mtoe)	SMF	PP	NG	HY	NU	NS	FW	VAW	CEC	NCEC	PEC
1978											
(15) Consumption	3,2	16,8	0,6	6,4	-	-	84,0	11,4	27,0	95,4	122,4
(16) Production	3,0	113,3	0,6	6,4	-	-	84,0	11,4	123,3	95,4	218,7
(17) + Imports	0,2	6,8	-						7,0		7,0
(18) - Exports	-	97,5	-						97,5		97,5
2000											
(19) Consumption	10	54	10	20	1	8	133	23	103	156	259
(20) Production	8	180	10	20	1	8	133	23	227	156	383
(21) + Imports	2	15	-						17		17
(22) - Exports	-	126	-						126		126
2020											
(23) Consumption	20	116	30	70	10	24	157	36	270	193	463
(24) Production	15	250	30	70	10	24	157	36	399	193	492
(25) + Imports	5	25	-						30		30
(26) - Exports	-	129	-						129		129

(*) of which balance factor BF : 1978 = 5,8 Mtoe ; 2000 = 15 Mtoe ; 2020 = 30 Mtoe.

// FORECAST TABLES //

Table   : 7
Appendix : 10

SCENARIO : I "NORMATIVE-COOPERATION"

REGION : /AFRICA SOUTH OF THE SAHARA/ (R6*)
(including South Africa)   06/82 (3)

	1960-1978 (%)	1978	1978-2000 (%)	2000	2000-2020 (%)	2020
(1) POPULATION (M.inhb)	2,6	347,8	3,1	681	2,7	1152
(2) ECONOMIC GROWTH						
(3) - total (G $ 78)	4,3	166,1	4,5	437	3,5	870
(4) - per cap. ($ 78)	1,7	478	1,3	642	0,8	755
(5) ENERGY CONSUMPTION						
(6) Elasticities						
(7) . commercial /p.c.	1,00		1,69		3,13	
(8) . total	0,74		0,84		1,11	
(9) Total (Mtoe)	3,2	176,6	3,8	403	3,9	870
(10) . commercial	4,5	77,5	5,4	244	5,2	675
(11) . non-commercial	2,5	99,1	2,2	159	1,0	195
(12) Per capita (toe)	0,6	0,508	0,7	0,592	1,2	0,755
(13) . commercial	1,7	0,223	2,7	0,358	2,5	0,586
(14) . non-commercial	-0,1	0,285	-0,9	0,234	-1,6	0,169

(Mtoe)	SMF	PP	NG	HY	NU	NS	FW	VAW	CEC	NCEC	PEC
1978											
(15) Consumption	42,6	27,4	0,6	6,9	-	-	86,5	12,6	77,5	99,1	176,6
(16) Production	53,0(*)	113,3(*)	0,6	6,9	-	-	86,5	12,6	173,8	99,1	272,9
(17) + Imports	0,2	22,4	-						22,6		22,6
(18) - Exports	9,5	97,5	-						107,0		107,0
2000											
(19) Consumption	130	64	10	21	6	13	135	24	244	159	403
(20) Production	188 (*)	180 (*)	10	21	6	13	135	24	418	159	577
(21) + Imports	2	30	-						32		32
(22) - Exports	55	126	-						181		181
2020											
(23) Consumption	383	126	30	72	25	39	158	37	675	195	870
(24) Production	468 (*)	250 (*)	30	72	25	39	158	37	884	195	1079
(25) + Imports	5	40	-						45		45
(26) - Exports	80	129	-						209		209

(*) of which balance factor BF : 1978 = 11,9 Mtoe ; 2000 = 25 Mtoe ; 2020 = 45 Mtoe.

of which { SMF 1,1 ; PP 10,8 }   of which { SMF 5 ; PP 20 }   of which { SMF 10 ; PP 35 }

// FORECAST TABLES //

SCENARIO :   I "NORMATIVE-COOPERATION"

REGION :   /SOUTH ASIA/ (R7)

06/1982 (2)

	1960-1978 (%)	1978	1978-2000 (%)	2000	2000-2020 (%)	2020
(1) POPULATION (M.inhb)	2,3	848,6	1,9	1284	1,25	1644
(2) ECONOMIC GROWTH						
(3) - total (G $ 78)	3,8	149,9	4,1	364,7	3,5	721,7
(4) - per cap. ($ 78)	1,4	177	2,2	284	2,2	439
(5) ENERGY CONSUMPTION						
(6) Elasticities						
(7) . commercial /p.c.	2,00		1,43		1,00	
(8) . total	0,87		0,85		0,77	
(9) Total (Mtoe)	3,3	214,7	3,5	461	2,7	783
(10) . commercial	5,2	96,3	5,1	288	3,5	569
(11) . non-commercial	2,2	118,4	1,7	173	1,1	214
(12) Per capita (toe)	1,0	0,252	1,6	0,359	1,4	0,476
(13) . commercial	2,8	0,113	3,15	0,224	2,2	0,346
(14) . non-commercial	-0,15	0,139	-0,15	0,135	-0,15	0,130

(Mtoe)	SMF	PP	NG	HY	NU	NS	FW	VAW	CEC	NCEC	PEC
/1978/											
(15) Consumption	48,2	29,3	5,4	12,8	0,6	-	71,2	47,2	96,3	118,4	214,7
(16) Production	49,2(*)	12,4	8,0	12,8	0,6	-	71,2	47,2	83,0	118,4	201,4
(17) + Imports	-	21,7(*)	-						21,7	118,4	21,7
(18) - Exports	-	-							2,6		2,6
/2000/											
(19) Consumption	129	85	30	27	8	9	95	78	288	173	461
(20) Production	132 (*)	55 (*)	40 (*)	27	8	9	95	78	271	173	444
(21) + Imports	-	42	-						42		42
(22) - Exports	1	-	5						6		6
/2020/											
(23) Consumption	246	128	60	54	24	57	107	107	569	214	783
(24) Production	254 (*)	85 (*)	80 (*)	54	24	57	107	107	554	214	768
(25) + Imports	-	61	-						61		61
(26) - Exports	4	-	10						14		14

(*) of which balance factor BF : 1978 = 5,8 Mtoe ; 2000 = 19 Mtoe ; 2020 = 32 Mtoe.
of which {1 SMF / 4,8 PP} ; of which {2 SMF / 12 PP / 5 NG} ; of which {4 SMF / 18 PP / 10 NG}

// FORECAST TABLES //

SCENARIO :   I "NORMATIVE-COOPERATION"

REGION :   /SOUTH-EAST ASIA/ (R8)

06/82 (2)

	1960-1978 (%)	1978	1978-2000 (%)	2000	2000-2020 (%)	2020
(1) POPULATION (M.inhb)	2,4	337,6	1,8	503	1,15	631
(2) ECONOMIC GROWTH						
(3) - total (G $ 78)	7,2	220,2	6,2	824	4,2	1848
(4) - per cap. ($ 78)	4,7	652	4,3	1639	3,0	2960
(5) ENERGY CONSUMPTION						
(6) Elasticities						
(7) . commercial /p.c.	1,37		1,07		0,73	
(8) . total	0,72		0,74		0,69	
(9) Total (Mtoe)	5,2	218,7	4,6	584	2,9	1032
(10) . commercial	8,9	118,1	6,5	472	3,4	920
(11) . non-commercial	2,6	100,6	0,5	112	0	112
(12) Per capita (toe)	2,7	0,648	2,7	1,161	1,7	1,635
(13) . commercial	6,4	0,350	4,6	0,938	2,2	1,458
(14) . non-commercial	0,2	0,298	-1,3	0,223	-1,1	0,177

(Mtoe)	SMF	PP	NG	HY	NU	NS	FW	VAW	CEC	NCEC	PEC
/1978/											
(15) Consumption	13,9	92,4	7,2	4,6	-	-	62,5	38,1	118,1	100,6	218,7
(16) Production	11,8	101,1(*)	18,0	4,6	-	-	62,5	38,1	135,5	100,6	236,1
(17) + Imports	2,6	85,2	-						87,8		87,8
(18) - Exports	0,5	73,5	10,8						84,8		84,8
/2000/											
(19) Consumption	110	247	50	20	35	10	65	47	472	112	584
(20) Production	70	150 (*)	100	20	35	10	65	47	385	112	497
(21) + Imports	40	127	-						167		167
(22) - Exports	-	-	50						50		50
/2020/											
(23) Consumption	260	310	150	80	80	40	55	57	920	112	1032
(24) Production	130	120	200	80	80	40	55	57	650	112	762
(25) + Imports	130	225 (*)	-						355		355
(26) - Exports	-	-	50						50		50

(*) of which balance factor BF : 1978 = 20,4 Mtoe ; 2000 = 30 Mtoe ; 2020 = 35 Mtoe.

/ / FORECAST TABLES / /

Table   : 10
Appendix : 10

SCENARIO :   I "NORMATIVE-COOPERATION"

REGION :   /CENTRALLY PLANNED ASIAN COUNTRIES/ (R9)   03/82 (3)

(Mtoe)	1960-1978 (%)	1978	1978-2000 (%)	2000	2000-2020 (%)	2020
(1) POPULATION (M.inhb)	2,1	1032	1,25	1359	0,75	1576
(2) ECONOMIC GROWTH						
(3) - total (G $ 78)	5,3	246,8	3,8	558	4,0	1217
(4) - per cap. ($ 78)	3,2	239	2,5	411	3,2	772
(5) ENERGY CONSUMPTION						
(6) Elasticities						
(7) . commercial /p.c.	1,22		0,8		0,5	
(8) . total	0,83		0,61		0,38	
(9) Total (Mtoe)	4,4	638	2,3	1063	1,5	1440
(10) . commercial	6,1	419	3,3	859	2,3	1360
(11) . non-commercial	2,2	219	-0,3	204	-5,3	80
(12) Per capita (toe)	2,3	0,62	1,1	0,78	0,8	0,91
(13) . commercial	3,9	0,41	2,0	0,63	1,4	0,86
(14) . non-commercial	0,1	0,21	-1,5	0,15	-5,3	0,05

(Mtoe)	SNF	PP	NG	HY	NU	NS	FW	VAW	CEC	NCEC	PEC
/1978/											
(15) Consumption	305,8	88,8	10	14,4	-	-	126	93	419	219	638
(16) Production	307	104 (*)	12 (*)	14	-	-	126	93	437	219	656
(17) + Imports	-	1	-						1		1
(18) - Exports	1	13	-						14		14
/2000/											
(19) Consumption	510	209	77	50	7	6	122	82	859	204	1063
(20) Production	530	315 (*)	85 (*)	50	7	6	122	82	993	204	1197
(21) + Imports	-	3	-						3		3
(22) - Exports	20	100	-						120		120
/2020/											
(23) Consumption	700	290	160	150	30	30	50	30	1360	80	1440
(24) Production	780	400 (*)	300	150	30	30	50	30	1690	80	1770
(25) + Imports	-	5	-						5		5
(26) - Exports	80	100	100						280		280

(*) of which balance factor BF : 1978 = 5 Mtoe of which { 2 NG / 3 PP ; 2000 = 17 Mtoe of which { 8 NG / 9 PP ; 2020 = 55 Mtoe. of which { 40 NG / 15 PP

/ / FORECAST TABLES / /

Table   : 11
Appendix : 10

SCENARIO :   I "NORMATIVE-COOPERATION"

REGION :   /LATIN AMERICA/ (R10)   08/82 (4)

(Mtoe)	1960-1978 (%)	1978	1978-2000 (%)	2000	2000-2020 (%)	2020
(1) POPULATION (M.inhb)	2,8	344,5	2,3	564	1,8	802
(2) ECONOMIC GROWTH						
(3) - total (G $ 78)	5,6	482,8	5,8	1685,4	4,5	4082
(4) - per cap. ($ 78)	2,8	1402	3,5	2988	2,7	5091
(5) ENERGY CONSUMPTION						
(6) Elasticities						
(7) . commercial /p.c.	1,29		0,94		0,81	
(8) . total	0,80		0,86		0,84	
(9) Total (Mtoe)	4,5	356,2	5,0	1036	3,8	2182
(10) . commercial	6,5	288,7	5,6	966	4,0	2122
(11) . non-commercial	0	67,5	0,2	70	-0,8	60
(12) Per capita (toe)	1,7	1,034	2,6	1,836	2,0	2,721
(13) . commercial	3,6	0,838	3,3	1,712	2,2	2,646
(14) . non-commercial	-2,7	0,196	-2,1	0,124	-2,5	0,075

(Mtoe)	SNF	PP	NG	HY	NU	NS	FW	VAW	CEC	NCEC	PEC
/1978/											
(15) Consumption	12,2	167,9	50,5	50,8	1,8	5,5	55,3	12,2	288,7	67,5	356,2
(16) Production	7,8	240,3	52,4	50,8	1,8	5,8	55,3	13,2	358,9	68,5	427,4
(17) + Imports	4,7	66	-						70,7		70,7
(18) - Exports	-	75	0,2						75,2		75,2
/2000/											
(19) Consumption	62	472	160	155	32	85	49	21	966	70	1036
(20) Production	55	541	189	155	32	90	49	21	1062	70	1132
(21) + Imports	10	40	2						52		52
(22) - Exports	3	40	25						68		68
/2020/											
(23) Consumption	300	732	280	437	145	228	30	30	2122	60	2182
(24) Production	290	822	335	437	145	240	30	30	2269	60	2329
(25) + Imports	20	20	-						40		40
(26) - Exports	10	20	40						70		70

(*) of which balance factor BF : 1978 = 64,6 Mtoe ; 2000 = 80 Mtoe ; 2020 = 117 Mtoe.

/ FORECAST TABLES /

Table : 12
Appendix : 10

SCENARIO : II "INCREASED TENSIONS"

REGION : /NORTH AMERICA/ (R1)

	1960-1978 (%)	1978	1978-2000 (%)	2000	2000-2020 (%)	2020
(1) POPULATION (M.inhb)	1,15	245	0,86	296	0,82	349
(2) ECONOMIC GROWTH						
(3) - total (G $ 78)	4,0	2348	2,0	3633	1,55	4941
(4) - per cap. ($ 78)	2,83	9584	1,1	12274	0,72	14158
(5) ENERGY CONSUMPTION						
(6) Elasticities						
(7) . commercial /p.c.						
(8) . total	0,78		0,33		0,74	
(9) Total (Mtoe)	3,1	2043	0,81	2439	1,17	3080
(10) . commercial	3,2	2008	0,72	2350	1,19	2977
(11) . non-commercial	-0,7	35	4,33	89	0,73	103
(12) Per capita (toe)	2,0	8,3	0,47	8,2	0,35	8,8
(13) . commercial	2,1	8,2	0,36	7,9	0,37	8,5
(14) . non-commercial	2,0	0,1	5,12	0,3	0	0,3

05/82 (2)*

(Mtoe)	SMF	PP	NG	HY	NU	NS	FW	VAW	CEC	NCEC	PEC
1978											
(15) Consumption	358	914	528	129	77	2	30	5	2008	35	2043
(16) Production	380	574 (*)	527	129	77	2	30	5	1689	35	1724
(17) + Imports	5	343	2						350		350
(18) - Exports	27	2	1						30		30
2000											
(19) Consumption	810	514	487	174	273	92	81	8	2350	89	2439
(20) Production	966	573(*)	475	174	273	92	81	8	2553	89	2642
(21) + Imports	0	120	12						132		132
(22) - Exports	156	2	-						158		158
2020											
(23) Consumption	1486	209	404	189	398	291	91	12	2977	103	3080
(24) Production	1654	443(*)	404	189	398	291	91	12	3379	103	3482
(25) + Imports	0	36							36		36
(26) - Exports	168	2	-						170		170

(*) of which balance factor BF : 1978 = 1 Mtoe ; 2000 = 177 Mtoe ; 2020 = 268 Mtoe.

/ FORECAST TABLES /

Table : 13
Appendix : 10

SCENARIO : II "INCREASED TENSIONS"

REGION : /WESTERN EUROPE/ (R2*)

	1960-1978 (%)	1978	1978-2000 (%)	2000	2000-2020 (%)	2020
(1) POPULATION (M.inhb)	0,87	417,2	0,5	463	0,3	489
(2) ECONOMIC GROWTH						
(3) - total (G $ 78)	3,94	2596,0	2,05	4059	1,5	5505
(4) - per cap. ($ 78)	3,05	6222	1,6	8767	1,25	11258
(5) ENERGY CONSUMPTION						
(6) Elasticities						
(7) . commercial /p.c.	0,95		0,71		0,73	
(8) . total	0,91		0,56		0,64	
(9) Total (Mtoe)	3,6	1226,3	1,45	1679	1,1	2090
(10) . commercial	3,8	1196,5	1,4	1625	1,1	2021
(11) . non-commercial	-1,3	29,8	2,7	54	1,2	69
(12) Per capita (toe)	2,7	2,94	0,95	3,63	0,8	4,27
(13) . commercial	2,9	2,87	0,9	3,51	0,8	4,13
(14) . non-commercial	-2,2	0,07	2,5	0,12	0,8	0,14

07/82 (4)

(Mtoe)	SMF	PP	NG	HY	NU	NS	FW	VAW	CEC	NCEC	PEC
1978											
(15) Consumption	260,3	626,8	173,1	96,6	39,7	-	15,6	14,2	1196,5	29,8	1226,3
(16) Production	220,2	91,6	151,6	96,6	39,7	-	15,6	14,2	599,7	29,8	629,5
(17) + Imports	41,0	656,0(*)	21,5						718,5		718,5
(18) - Exports	0,9	54,0							54,9		54,9
2000											
(19) Consumption	421	588	230	130	211	45	24	30	1625	54	1679
(20) Production	300	166	130	130	211	45	24	30	982	54	1036
(21) + Imports	121	511 (*)	100						732		732
(22) - Exports	-	40	-						40		40
2020											
(23) Consumption	605	418	232	165	505	95	25	44	2020	69	2089
(24) Production	345	132	102	165	505	95	25	44	1344	69	1413
(25) + Imports	260	346 (*)	130						736		736
(26) - Exports	-	20	-						20		20

(*) of which balance factor BF : 1978 = 66,8 Mtoe ; 2000 = 49 Mtce ; 2020 = 40 Mtoe.

// FORECAST TABLES //

SCENARIO :   II "INCREASED TENSIONS"

REGION :   /PACIFIC INDUSTRIALIZED COUNTRIES/ (R3)

06/82 (3)

	1960-1978 (%)	1978	1978-2000 (%)	2000	2000-2020 (%)	2020
(1) POPULATION (M.inhb)	1,2	132,3	0,55	149,5	0,07	151,7
(2) ECONOMIC GROWTH						
(3) – total (G $ 78)	8,8	1017	3	1949	1,5	2625
(4) – per cap. ($ 78)	7,5	7680	2,4	13035	1,4	17300
(5) ENERGY CONSUMPTION						
(6) Elasticities						
(7) . commercial /p.c.	0,84		0,58		0,64	
(8) . total	0,67		0,67		0,67	
(9) Total (Mtoe)	5,9	397	2,0	615	1,0	750
(10) . commercial	6,3	395	2,0	611	1,0	744
(11) . non-commercial	-3,2	2	3,2	4	2,0	6
(12) Per capita (toe)	4,7	3,00	1,4	4,11	0,9	4,94
(13) . commercial	5,0	2,98	1,4	4,08	0,9	4,90
(14) . non-commercial	4,4	0,02	0	0,03	2,0	0,04

(Mtoe)	SMF	PP	NG	HY	NU	NS	FW	VAW	CEC	NCEC	PEC
1978/											
(15) Consumption	82,8	247,3	23,6	26,8	14,5	–	0,5	1,5	395,0	2,0	397,0
(16) Production	71,1	24,6	9,9	26,8	14,5	–	0,5	1,5	146,9	2,0	148,9
(17) + Imports	18,7	251,8(*)	13,7						284,2		284,2
(18) – Exports	7,0	–	–						7,0		7,0
2000/											
(19) Consumption	190	208	90	30	82	11	1	3	611	4	615
(20) Production	165	32	25	30	82	11	1	3	345	4	349
(21) + Imports	90	201 (*)	65						356		356
(22) – Exports	65	–	–						65		65
2020/											
(23) Consumption	250	160	140	35	130	29	2	4	744	6	750
(24) Production	240	45	50	35	130	29	2	4	529	6	535
(25) + Imports	110	136 (*)	90						336		336
(26) – Exports	100	–	–						100		100

(*) of which balance factor BF : 1978 = 29,1 Mtoe ; 2000 = 25   Mtoe ; 2020 = 21   Mtoe.

// FORECAST TABLES //

SCENARIO :   II "INCREASED TENSIONS"

REGION :   /EASTERN EUROPE/ (R4)

05/1982 (3)

	1960-1978 (%)	1978	1978-2000 (%)	2000	2000-2020 (%)	2020
(1) POPULATION (M.inhb)	0,95	372,4	0,55	420	0,5	462
(2) ECONOMIC GROWTH						
(3) – total (G $ 78)	6,1					
(4) – per cap. ($ 78)	5,1					
(5) ENERGY CONSUMPTION						
(6) Elasticities						
(7) . commercial /p.c.						
(8) . total						
(9) Total (Mtoe)	4,4	1413	1,9	2155	1,3	2775
(10) . commercial	4,6	1367	2,0	2115	1,3	2740
(11) . non-commercial	0	46	-0,6	40	-0,7	35
(12) Per capita (toe)	3,4	3,79	1,4	5,13	0,8	6,00
(13) . commercial	3,6	3,67	1,4	5,03	0,8	5,93
(14) . non-commercial	-1,3	0,12	-0,8	0,10	-1,8	0,07

(Mtoe)	SMF	PP	NG	HY	NU	NS	FW	VAW	CEC	NCEC	PEC
1978/											
(15) Consumption	574,6	403,9	334,4	42,2	11,7	–	30,2	15,8	1366,8	46	1412,8
(16) Production	599,2	574,6(*)	343,7	42,2	11,7	–	30,2	15,8	1571,4	46	1617,4
(17) + Imports	1,3	26,0	8,1						35,4		35,4
(18) – Exports	25,9	57,9	17,3						101,1		101,1
2000/											
(19) Consumption	635	545	670	65	180	20	25	15	2115	40	2155
(20) Production	665	520 (*)	720	65	180	20	25	15	2170	40	2210
(21) + Imports	–	70	5						75		75
(22) – Exports	30	–	55						85		85
2020/											
(23) Consumption	780	420	940	100	450	50	25	10	2740	35	2775
(24) Production	800	400 (*)	1000	100	450	50	25	10	2800	35	2835
(25) + Imports	–	50	10						60		60
(26) – Exports	20	–	70						90		90

(*) of which balance factor BF : 1978 = 138,9 Mtoe ; 2000 = 45   Mtoe ; 2020 = 30   Mtoe.

// FORECAST TABLES //

Table : 16
Appendix : 10

SCENARIO : II "INCREASED TENSIONS"

REGION : /NORTH AFRICA / MIDDLE EAST/ (R5)

06/1982 (2)

	1960-1978 (%)	1978	1978-2000 (%)	2000	2000-2020 (%)	2020
(1) POPULATION (M.inhb)	2,8	183,8	2,55	320	1,9	465
(2) ECONOMIC GROWTH						
(3) - total (G $ 78)	9,1	291,9	4,6	779	3,6	1577
(4) - per cap. ($ 78)	6,1	1588	2,0	2434	1,7	3391
(5) ENERGY CONSUMPTION						
(6) Elasticities						
(7) . commercial /p.c.	1,10					
(8) . total	0,89					
(9) Total (Mtoe)	8,1	134,5	4,8	377	3,1	696
(10) . commercial	9,7	116,5	5,2	352	3,2	666
(11) . non-commercial	2,8	18,0	1,5	25	0,9	30
(12) Per capita (toe)	5,2	0,732	2,2	1,178	1,2	1,497
(13) . commercial	6,7	0,634	2,5	1,100	1,3	1,432
(14) . non-commercial	0	0,098	-1,0	0,078	-0,9	0,065

(Mtoe)	SMF	PP	NG	HY	NU	NS	FW	VAW	CEC	NCEC	PEC
/1978/											
(15) Consumption	2,3	79,2	29,2	5,8	-	-	7,7	10,3	116,5	18,0	134,5
(16) Production	1,1	1182,4(*)	58,6(*)	5,8			7,7	10,3	1247,9	18,0	1265,9
(17) + Imports	1,2	-	-						1,2		1,2
(18) - Exports	-	1002,1	27,3						1029,4		1029,4
/2000/											
(19) Consumption	7	218	104	13	4	6	10	15	352	25	377
(20) Production	3	1070 (*)	190 (*)	13	4	6	10	15	1286	25	1311
(21) + Imports	4	-	-						4		4
(22) - Exports	-	757	75						832		832
/2020/											
(23) Consumption	41	287	246	20	33	39	11	19	666	30	696
(24) Production	6	990 (*)	350 (*)	20	33	39	11	19	1438	30	1468
(25) + Imports	35	-	-						35		35
(26) - Exports	-	548	83						631		631

(*) of which balance factor   BF : 1978 = 103,2   Mtoe ; 2000 = 106   Mtoe ; 2020 = 176   Mtoe.

// FORECAST TABLES //

Table : 17
Appendix : 10

SCENARIO : II "INCREASED TENSIONS"

REGION : /AFRICA SOUTH OF THE SAHARA/ (R6)
(excluding South Africa)

03/1982 (2)

	1960-1978 (%)	1978	1978-2000 (%)	2000	2000-2020 (%)	2020
(1) POPULATION (M.inhb)	2,6	320,1	3,1	630	2,7	1071
(2) ECONOMIC GROWTH						
(3) - total (G $ 78)	4,0	122,3	4,0	290	3,0	524
(4) - per cap. ($ 78)	1,4	382	0,8	460	0,3	489
(5) ENERGY CONSUMPTION						
(6) Elasticities						
(7) . commercial /p.c.	2,01					
(8) . total	0,75					
(9) Total (Mtoe)	3,0	122,4	3,2	244	2,7	414
(10) . commercial	5,5	27,0	4,4	69	4,7	174
(11) . non-commercial	2,5	95,4	2,8	175	1,6	240
(12) Per capita (toe)	0,4	0,38	0,1	0,39	0	0,39
(13) . commercial	2,9	0,08	1,2	0,11	1,9	0,16
(14) . non-commercial	0	0,30	-0,3	0,28	-1,1	0,23

(Mtoe)	SMF	PP	NG	HY	NU	NS	FW	VAW	CEC	NCEC	PEC
/1978/											
(15) Consumption	3,2	16,8	0,6	6,4	-	-	84,0	11,4	27,0	95,4	122,4
(16) Production	3,0	113,3	0,6	6,4	-	-	84,0	11,4	123,3	95,4	218,7
(17) + Imports	0,2	6,8	-						7,0		7,0
(18) - Exports	-	97,5	-						97,5		97,5
/2000/											
(19) Consumption	8	30	5	15	1	10	150	25	69	175	244
(20) Production	6	150	5	15	1	10	150	25	187	175	362
(21) + Imports	2	10							12		12
(22) - Exports	-	120							120		120
/2020/											
(23) Consumption	16	63	15	45	5	30	200	40	174	240	414
(24) Production	12	180	15	45	5	30	200	40	287	240	527
(25) · Imports	4	20							24		24
(26) - Exports	-	122							122		122

(*) of which balance factor   BF : 1978 = 5,8   Mtoe ; 2000 = 10   Mtoe ; 2020 = 15   Mtoe.

// FORECAST TABLES //

SCENARIO :   II "INCREASED TENSIONS"

REGION :   /AFRICA SOUTH OF THE SAHARA/ (R6*)
            (including South Africa)

06/82 (3)

	1960-1978 (%)	1978	1978-2000 (%)	2000	2000-2020 (%)	2020
(1) POPULATION (M.inhb)	2,6	347,8	3,1	681	2,7	1152
(2) ECONOMIC GROWTH						
(3) - total (G $ 78)	4,3	166,1	4,0	394	3,0	712
(4) - per cap. ($ 78)	1,7	478	0,9	578	0,3	618
(5) ENERGY CONSUMPTION						
(6) Elasticities						
(7) . commercial /p.c.	1,00		0,89		2,00	
(8) . total	0,74		0,83		1,50	
(9) Total (Mtoe)	3,2	176,6	3,3	360	3,2	682
(10) . commercial	4,5	77,5	3,9	181	4,5	439
(11) . non-commercial	2,5	99,1	2,7	179	1,5	243
(12) Per capita (toe)	0,6	0,508	0,2	0,529	0,6	0,592
(13) . commercial	1,7	0,223	0,8	0,266	1,8	0,381
(14) . non-commercial	-0,1	0,285	-0,4	0,263	-1,1	0,211

(Mtoe)	SMF	PP	NG	HY	NU	NS	FW	VAW	CEC	NCEC	PEC
/1978/											
(15) Consumption	42,6	27,4	0,6	6,9	–	–	86,5	12,6	77,5	99,1	176,6
(16) Production	53,0(*)	113,3(*)	0,6	6,9	–	–	86,5	12,8	173,8	99,1	272,9
(17) + Imports	0,2	22,4	–						22,6		22,6
(18) - Exports	9,5	97,5	–						107,0		107,0
/2000/											
(19) Consumption	104	37	5	16	6	13	152	27	181	179	360
(20) Production	146(*)	150(*)	5	16	6	13	152	27	336	179	515
(21) + Imports	2	20	–						22		22
(22) - Exports	40	120							160		160
/2020/											
(23) Consumption	260	63	15	46	15	40	202	41	439	243	682
(24) Production	314(*)	180(*)	15	46	15	40	202	41	610	243	853
(25) + Imports	4	20	–						24		24
(26) - Exports	50	122	–						172		172

(*) of which balance factor BF : 1978 = 11,9 Mtoe ; 2000 = 17 Mtoe ; 2020 = 23 Mtoe.
of which { SMF 1,1 / PP 10,8   of which { SMF 4 / PP 13   of which { SMF 8 / PP 15

// FORECAST TABLES //

SCENARIO :   II "INCREASED TENSIONS"

REGION :   /SOUTH ASIA/ (R7)

06/ 1982 (2)

	1960-1978 (%)	1978	1978-2000 (%)	2000	2000-2020 (%)	2020
(1) POPULATION (M.inhb)	2,3	848,6	1,9	1284	1,25	1644
(2) ECONOMIC GROWTH						
(3) - total (G $ 78)	3,8	149,9	3,1	294,1	2,45	506,3
(4) - per cap. ($ 78)	1,4	177	1,2	229	1,2	308
(5) ENERGY CONSUMPTION						
(6) Elasticities						
(7) . commercial /p.c.	2,00		1,43		1,00	
(8) . total	0,87		0,87		0,77	
(9) Total (Mtoe)	3,3	214,7	2,7	389	1,9	570
(10) . commercial	5,2	96,3	3,6	211	2,4	342
(11) . non-commercial	2,2	118,4	1,9	178	1,2	228
(12) Per capita (toe)	1,0	0,252	0,8	0,303	0,7	0,347
(13) . commercial	2,8	0,113	1,7	0,164	1,2	0,208
(14) . non-commercial	-0,15	0,139	0	0,139	0	0,139

(Mtoe)	SMF	PP	NG	HY	NU	NS	FW	VAW	CEC	NCEC	PEC
/1978/											
(15) Consumption	48,2	29,3	5,4	12,8	0,6	–	71,2	47,2	96,3	118,4	214,7
(16) Production	49,2(*)	12,4	8,0	12,8	0,6	–	71,2	47,2	83,0	118,4	201,4
(17) + Imports	–	21,7(*)	–						21,7		21,7
(18) - Exports	–	–	2,6						2,6		2,6
/2000/											
(19) Consumption	102	60	17	20	6	6	98	80	211	178	389
(20) Production	104 (*)	40 (*)	27 (*)	20	6	6	98	80	203	178	381
(21) + Imports	–	30	–						30		30
(22) - Exports	–	–	5						5		5
/2020/											
(23) Consumption	147	75	34	40	12	34	114	114	342	228	570
(24) Production	151 (*)	50 (*)	54 (*)	40	12	34	114	114	341	228	569
(25) + Imports	–	37	–						37		37
(26) - Exports	–	–	10						10		10

(*) of which balance factor BF : 1978 = 5,8 Mtoe ; 2000 = 17 Mtoe ; 2020 = 26 Mtoe.
of which { 1 SMF / 4,8 PP   of which { 2 SMF / 10 PP / 5 NG   of which { 4 SMF / 12 PP / 10 NG

// FORECAST TABLES //

SCENARIO :   II "INCREASED TENSIONS"

REGION : /SOUTH-EAST ASIA/ (R8)

Table   : 20
Appendix : 10

	1960-1978 (%)	1978	1978-2000 (%)	2000	2000-2020 (%)	2020	06/82 (2)*
(1) POPULATION (M.inhb)	2,4	337,6	1,8	503	1,15	631	
(2) ECONOMIC GROWTH							
(3) - total (G $ 78)	7,2	220,2	4,5	580	3,0	1048	
(4) - per cap. ($ 78)	4,7	652	2,6	1153	1,85	1661	
(5) ENERGY CONSUMPTION							
(6) Elasticities							
(7) . commercial /p.c.	1,37		1,00		0,68		
(8) . total	0,72		0,71		0,63		
(9) Total (Mtoe)	5,2	218,7	3,2	436	1,9	638	
(10) . commercial	8,9	118,1	4,5	311	2,4	500	
(11) . non-commercial	2,6	100,6	1	125	0,5	138	
(12) Per capita (toe)	2,7	0,648	1,3	0,866	0,8	1,011	
(13) . commercial	6,4	0,350	2,6	0,618	1,25	0,792	
(14) . non-commercial	0,2	0,298	-0,8	0,248	-0,6	0,219	

(Mtoe)	SNF	PP	NG	HY	NU	NS	FW	VAW	CEC	NCEC	PEC
/1978/											
(15) Consumption	13,9	92,4	7,2	4,6	-	-	62,5	38,1	118,1	100,6	218,7
(16) Production	11,8	101,1(*)	18,0	4,6	-	-	62,5	38,1	195,5	100,6	236,1
(17) + Imports	2,6	85,2	-						87,8		87,8
(18) - Exports	0,5	73,5	10,8						84,8		84,8
/2000/											
(19) Consumption	100	132	30	17	20	12	70	55	311	125	436
(20) Production	80	85 (*)	65	17	20	12	70	55	279	125	404
(21) + Imports	20	67	-						87		87
(22) - Exports	-	-	35						35		35
/2020/											
(23) Consumption	170	80	100	60	40	50	65	73	500	138	638
(24) Production	150	50 (*)	130	60	40	50	65	73	480	138	618
(25) + Imports	20	45	-						65		65
(26) - Exports	-	-	30						30		30

(*) of which balance factor   BF : 1978 =   20,4 Mtoe ; 2000 =   20   Mtoe ; 2020 =   15   Mtoe.

// FORECAST TABLES //

SCENARIO :   II "INCREASED TENSIONS"

REGION : /CENTRALLY PLANNED ASIAN CONTRIES/ (R9)

Table   : 21
Appendix : 10

	1960-1978 (%)	1978	1978-2000 (%)	2000	2000-2020 (%)	2020	03/82 (3)
(1) POPULATION (M.inhb)	2,1	1032	1,25	1359	0,75	1576	
(2) ECONOMIC GROWTH							
(3) - total (G $ 78)	5,3	246,8	1,35	332	2,35	528	
(4) - per cap. ($ 78)	3,2	239	0,1	244	1,6	335	
(5) ENERGY CONSUMPTION							
(6) Elasticities							
(7) . commercial /p.c.	1,22		1,0		0,8		
(8) . total	0,83		0,93		0,38		
(9) Total (Mtoe)	4,4	638	1,25	838	0,9	1006	
(10) . commercial	6,1	419	1,4	566	2,0	846	
(11) . non-commercial	2,2	219	1,0	272	-2,6	160	
(12) Per capita (toe)	2,3	0,62	0	0,62	0,2	0,64	
(13) . commercial	3,9	0,41	0,1	0,42	1,3	0,54	
(14) . non-commercial	0,1	0,21	-0,2	0,20	-3,4	0,10	

(Mtoe)	SNF	PP	NG	HY	NU	NS	FW	VAW	CEC	NCEC	PEC
/1978/											
(15) Consumption	305,8	88,8	10	14,4	-	-	126	93	419	219	638
(16) Production	307	104 (*)	12 (*)	14	-	-	126	93	437	219	656
(17) + Imports	-	1	-						1		1
(18) - Exports	1	13							14		14
/2000/											
(19) Consumption	375	128	28	29	3	3	190	82	566	272	838
(20) Production	385	158 (*)	32 (*)	29	3	3	190	82	610	272	882
(21) + Imports	-	2	-						2		2
(22) - Exports	10	25							35		35
/2020/											
(23) Consumption	480	196	90	60	10	10	110	50	846	160	1006
(24) Production	510	200 (*)	100 (*)	60	10	10	110	50	890	160	1050
(25) + Imports	-	5	-						5		5
(26) - Exports	30	-	-						30		30

(*) of which balance factor   BF : 1978 =   5   Mtoe ; 2000 =   11   Mtoe ; 2020 =   19   Mtoe.
                                    of which { 2 NG   of which { 4 NG   of which { 10 NG
                                             { 3 PP   { 7 PP   { 9 PP

// FORECAST TABLES //

Table   : 22
Appendix : 10

SCENARIO :   II "INCREASED TENSIONS"

REGION :   /LATIN AMERICA/ (R10)

08/82 (4)

	1960-1978 (%)	1978	1978-2000 (%)	2000	2000-2020 (%)	2020
(1) POPULATION (M.inhb)	2,8	344,5	2,3	564	1,8	802
(2) ECONOMIC GROWTH						
(3) - total (G $ 78)	5,6	482,8	4,3	1222	3,5	2435
(4) - per cap. ($ 78)	2,8	1402	2,0	2167	1,7	3036
(5) ENERGY CONSUMPTION						
(6) Elasticities						
(7) . commercial /p.c.	1,29		1,0		0,88	
(8) . total	0,80		0,88		0,86	
(9) Total (Mtoe)	4,5	356,2	3,8	816	3,0	1480
(10) . commercial	6,5	288,7	4,3	731	3,3	1400
(11) . non-commercial	0	67,5	1,1	85	-0,3	80
(12) Per capita (toe)	1,7	1,034	1,5	1,447	1,2	1,846
(13) . commercial	3,6	0,838	2,0	1,296	1,5	1,746
(14) . non-commercial	-2,7	0,196	-1,2	0,151	-2,0	0,100

(Mtoe)	SMF	PP	NG	HY	NU	NS	FW	VAW	CEC	NCEC	PEC
/1978/											
(15) Consumption	12,2	167,9	50,5	50,8	1,8	5,5	55,3	12,2	288,7	67,5	356,2
(16) Production	7,8	240,3	52,4	50,8	1,8	5,8	55,3	13,2	358,9	68,5	427,4
(17) + Imports	4,7	65	-						70,7		70,7
(18) - Exports	-	75	0,2						75,2		75,2
/2000/											
(19) Consumption	44	358	120	128	19	62	60	25	731	85	816
(20) Production	40	400	144	128	19	65	60	25	796	85	881
(21) + Imports	7	30	-						37		37
(22) - Exports	3	30	20						53		53
/2020/											
(23) Consumption	170	520	200	288	60	162	40	40	1400	80	1480
(24) Production	170	562	240	288	60	170	40	40	1490	80	1570
(25) + Imports	10	15	-						25		25
(26) - Exports	10	15	30						55		55

(*) of which balance factor BF : 1978 = 64,6 Mtoe ; 2000 = 49 Mtoe ; 2020 = 60 Mtoe.

Annex 11

ENERGY CONSUMPTIONS FORECASTS

Table   : 1
Appendix : 11

ENERGY CONSUMPTION FORECASTS (Mtoe)

- 2000 (I) -

(Mtoe)	SMF	O	NG	HY	NU	NS	FW	VAW	CEC	NCEC	PEC
R1	898	552	526	179	288	111	84	10	2554	94	2648
R2	477	629	281	134	243	55	23	29	1819	52	1871
R3	200	225	96	35	109	15	1	2	680	3	683
R4	715	585	770	70	240	25	25	10	2405	35	2440
R5	10	335	155	17	9	7	9	13	533	22	555
R6	130	64	10	21	6	13	135	24	244	159	403
R7	129	85	30	27	8	9	95	78	288	173	461
R8	110	247	50	20	35	10	65	47	472	112	584
R9	510	209	77	50	7	6	122	82	859	204	1063
R10	62	472	160	155	32	85	49	21	966	70	1036
WIC	1695	1416	903	349	645	186	110	42	5194	152	5346
IC	2410	2001	1673	419	885	211	135	52	7599	187	7786
TW	831	1402	482	289	92	125	473	264	3221	737	3958
TW-R9	321	1193	405	239	85	119	351	182	2362	533	2895
WORLD	3241	3403	2155	708	977	336	608	316	10820	924	11744

Table   : 2
Appendix : 11

ENERGY CONSUMPTION FORECASTS (Mtoe)

- 2000 (II) -

(Mtoe)	SMF	O	NG	HY	NU	NS	FW	VAW	CEC	NCEC	PEC
R1	810	514	487	174	273	92	81	8	2350	89	2439
R2	421	588	230	130	211	45	24	30	1625	54	1679
R3	190	208	90	30	82	11	1	3	611	4	615
R4	635	545	670	65	180	20	25	15	2115	40	2155
R5	7	218	104	13	4	6	10	15	352	25	377
R6	104	37	5	16	6	13	152	27	181	179	360
R7	102	60	17	20	6	6	98	80	211	178	389
R8	100	132	30	17	20	12	70	55	311	125	436
R9	375	128	28	29	3	3	190	82	566	272	838
R10	44	358	120	128	19	62	60	25	731	85	816
WIC	1517	1317	807	335	571	151	108	43	4698	151	4849
IC	2152	1862	1477	400	751	171	133	58	6813	191	7004
TW	636	926	304	222	53	99	578	282	2240	860	3100
TW-R9	261	798	276	193	50	96	388	200	1674	588	2262
WORLD	2788	2788	1781	622	804	270	711	340	9053	1051	10104

Table   : 1
Appendix : 11

ENERGY CONSUMPTION FORECASTS (Mtoe)

- 2020 (I) -

(Mtoe)	SMF	O	NG	HY	NU	NS	FW	VAW	CEC	NCEC	PEC
R1	1774	234	460	199	484	318	91	15	3469	106	3575
R2	730	446	303	171	615	115	24	42	2380	66	2446
R3	290	185	165	40	200	36	1	3	916	4	920
R4	930	530	1100	130	600	80	20	10	3370	30	3400
R5	66	595	449	26	95	55	9	14	1286	23	1309
R6	383	126	30	72	25	39	158	37	675	195	870
R7	246	128	60	54	24	57	107	107	569	214	783
R8	260	310	150	80	80	40	55	57	920	112	1032
R9	700	290	160	150	30	30	50	30	1360	80	1440
R10	300	732	280	437	145	228	30	30	2122	60	2182
WIC	3157	875	928	412	1314	484	117	61	7170	178	7348
IC	4087	1405	2028	542	1914	564	137	71	10540	208	10748
TW	1592	2171	1129	817	384	434	408	274	6527	682	7209
TW-R9	892	1881	969	667	354	404	358	244	5167	602	5769
WORLD	5679	3576	3157	1359	2298	998	545	345	17067	890	17957

Table : 4
Appendix : 11

ENERGY CONSUMPTION FORECASTS (Mtoe)

- 2020 (II) -

(Mtoe)	SMF	O	NG	HY	NU	NS	FW	VAW	CEC	NCEC	PEC
R1	1486	209	404	189	398	291	91	12	2977	103	3080
R2	605	418	232	165	505	95	25	44	2020	69	2089
R3	250	160	140	35	130	29	2	4	744	6	750
R4	780	420	940	100	450	50	25	10	2740	35	2775
R5	41	287	246	20	33	39	11	19	666	30	696
R6	260	63	15	46	15	40	202	41	439	243	682
R7	147	75	34	40	12	34	114	114	342	228	570
R8	170	80	100	60	40	50	65	73	500	138	638
R9	480	196	90	60	10	10	110	50	846	160	1006
R10	170	520	200	288	60	162	40	40	1400	80	1480
WIC	2585	787	776	390	1043	425	120	61	6006	181	6187
IC	3365	1207	1716	490	1493	475	145	71	8746	216	8962
TW	1024	1221	685	513	160	325	540	336	3928	876	4804
TW-R9	544	1025	595	453	150	315	430	286	3082	716	3798
WORLD	4389	2428	2401	1003	1653	800	685	407	12674	1092	13766

Table : 5
Appendix : 11

ENERGY CONSUMPTION FORECASTS

REGIONAL STRUCTURE %
- 2000 (I) -

(%)	SMF	O	NG	HY	NU	NS	FW	VAW	CEC	NCEC	PEC
R1	34	21	20	7	11	4	3	Ɛ	97	3	100
R2	25	34	15	7	13	3	1	2	97	3	100
R3	29	33	14	5	16	2	Ɛ	1	99	1	100
R4	29	24	32	3	10	1	1	Ɛ	99	1	100
R5	2	60	28	3	2	1	2	2	96	4	100
R6	32	16	3	5	2	3	33	6	61	39	100
R7	28	18	6	6	2	2	21	17	62	38	100
R8	19	42	9	3	6	2	11	8	81	19	100
R9	48	20	7	5	1	Ɛ	11	8	81	19	100
R10	6	46	15	15	3	8	5	2	93	7	100
WIC	32	26	17	7	12	3	2	1	97	3	100
IC	31	26	21	5	11	3	2	1	97	3	100
TW	21	36	12	7	2	3	12	7	81	19	100
TW-R9	11	41	14	8	3	4	12	7	81	19	100
WORLD	28	29	18	6	8	3	5	3	92	8	100

Table   6
Appendix : 11

ENERGY CONSUMPTION FORECASTS

REGIONAL STRUCTURE %

- 2000 (II) -

(%)	SMF	O	NG	HY	NU	NS	FW	VAW	CEC	NCEC	PEC
R1	33	21	20	7	11	4	3	1	96	4	100
R2	25	35	14	8	12	3	1	2	97	3	100
R3	31	34	14	5	13	2	ε	1	99	1	100
R4	30	25	31	3	8	1	1	1	98	2	100
R5	2	58	27	3	1	2	3	4	93	7	100
R6	29	10	1	4	2	4	42	8	50	50	100
R7	26	15	4	5	2	2	25	21	54	46	100
R8	23	30	7	4	4	3	16	13	71	29	100
R9	45	15	3	4	ε	ε	23	10	67	33	100
R10	5	44	15	16	2	8	7	3	90	10	100
WIC	31	27	17	7	12	3	2	1	97	3	100
IC	31	26	21	6	11	2	2	1	97	3	100
TW	20	30	10	7	2	3	19	9	72	28	100
TW-R9	12	35	12	9	2	4	17	9	74	26	100
WORLD	28	28	17	6	8	3	7	3	90	10	100

ENERGY CONSUMPTION FORECASTS

REGIONAL STRUCTURE %

- 2020 (I) -

(%)	SMF	O	NG	HY	NU	NS	FW	VAW	CEC	NCEC	PEC
R1	49	7	13	6	13	9	3	ε	97	3	100
R2	30	18	12	7	25	5	1	2	97	3	100
R3	32	20	18	4	22	4	ε	ε	100	ε	100
R4	27	16	32	4	18	2	1	ε	99	1	100
R5	5	46	34	2	7	4	1	1	98	2	100
R6	44	15	4	8	3	4	18	4	78	22	100
R7	31	16	8	7	3	7	14	14	72	28	100
R8	25	30	14	8	8	4	5	6	89	11	100
R9	49	20	11	10	2	2	4	2	94	6	100
R10	14	34	13	20	7	10	1	1	98	2	100
WIC	43	12	12	6	18	6	2	1	97	3	100
IC	38	13	19	5	18	5	1	1	98	2	100
TW	22	30	16	11	5	6	6	4	90	10	100
TW-R9	15	33	17	12	6	7	6	4	90	10	100
WORLD	32	20	17	7	13	6	3	2	95	5	100

ENERGY CONSUMPTION FORECASTS

REGIONAL STRUCTURE %

- 2020 (II) -

(%)	SMF	O	NG	HY	NU	NS	FW	VAW	CEC	NCEC	PEC
R1	48	7	13	6	13	10	3	ε	97	3	100
R2	29	20	11	8	24	5	1	2	97	3	100
R3	33	21	19	5	17	4	ε	1	99	1	100
R4	28	15	34	4	16	2	1	ε	99	1	100
R5	6	41	35	3	5	5	2	3	95	5	100
R6	38	9	2	7	2	6	30	6	64	36	100
R7	26	13	6	7	2	6	20	20	60	40	100
R8	27	13	16	9	6	8	10	11	79	21	100
R9	48	19	9	6	1	1	11	5	84	16	100
R10	11	35	14	19	4	11	3	3	94	6	100
WIC	42	13	12	6	17	7	2	1	97	3	100
IC	38	13	19	5	17	5	2	1	97	3	100
TW	21	26	14	11	3	7	11	7	82	18	100
TW-R9	14	27	16	12	4	8	11	8	81	19	100
WORLD	32	18	17	7	12	6	5	3	92	8	100

ENERGY CONSUMPTION FORECASTS

FUEL STRUCTURE %

- 2000 (I) -

(%)	SMF	O	NG	HY	NU	NS	FW	VAW	CEC	NCEC	PEC
R1	28	16	25	25	30	33	14	3	24	10	22
R2	15	18	13	19	25	16	4	9	17	6	16
R3	6	7	4	5	11	5	ε	1	6	ε	6
R4	22	17	36	10	24	8	4	3	22	4	21
R5	ε	10	7	2	1	2	2	4	5	2	5
R6	4	2	1	3	1	4	22	7	2	17	3
R7	4	3	1	4	1	3	15	25	3	19	4
R8	3	7	2	3	3	3	11	15	4	12	5
R9	16	6	4	7	1	2	20	26	8	22	9
R10	2	14	7	22	3	25	8	7	9	8	9
WIC	52	42	42	49	66	55	18	13	48	16	45
IC	74	59	78	59	90	63	22	16	70	20	66
TW	26	41	22	41	10	37	78	84	30	80	34
TW-R9	10	35	18	34	9	35	58	58	22	68	25
WORLD	100	100	100	100	100	100	100	100	100	100	100

ENERGY CONSUMPTION FORECASTS

FUEL STRUCTURE %

- 2000 (II) -

(%)	SMF	O	NG	HY	NU	NS	FW	VAW	CEC	NCEC	PEC
R1	29	18	27	28	34	34	11	2	26	9	24
R2	15	21	13	21	26	17	3	9	18	5	17
R3	7	7	5	5	10	4	ϵ	1	7	ϵ	6
R4	23	20	38	10	22	7	4	4	23	4	21
R5	ϵ	8	6	2	1	2	1	4	4	2	4
R6	4	1	ϵ	3	1	5	21	8	2	17	4
R7	4	2	1	3	1	2	14	24	2	17	4
R8	3	5	2	3	2	5	10	16	4	12	4
R9	13	5	2	5	1	1	27	24	6	26	8
R10	2	13	6	20	2	23	9	8	8	8	8
WIC	54	47	45	54	71	56	15	13	52	14	48
IC	77	67	83	64	93	63	19	17	75	18	69
TW	23	33	17	36	7	37	81	83	25	82	31
TW-R9	10	28	15	31	6	36	54	59	19	56	23
WORLD	100	100	100	100	100	100	100	100	100	100	100

ENERGY CONSUMPTION FORECASTS

FUEL STRUCTURE %

- 2020 (I) -

(%)	SMF	O	NG	HY	NU	NS	FW	VAW	CEC	NCEC	PEC
R1	31	7	14	15	21	32	17	4	20	12	20
R2	13	12	10	12	27	11	4	12	14	8	14
R3	5	5	5	3	9	4	ϵ	1	5	ϵ	5
R4	16	15	35	10	26	8	4	3	20	3	19
R5	12	17	14	2	4	5	2	4	8	2	7
R6	7	4	1	5	1	4	29	11	4	22	5
R7	4	4	2	4	1	6	20	31	3	24	4
R8	5	8	5	6	4	4	10	16	5	13	6
R9	12	8	5	11	1	3	9	9	8	9	8
R10	5	20	9	32	6	23	5	9	13	7	12
WIC	56	24	29	30	57	48	21	18	42	20	41
IC	72	39	64	40	83	56	25	21	62	23	60
TW	28	61	36	60	17	44	75	79	38	77	40
TW-R9	16	53	31	49	16	41	66	70	30	68	32
WORLD	100	100	100	100	100	100	100	100	100	100	100

ENERGY CONSUMPTION FORECASTS

FUEL STRUCTURE %

- 2020 (II) -

(%)	SMF	O	NG	HY	NU	NS	FW	VAW	CEC	NCEC	PEC
R1	34	9	17	19	24	36	13	3	23	9	22
R2	14	17	10	16	30	12	4	11	16	6	15
R3	5	7	6	3	8	4	E	1	6	1	6
R4	18	17	39	10	27	6	4	2	22	3	20
R5	1	12	10	2	2	5	2	5	5	3	5
R6	6	3	1	5	1	5	29	10	3	22	5
R7	3	3	1	4	1	5	17	28	3	21	4
R8	4	3	4	6	2	6	9	18	4	13	5
R9	11	8	4	6	1	1	16	12	7	15	7
R10	4	21	8	29	4	20	6	10	11	7	11
WIC	59	33	32	39	63	53	17	15	47	17	45
IC	77	50	71	49	90	59	21	17	69	20	65
TW	23	50	39	51	10	41	79	83	31	80	35
TW-R9	12	42	35	45	9	40	63	71	24	65	28
WORLD	100	100	100	100	100	100	100	100	100	100	100

ENERGY CONSUMPTIONS

- AVERAGE ANNUAL RATES OF GROWTH (%) -

- 1978 - 2000 (I) -

(%)	SMF	O	NG	HY	NU	NS	FW	VAW	CEC	NCEC	PEC
R1	4,3	-2,3	0	1,5	6,2	-	4,8	3,2	1,1	4,6	1,2
R2	2,8	0	2,2	1,5	8,5	-	1,7	3,4	1,9	2,5	1,9
R3	4,1	-0,4	6,5	1,2	9,8	-	0	3,2	2,5	1,9	2,5
R4	1,0	1,7	3,9	2,3	14,6	-	-0,8	-2,1	2,6	-1,2	2,5
R5	7,6	6,8	7,9	4,8	-	-	0,5	1,2	7,2	0,9	6,7
R6	5,2	4,0	11,0	5,1	-	-	2,1	2,8	5,3	2,2	3,8
R7	4,6	5,0	8,5	3,4	10,0	-	1,3	2,3	5,1	1,8	3,5
R8	9,8	4,6	9,3	6,5	-	-	0,1	1,0	6,5	0,5	4,6
R9	2,3	4,0	9,7	6,0	-	-	-0,2	-0,6	3,3	-0,3	2,3
R10	7,7	4,8	5,4	5,2	13,4	12,8	-0,5	2,6	5,6	0,2	5,0
WIC	3,8	-1,1	1,0	1,5	7,5	-	3,7	3,2	1,6	3,6	1,7
IC	2,8	-0,4	2,1	1,6	8,6	-	2,5	1,6	1,9	2,2	1,9
TW	3,5	5,1	7,3	5,2	16,8	-	0,7	1,0	5,2	0,8	4,0
TW - R9	6,5	5,3	7,0	5,0	16,4	-	1,0	1,9	6,1	1,3	4,7
WORLD	3,0	1,1	2,9	2,7	9,0	18,5	1,0	1,1	2,7	1,0	2,5

ENERGY CONSUMPTIONS

Table : 14
Appendix : 11

- AVERAGE ANNUAL RATES OF GROWTH (%) -
- 1978 - 2000 (II) -

(%)	SMF	O	NG	HY	NU	NS	FW	VAW	CEC	NCEC	PEC
R1	3,8	-2,6	-0,4	1,4	5,9	-	4,6	2,2	0,7	4,3	0,8
R2	2,2	-0,3	1,3	1,3	7,9	-	1,9	3,5	1,4	2,7	1,4
R3	3,8	-0,8	6,2	0,5	8,4	-	0	5,1	2,0	3,2	2,0
R4	0,5	1,4	3,2	2,0	13,1	-	-0,8	0	2,0	-0,6	1,9
R5	5,8	4,7	6,0	3,6	-	-	1,0	1,9	5,2	1,5	4,8
R6	4,1	1,4	7,6	3,8	-	-	2,6	3,4	3,9	2,7	3,3
R7	3,5	3,4	5,7	2,0	8,5	-	1,5	2,4	3,6	1,9	2,8
R8	9,3	1,7	6,8	5,7	-	-	0,5	1,7	4,5	1,0	3,2
R9	0,9	1,7	4,8	3,4	-	-	1,9	-0,5	1,4	1,0	1,2
R10	6,1	3,5	4,1	4,3	10,8	11,2	0,4	3,4	4,3	1,1	3,8
WIC	3,3	-1,4	0,5	1,3	6,9	-	3,7	3,3	1,2	3,6	1,2
IC	2,3	-0,8	1,5	1,4	7,8	-	2,4	2,1	1,4	2,3	1,4
TW	2,3	3,1	5,1	3,9	13,9	-	1,6	1,3	3,4	1,5	2,8
TW - R9	5,5	3,4	5,1	4,0	13,6	-	1,5	2,4	4,4	1,8	3,6
WORLD	2,3	0,2	2,0	2,1	8,1	17,3	1,7	1,4	1,8	1,6	1,8

ENERGY CONSUMPTIONS

Table : 15
Appendix : 11

- AVERAGE ANNUAL RATES OF GROWTH (%) -
- 2000 - 2020 (I) -

(%)	SMF	O	NG	HY	NU	NS	FW	VAW	CEC	NCEC	PEC
R1	3,5	-4,2	-0,7	0,5	2,6	5,4	0,4	2,0	1,5	0,6	1,5
R2	2,2	-1,7	0,4	1,2	4,8	3,8	0,2	1,9	1,4	1,2	1,3
R3	1,9	-1,0	2,7	0,7	3,1	4,5	0	2,0	1,5	1,4	1,5
R4	1,3	-0,5	1,8	3,1	4,7	6,0	-1,1	0	1,7	-0,8	1,7
R5	9,9	2,9	5,5	2,1	12,5	10,9	0	0,4	4,5	0,2	4,4
R6	5,6	3,4	5,6	6,3	7,4	5,6	0,8	2,2	5,2	1,0	3,9
R7	3,3	2,1	3,5	3,5	5,6	9,7	0,6	1,6	3,5	1,1	2,7
R8	4,4	1,1	5,6	7,2	4,2	7,2	-0,8	1,0	3,4	0	2,9
R9	1,6	1,7	3,7	5,6	7,5	8,4	-4,4	-4,9	2,3	-4,6	1,5
R10	8,2	2,2	2,8	5,3	7,8	5,1	-2,4	1,8	4,0	-0,8	3,8
WIC	3,2	-2,4	0,1	0,8	3,6	4,9	0,3	1,9	1,6	0,8	1,6
IC	2,7	-1,8	1,0	1,3	3,9	5,0	0,1	1,6	1,6	0,5	1,6
TW	3,3	2,2	4,3	5,3	7,4	6,4	-0,7	0,2	3,6	-0,4	3,0
TW - R9	5,2	2,3	4,5	5,3	7,4	6,3	0,1	1,5	4,0	0,6	3,5
WORLD	2,8	0,2	1,9	3,3	4,4	5,6	-0,5	0,4	2,3	-0,2	2,1

ENERGY CONSUMPTIONS

- AVERAGE ANNUAL RATES OF GROWTH (%) -
- 2000 - 2020 (II) -

(%)	SMF	O	NG	HY	NU	NS	FW	VAW	CEC	NCEC	PEC
R1	3,1	-4,4	-0,9	0,4	1,9	5,9	0,6	2,0	1,2	0,7	1,2
R2	1,8	-1,7	0	1,2	4,5	3,8	0,2	1,9	1,1	1,2	1,1
R3	1,4	-1,3	2,2	0,8	2,3	5,0	3,5	1,4	1,0	2,0	1,0
R4	1,0	-1,3	1,7	2,2	4,7	4,7	0	-2,0	1,3	-0,7	1,3
R5	9,2	1,4	4,4	2,2	11,1	9,8	0,5	1,2	3,2	0,9	3,1
R6	4,7	2,7	5,6	5,4	4,7	5,8	1,4	2,1	4,5	1,5	3,2
R7	1,8	1,1	3,5	3,5	3,5	9,1	0,8	1,8	2,4	1,2	1,9
R8	2,7	-2,5	6,2	6,5	3,5	7,4	-0,4	1,4	2,4	0,5	1,9
R9	1,2	2,2	6,0	3,7	6,2	6,2	-2,7	-2,4	2,0	-2,6	0,9
R10	7,0	1,9	2,6	4,1	5,9	4,9	-2,0	2,4	3,3	-0,3	3,0
WIC	2,7	-2,5	-0,2	0,8	3,1	5,3	0,5	1,8	1,2	0,9	1,2
IC	2,3	-2,1	0,8	1,0	3,5	5,2	0,4	1,0	1,3	0,6	1,2
TW	2,4	1,4	4,1	4,3	5,7	6,1	-0,3	0,9	2,8	0,1	2,2
TW - R9	3,7	1,3	3,9	4,4	5,6	6,1	0,5	1,8	3,1	1,0	2,6
WORLD	2,3	-0,7	1,5	2,4	3,7	5,6	-0,2	0,9	1,7	0,2	1,6

ENERGY CONSUMPTION PER CAPITA

COMMERCIAL AND NON-COMMERCIAL

(toe/p.c.)	COMMERCIAL CONSUMPTIONS						NON-COMMERCIAL CONSUMPTIONS					
	1960	1978	2000		2020		1960	1978	2000		2020	
			I	II	I	II			I	II	I	II
R1	5,63	8,20	8,63	7,94	9,94	8,53	0,20	0,14	0,32	0,30	0,30	0,30
R2	1,71	2,87	3,93	3,51	4,87	4,13	0,11	0,07	0,11	0,12	0,13	0,14
R3	1,06	2,98	4,56	4,10	6,02	4,89	0,03	0,02	0,02	0,03	0,03	0,04
R4	1,69	3,67	5,73	5,03	7,29	5,93	0,15	0,12	0,08	0,10	0,07	0,08
R5	0,20	0,63	1,66	1,10	2,77	1,43	0,10	0,10	0,07	0,08	0,05	0,07
R6	0,17	0,22	0,36	0,27	0,59	0,38	0,29	0,29	0,23	0,26	0,17	0,21
R7	0,07	0,11	0,22	0,16	0,35	0,21	0,14	0,14	0,14	0,14	0,13	0,14
R8	0,12	0,35	0,94	0,62	1,46	0,79	0,28	0,30	0,22	0,25	0,18	0,22
R9	0,20	0,41	0,63	0,42	0,86	0,54	0,21	0,21	0,15	0,20	0,05	0,10
R10	0,44	0,84	1,71	1,30	2,65	1,75	0,32	0,19	0,13	0,15	0,07	0,10
WIC	2,77	4,44	5,41	4,90	6,69	5,61	0,12	0,09	0,16	0,16	0,17	0,17
IC	2,43	4,20	5,51	4,94	6,88	5,71	0,13	0,10	0,14	0,14	0,13	0,14
TW	0,17	0,35	0,69	0,48	1,05	0,64	0,21	0,20	0,16	0,19	0,11	0,14
TW - R9	0,15	0,31	0,72	0,51	1,12	0,67	0,21	0,20	0,16	0,18	0,13	0,15
WORLD	0,91	1,43	1,79	1,50	2,21	1,64	0,19	0,17	0,15	0,17	0,12	0,14

Table   : 18
Appendix : 11

ENERGY CONSUMPTION PER CAPITA

- TOTAL AND RATES OF GROWTH -

	TOTAL CONSUMPTION (toe/p.c.)						RATES OF GROWTH (% /year)				
			2000		2020			1978-2000		2000-2020	
	1960	1978	I	II	I	II	1960-1978	I	II	I	II
R1	5,83	8,34	8,95	8,24	10,24	8,83	2,0	0,3	-0,1	0,7	0,3
R2	1,82	2,94	4,04	3,63	5,00	4,27	2,7	1,5	1,0	1,1	0,8
R3	1,09	3,00	4,58	4,13	6,05	4,93	5,8	1,9	1,5	1,4	0,9
R4	1,84	3,79	5,81	5,13	7,36	6,01	4,1	2,0	1,4	1,2	0,8
R5	0,30	0,73	1,73	1,18	2,82	1,50	5,1	4,0	2,2	2,5	1,2
R6	0,46	0,51	0,59	0,53	0,76	0,59	0,6	0,7	0,2	1,3	0,5
R7	0,21	0,25	0,36	0,30	0,48	0,35	1,0	1,7	0,8	1,4	0,8
R8	0,40	0,65	1,16	0,87	1,64	1,01	2,7	2,7	1,3	1,7	0,7
R9	0,41	0,62	0,78	0,62	0,91	0,64	2,3	1,0	0	0,8	0,2
R10	0,76	1,03	1,84	1,45	2,72	1,85	1,7	2,7	1,6	2,0	1,2
WIC	2,89	4,53	5,57	5,06	6,86	5,78	2,5	0,9	0,5	1,0	0,7
IC	2,56	4,30	5,65	5,08	7,01	5,85	2,9	1,2	0,8	1,1	0,7
TW	0,38	0,55	0,85	0,67	1,16	0,78	2,1	2,0	0,9	1,6	0,8
TW-R9	0,36	0,51	0,88	0,69	1,25	0,82	2,0	2,5	1,4	1,8	0,9
WORLD	1,10	1,60	1,94	1,67	2,33	1,78	2,1	0,9	0,2	0,9	0,3

ENERGY PRODUCTION FORECASTS

ENERGY PRODUCTION FORECASTS (Mtoe)

Table : 1
Appendix : 12

- 2000 (I) -

(Mteo)	SMF	O	NG	HY	NU	NS	FW	VAW	CEP	NCEP	PEP
R1	1067	570	502	179	288	111	84	10	2717	94	2811
R2	305	158	121	134	243	55	23	29	1016	52	1068
R3	180	35	30	35	109	15	1	2	404	3	407
R4	790	595	835	70	240	25	25	10	2555	35	2590
R5	5	1340	261	17	9	7	9	13	1639	22	1661
R6	188	180	10	21	6	13	135	24	418	159	577
R7	132	55	40	27	8	9	95	78	271	173	444
R8	70	150	100	20	35	10	65	47	385	112	497
R9	530	315	85	50	7	6	122	82	993	204	1197
R10	55	541	189	155	32	90	49	21	1062	70	1132
WIC	1732	763	653	349	645	186	110	42	4328	152	4480
IC	2522	1358	1488	419	885	211	135	52	6883	187	7070
TW	800	2581	685	289	92	130	473	264	4577	737	5314
TW-R9	270	2266	600	239	85	124	351	182	3584	533	4117
WORLD	3322	3939	2173	708	977	341	608	316	11460	924	12384

ENERGY PRODUCTION FORECASTS (Mtoe)

- 2000 (II) -

(Mtoe)	SMF	O	NG	HY	NU	NS	FW	VAW	CEP	NCEP	PEP
R1	966	573	475	174	273	92	81	8	2553	89	2642
R2	300	166	130	130	211	45	24	30	982	54	1036
R3	165	32	25	30	82	11	1	3	345	4	349
R4	665	520	720	65	180	20	25	15	2170	40	2210
R5	3	1070	190	13	4	6	10	15	1286	25	1311
R6	146	150	5	16	6	13	152	27	336	179	515
R7	104	40	27	20	6	6	98	80	203	178	381
R8	80	85	65	17	20	12	70	55	279	125	404
R9	385	158	32	29	3	3	190	82	610	272	882
R10	40	400	144	128	19	65	60	25	796	85	881
WIC	1571	771	630	335	571	151	108	43	4029	151	4180
IC	2236	1291	1350	400	751	171	133	58	6199	191	6390
TW	618	1903	463	222	53	102	578	282	3361	860	4221
TW-R9	233	1745	431	193	50	99	388	200	2751	588	3339
WORLD	2854	3194	1813	622	804	273	711	340	9560	1051	10611

ENERGY PRODUCTION FORECASTS (Mtoe)

- 2020 (I) -

(Mtoe)	SMF	O	NG	HY	NU	NS	FW	VAW	CEP	NCEP	PEP
R1	1958	484	460	199	484	318	91	15	3903	106	4009
R2	360	115	103	171	615	115	24	42	1479	66	1545
R3	290	50	60	40	200	36	1	3	676	4	680
R4	1000	520	1200	130	600	80	20	10	3530	30	3560
R5	8	1310	515	26	95	55	9	14	2009	23	2032
R6	468	250	30	72	25	39	158	37	884	195	1079
R7	254	85	80	54	24	57	107	107	554	214	768
R8	130	120	200	80	80	40	55	57	650	112	762
R9	780	400	300	150	30	30	50	30	1690	80	1770
R10	290	822	335	437	145	240	30	30	2269	60	2329
WIC	3061	649	623	412	1314	484	117	61	6543	178	6721
IC	4061	1169	1823	542	1914	564	137	71	10073	208	10281
TW	1477	2987	1460	817	384	446	408	274	7571	682	8253
TW-R9	697	2587	1160	667	354	416	358	244	5881	602	6483
WORLD	5538	4156	3283	1359	2298	1010	545	345	17644	890	18534

ENERGY PRODUCTION FORECASTS (Mtoe)

- 2020 (II) -

(Mtoe)	SMF	O	NG	HY	NU	NS	FW	VAW	CEP	NCEP	PEP
R1	1654	443	404	189	398	291	91	12	3379	103	3482
R2	345	132	102	165	505	95	25	44	1344	69	1413
R3	240	45	50	35	130	29	2	4	529	6	535
R4	800	400	1000	100	450	50	25	10	2800	35	2835
R5	6	990	350	20	33	39	11	19	1438	30	1468
R6	314	180	15	46	15	40	202	41	610	243	853
R7	151	50	54	40	12	34	114	114	341	228	569
R8	150	50	130	60	40	50	65	73	480	138	618
R9	510	200	100	60	10	10	110	50	890	160	1050
R10	170	562	240	288	60	170	40	40	1490	80	1570
WIC	2541	620	556	390	1043	425	120	61	5575	181	5756
IC	3341	1020	1556	490	1493	475	145	71	8375	216	8591
TW	999	2032	889	513	160	333	540	336	4926	876	5802
TW-R9	489	1832	789	453	150	323	430	286	4036	716	4752
WORLD	4340	3052	2445	1003	1653	808	685	407	13301	1092	14393

ENERGY PRODUCTION FORECASTS

REGIONAL STRUCTURE %

- 2000 (I) -

(%)	SMF	O	NG	HY	NU	NS	FW	VAW	CEP	NCEP	PEP
R1	38	20	18	7	10	4	3	E	97	3	100
R2	28	15	11	13	23	5	2	3	95	5	100
R3	43	9	7	9	27	4	E	1	99	1	100
R4	31	23	32	3	9	1	1	E	99	1	100
R5	E	80	16	1	1	E	1	1	98	2	100
R6	32	31	2	4	1	2	23	4	72	28	100
R7	29	12	9	6	2	3	21	18	61	39	100
R8	14	30	20	4	7	2	13	10	77	23	100
R9	44	26	7	4	1	1	10	7	83	17	100
R10	5	48	16	14	3	8	4	2	94	6	100
WIC	39	17	15	8	14	4	2	1	97	3	100
IC	36	19	21	6	12	3	2	1	97	3	100
TW	15	48	13	5	2	3	9	5	86	14	100
TW-R9	6	55	15	6	2	3	9	4	87	13	100
WORLD	27	32	17	6	8	3	5	2	93	7	100

ENERGY PRODUCTION FORECASTS

REGIONAL STRUCTURE %

- 2000 (II) -

(%)	SMF	O	NG	HY	NU	NS	FW	VAW	CEP	NCEP	PEP
R1	37	22	18	7	10	3	3	ε	97	3	100
R2	29	16	13	13	20	4	2	3	95	5	100
R3	47	9	8	9	23	3	ε	1	99	1	100
R4	30	24	32	3	8	1	1	1	98	2	100
R5	ε	82	14	1	ε	1	1	1	98	2	100
R6	29	29	1	3	1	2	30	5	65	35	100
R7	27	10	7	5	2	2	26	21	53	47	100
R8	20	21	16	4	5	3	17	14	69	31	100
R9	44	18	4	3	ε	ε	22	9	69	31	100
R10	5	45	16	15	2	7	7	3	90	10	100
WIC	37	18	15	8	14	4	3	1	96	4	100
IC	35	20	21	6	12	3	2	1	97	3	100
TW	15	45	11	5	1	2	14	7	79	21	100
TW-R9	7	52	13	6	1	3	12	6	82	18	100
WORLD	27	30	17	6	7	3	7	3	90	10	100

ENERGY PRODUCTION FORECASTS

REGIONAL STRUCTURE %

- 2020 (I) -

(%)	SMF	O	NG	HY	NU	NS	FW	VAW	CEP	NCEP	PEP
R1	49	12	12	5	12	8	2	ε	98	2	100
R2	23	7	7	11	40	7	2	3	95	5	100
R3	43	7	9	6	29	5	ε	1	99	1	100
R4	28	14	34	4	17	2	1	ε	99	1	100
R5	ε	65	25	1	5	3	ε	1	99	1	100
R6	43	23	3	7	2	4	15	3	82	18	100
R7	33	11	10	7	3	8	14	14	72	28	100
R8	17	16	26	10	10	6	7	8	85	15	100
R9	44	22	17	8	2	2	3	2	95	5	100
R10	13	35	14	19	6	11	1	1	98	2	100
WIC	45	10	9	6	20	7	2	1	97	3	100
IC	40	11	18	5	19	5	1	1	98	2	100
TW	18	36	18	10	5	5	5	3	92	8	100
TW-R9	11	40	18	10	5	6	6	4	90	10	100
WORLD	30	22	18	7	12	6	3	2	95	5	100

ENERGY PRODUCTION FORECASTS

REGIONAL STRUCTURE %

- 2020 (II) -

Table : 8
Appendix . 12

(%)	SMF	O	NG	HY	NU	\>	FV	VAW	CEP	NCEP	PEP
R1	48	13	12	5	11	8	3	E	97	3	100
R2	24	9	7	12	36	7	2	3	95	5	100
R3	45	9	9	7	24	5	E	1	99	1	100
R4	28	14	35	4	16	:	1	E	99	1	100
R5	E	68	24	1	2	3	1	1	98	2	100
R6	37	21	2	5	2	5	23	5	72	28	100
R7	27	9	9	7	2	6	20	20	60	40	100
R8	24	8	21	10	7	8	10	12	78	22	100
R9	49	19	9	6	1	1	10	5	85	15	100
R10	11	35	15	18	4	11	3	3	94	6	100
WIC	44	11	10	7	18	7	2	1	97	3	100
IC	39	12	18	6	17	5	2	1	97	3	100
TW	17	35	15	9	3	6	9	6	85	15	100
TW-R9	10	38	17	10	3	7	9	6	85	15	100
WORLD	30	21	17	7	11	6	5	3	92	8	100

ENERGY PRODUCTION FORECASTS

FUEL STRUCTURE %

- 2000 (I) -

Table : 9
Appendix : 12

(%)	SMF	O	NG	HY	NU	NS	FV	VAW	CEP	NCEP	PEP
R1	32	14	23	25	30	33	14	3	24	10	23
R2	9	4	6	19	25	16	4	9	9	6	9
R3	5	1	1	5	11	4	E	1	4	E	3
R4	24	15	38	10	24	7	4	3	22	4	21
R5	E	34	12	2	1	2	2	4	14	2	13
R6	6	4	1	3	1	4	22	7	4	17	5
R7	4	1	2	4	1	3	15	25	2	19	3
R8	2	4	4	3	3	3	11	15	3	12	4
R9	16	8	4	7	1	2	20	26	9	22	10
R10	2	14	9	22	3	26	8	7	9	8	9
WIC	52	19	30	49	66	55	18	13	38	16	36
IC	76	35	68	59	90	62	22	16	60	20	57
TW	24	65	32	41	10	38	78	84	40	80	43
TW-R9	8	57	28	34	9	36	58	58	31	68	33
WORLD	100	100	100	100	100	100	100	100	100	100	100

ENERGY PRODUCTION FORECASTS

Table : 10
Appendix : 12

FUEL STRUCTURE %

- 2000 (II) -

(%)	SMF	O	NG	HY	NU	NS	FW	VAW	CEP	NCEP	PEP
R1	34	18	26	28	34	34	11	2	27	9	25
R2	10	5	7	21	26	16	3	9	10	5	10
R3	6	1	1	5	10	4	ε	1	4	ε	3
R4	23	16	40	10	22	7	4	4	23	4	21
R5	ε	33	10	2	1	2	1	4	13	2	12
R6	5	5	ε	3	1	5	21	8	4	17	5
R7	4	1	2	3	1	2	14	24	2	17	4
R8	3	3	3	3	2	5	10	16	3	12	4
R9	14	5	2	5	1	1	27	24	6	26	8
R10	1	13	8	20	2	24	9	8	8	8	8
WIC	55	24	35	54	71	56	15	13	42	14	39
IC	78	40	75	64	93	63	19	17	65	18	60
TW	22	60	25	36	7	37	81	83	35	82	40
TW-R9	8	55	23	31	6	36	54	59	29	56	32
WORLD	100	100	100	100	100	100	100	100	100	100	100

ENERGY PRODUCTION FORECASTS

Table : 11
Appendix : 12

FUEL STRUCTURE %

- 2020 (I) -

(%)	SMF	O	NG	HY	NU	NS	FW	VAW	CEP	NCEP	PEP
R1	35	12	14	15	21	31	17	4	22	12	22
R2	7	3	3	12	27	11	4	12	8	8	8
R3	5	1	2	3	9	4	ε	1	4	ε	4
R4	18	12	37	10	26	8	4	3	20	3	19
R5	ε	31	16	2	4	5	2	4	11	2	11
R6	8	6	1	5	1	4	29	11	5	22	6
R7	5	2	2	4	1	6	20	31	3	24	4
R8	3	3	6	6	4	4	10	16	4	13	4
R9	14	10	9	11	1	3	9	9	10	9	10
R10	5	20	10	32	6	24	5	9	13	7	12
WIC	55	16	19	30	57	48	21	18	37	20	36
IC	73	28	56	40	83	56	25	21	57	23	55
TW	27	72	44	60	17	44	75	79	43	77	45
TW-R9	13	62	35	49	16	41	66	70	33	68	35
WORLD	100	100	100	100	100	100	100	100	100	100	100

ENERGY PRODUCTION FORECASTS

FUEL STRUCTURE %

- 2000 (II) -

(%)	SMF	O	NG	HY	NU	NS	FW	VAW	CEP	NCEP	PEP
R1	34	18	26	28	34	34	11	2	27	9	25
R2	10	5	7	21	26	16	3	9	10	5	10
R3	6	1	1	5	10	4	ε	1	4	ε	3
R4	23	16	40	10	22	7	4	4	23	4	21
R5	ε	33	10	2	1	2	1	4	13	2	12
R6	5	5	ε	3	1	5	21	8	4	17	5
R7	4	1	2	3	1	2	14	24	2	17	4
R8	3	3	3	3	2	5	10	16	3	12	4
R9	14	5	2	5	1	1	27	24	6	26	8
R10	1	13	8	20	2	24	9	8	8	8	8
WIC	55	24	35	54	71	56	15	13	42	14	39
IC	78	40	75	64	93	63	19	17	65	18	60
TW	22	60	25	36	7	37	81	83	35	82	40
TW-R9	8	55	23	31	6	36	54	59	29	56	32
WORLD	100	100	100	100	100	100	100	100	100	100	100

ENERGY EXCHANGES FORECASTS

INTER REGIONAL ENERGY EXCHANGE FORECASTS (Mtoe)

- 2000 (I) -

(Mtoe)	SMF			CO			NG			TOTAL		
	E	I	B	E	I	B	E	I	B	E	I	B
R1	169	-	169	2	162	-160	-	24	-24	171	186	-15
R2	-	172	-172	40	565	-525	-	160	-160	40	897	-857
R3	75	95	-20	-	218	-218	-	66	-66	75	379	-304
R4	75	-	75	-	70	-70	80	15	65	155	85	70
R5	-	5	-5	930	-	930	95	-	95	1025	5	1020
R6	55	2	53	126	30	96	-	-	-	181	32	149
R7	1	-	1	-	42	-42	5	-	5	6	42	-36
R8	-	40	-40	-	127	-127	50	-	50	50	167	-117
R9	20	-	20	100	3	97	-	-	-	120	3	117
R10	3	10	-7	40	40	-	25	2	23	68	52	16
WORLD	398	324	74	1238	1257	-19	255	267	-12	1891	1848	43

INTER REGIONAL ENERGY EXCHANGE FORECASTS (Mtoe)

- 2000 (II) -

(Mtoe)	SMF			CO			NG			TOTAL		
	E	I	B	E	I	B	E	I	B	E	I	B
R1	156	-	156	2	120	-118	-	12	-12	158	132	26
R2	-	121	-121	40	511	-471	-	100	-100	40	732	-692
R3	65	90	-25	-	201	-201	-	65	-65	65	356	-291
R4	30	-	30	-	70	-70	55	5	50	85	75	10
R5	-	4	-4	757	-	757	75	-	75	832	4	828
R6	40	2	38	120	20	100	-	-	-	160	22	138
R7	-	-	-	-	30	-30	5	-	5	5	30	-25
R8	-	20	-20	-	67	-67	35	-	35	35	87	-52
R9	10	-	10	25	2	23	-	-	-	35	2	33
R10	3	7	-4	30	30	0	20	-	20	53	37	16
WORLD	304	244	60	974	1051	-77	190	182	8	1468	1477	-9

Table : 3
Appendix : 13

INTER REGIONAL ENERGY EXCHANGE FORECASTS (Mtoe)

- 2020 (I) -

(Mtoe)	SMF			CO			NG			TOTAL		
	E	I	B	E	I	B	E	I	B	E	I	B
R1	184	-	184	2	51	-49	-	-	-	186	51	135
R2	-	370	-370	20	399	-379	-	200	-200	20	969	-949
R3	120	120	0	-	162	-162	-	105	-105	120	387	-267
R4	70	-	70	-	60	-60	120	20	100	190	80	110
R5	-	58	-58	665	-	665	41	-	41	706	58	648
R6	80	5	75	129	40	89	-	-	-	209	45	164
R7	4	-	4	-	61	-61	10	-	10	14	61	-47
R8	-	130	-130	-	225	-225	50	-	50	50	355	-305
R9	80	-	80	100	5	95	100	-	100	280	5	275
R10	10	20	-10	20	20	0	40	-	40	70	40	30
WORLD	548	703	-155	936	1023	-87	361	325	36	1845	2051	-206

Table : 4
Appendix : 13

INTER REGIONAL ENERGY EXCHANGE FORECASTS (Mtoe)

- 2020 (II) -

(Mtoe)	SMF			CO			NG			TOTAL		
	E	I	B	E	I	B	E	I	B	E	I	B
R1	168	-	168	2	36	-34	-	-	-	170	36	134
R2	-	260	-260	20	346	-326	-	130	-130	20	736	-716
R3	100	110	-10	-	136	-136	-	90	-90	100	336	-236
R4	20	-	20	-	50	-50	70	10	60	90	60	30
R5	-	35	-35	548	-	548	83	-	83	631	35	596
R6	50	4	46	122	20	102	-	-	-	172	24	148
R7	-	-	-	-	37	-37	10	-	10	10	37	-27
R8	-	20	-20	-	45	-45	30	-	30	30	65	-35
R9	30	-	30	-	5	-5	-	-	-	30	5	25
R10	10	10	0	15	15	0	30	-	30	55	25	30
WORLD	378	439	-61	707	690	17	223	230	-7	1308	1359	-51

INTER REGIONAL ENERGY EXCHANGE FORECASTS

REGIONAL STRUCTURE %

- 2000 (I) -

(%)	SMF			CO			NG			TOTAL		
	E	I	B	E	I	B	E	I	B	E	I	B
R1	99	-		1	87		-	13		100	100	
R2	-	19		100	63		-	18		100	100	
R3	100	25		-	58		-	17		100	100	
R4	48	-		-	82		52	18		100	100	
R5	-	100		91	-		9	-		100	100	
R6	30	6		70	94		-	-		100	100	
R7	17	-		-	100		83	-		100	100	
R8	-	24		-	76		100	-		100	100	
R9	17	-		83	-		-	-		100	100	
R10	4	19		59	77		37	4		100	100	
WORLD	21	18		65	68		14	14		100	100	

INTER REGIONAL ENERGY EXCHANGE FORECASTS

REGIONAL STRUCTURE %

- 2000 (II) -

(%)	SMF			CO			NG			TOTAL		
	E	I	B	E	I	B	E	I	B	E	I	B
R1	99	-		1	91		-	9		100	100	
R2	-	17		100	70		-	13		100	100	
R3	100	25		-	57		-	18		100	100	
R4	35	-		-	93		65	7		100	100	
R5	-	100		91	-		9	-		100	100	
R6	25	9		75	91		-	-		100	100	
R7	-	-		-	100		100	-		100	100	
R8	-	23		-	77		100	-		100	100	
R9	28	-		72	100		-	-		100	100	
R10	6	19		57	81		37	-		100	100	
WORLD	21	17		66	71		13	12		100	100	

Table   :   7
Appendix :   13

INTER REGIONAL ENERGY EXCHANGE FORECASTS

REGIONAL STRUCTURE %

- 2020 (I) -

(%)	SMF			CO			NG			TOTAL		
	E	I	B	E	I	B	E	I	B	E	I	B
R1	99	–		1	100		–	–		100	100	
R2	–	38		100	41		–	21		100	100	
R3	100	31		–	42		–	27		100	100	
R4	37	–		–	75		63	25		100	100	
R5	–	100		94	–		6	–		100	100	
R6	38	11		62	89		–	–		100	100	
R7	29	–		–	100		71	–		100	100	
R8	–	37		–	63		100	–		100	100	
R9	28	–		36	100		36	–		100	100	
R10	14	50		29	50		57	–		100	100	
WORLD	30	34		51	50		19	16		100	100	

Table   :   8
Appendix :   13

INTER REGIONAL ENERGY EXCHANGE FORECASTS

REGIONAL STRUCTURE %

- 2020 (II) -

(%)	SMF			CO			NG			TOTAL		
	E	I	B	E	I	B	E	I	B	E	I	B
R1	99	–		1	100		–	–		100	100	
R2	–	35		100	47		–	18		100	100	
R3	100	33		–	40		–	27		100	100	
R4	22	–		–	83		78	17		100	100	
R5	–	100		87	–		13	–		100	100	
R6	29	17		71	83		–	–		100	100	
R7	–	–		–	100		100	–		100	100	
R8	–	31		–	69		100	–		100	100	
R9	100	–		–	100		–	–		100	100	
R10	18	40		27	60		55	–		100	100	
WORLD	29	32		54	51		17	17		100	100	

Table   :   9
Appendix : 13

INTER REGIONAL ENERGY EXCHANGE FORECASTS

FUEL STRUCTURE %

- 2000 (I) -

(%)	SMF			CO			NG			TOTAL		
	E	I	B	E	I	B	E	I	B	E	I	B
R1	42	-		Ɛ	13		-	9		9	9	
R2	-	53		3	45		-	60		2	49	
R3	19	29		-	18		-	25		4	21	
R4	19	-		-	6		31	5		8	5	
R5	-	2		76	-		37	-		54	Ɛ	
R6	14	1		10	2		-	-		10	2	
R7	Ɛ	-		-	3		2	-		Ɛ	2	
R8	-	12		-	10		20	-		3	9	
R9	5	-		8	Ɛ		-	-		6	Ɛ	
R10	1	3		3	3		10	1		4	3	
WORLD	100	100		100	100		100	100		100	100	

Table   :   10
Appendix : 13

INTER REGIONAL ENERGY EXCHANGE FORECASTS

FUEL STRUCTURE %

- 2000 (II) -

(%)	SMF			CO			NG			TOTAL		
	E	I	B	E	I	B	E	I	B	E	I	B
R1	51	-		Ɛ	11		-	6		11	9	
R2	-	49		4	49		-	55		3	49	
R3	22	37		-	19		-	36		4	24	
R4	10	-		-	7		29	3		6	5	
R5	-	2		78	-		39	-		57	Ɛ	
R6	13	1		12	2		-	-		11	2	
R7	-	-		-	3		3	-		Ɛ	2	
R8	-	8		-	6		18	-		2	6	
R9	3	-		3	Ɛ		-	-		2	Ɛ	
R10	1	3		3	3		11	-		4	3	
WORLD	100	100		100	100		100	100		100	100	

INTER REGIONAL ENERGY EXCHANGE FORECASTS

FUEL STRUCTURE %

- 2020 (I) -

(%)	SMF			CO			NG			TOTAL		
	E	I	B	E	I	B	E	I	B	E	I	B
R1	32	-		ε	5		-	-		10	3	
R2	-	53		2	39		-	62		1	47	
R3	22	17		-	16		-	32		7	19	
R4	13	-		-	6		33	6		10	4	
R5	-	8		71	-		11	-		38	3	
R6	15	1		14	4		-	-		11	2	
R7	1	-		-	6		3	-		1	3	
R8	-	18		-	22		14	-		3	17	
R9	15	-		11	ε		28	-		15	ε	
R10	2	3		2	2		11	-		4	2	
WORLD	100	100		100	100		100	100		100	100	

INTER REGIONAL ENERGY EXCHANGE FORECASTS

FUEL STRUCTURE %

- 2020 (II) -

(%)	SMF			CO			NG			TOTAL		
	E	I	B	E	I	B	E	I	B	E	I	B
R1	45	-		ε	5		-	-		13	3	
R2	-	59		3	50		-	57		2	54	
R3	26	25		-	20		-	39		8	25	
R4	5	-		-	7		31	4		7	4	
R5	-	8		78	-		37	-		48	2	
R6	13	1		17	3		-	-		13	2	
R7	-	-		-	5		4	-		1	3	
R8	-	5		-	7		14	-		2	5	
R9	8	-		-	1		-	-		2	ε	
R10	3	2		2	2		14	-		4	2	
WORLD	100	100		100	100		100	100		100	100	

BALANCE FACTOR

Energy consumptions, as they were defined here for certain UN Conventions (in particular that of the *Yearbook of Energy Statistics* which provided the reference base of consumptions), particularly excluded non-energy uses and bunkers (which correspond to effective consumption outside the regional territory).

However, when we attempted to calculate not only consumption forecasts but also production forecasts, we had to operate a quantitative reconciliation between production and consumption (according to the definition adopted). For each of the energy sources, one therefore had a general equation of the following type:

$$\text{Production} = \text{Consumption} + \text{Exports} - \text{Imports} + \text{Balance Factor}$$
$$P \quad = \quad C + E - I + BF$$

One knows that the problem was simplified by making the assumption $P = C$ for all the primary sources of electricity (hydropower, nuclear, etc.) and for non-commercial sources. There thus remained in issue all fossil fuels (coal, oil and natural gas) and, possibly, new energies.

Apart from the possibility of interregional trade, one could ascertain the existence of a balance factor for each of these sources which could, if need be, settle the primary balances total.

This relatively heterogeneous balance factor could include very diverse elements:

- non energy uses of the resources;
- bunkers;
- stock variations.

All bunkers were attributed to oil, and stock variations were ignored in forecasts. As a result, the only potential non-energy uses which appear out of oil are those for coal (projects in South Africa and in India), for gas and for new energies (especially in Brazil, based on alcohol).

In the case of oil, an additional element occurred. The final consumption of oil is almost never a consumption of crude oil but one of refined products. However, the production of oil is one of crude oil. In the case of oil one therefore has:

$$PC = CRP + EC - IC + OBF$$

with OBF = $ERP + NEU + MAB \pm \Delta SCP \pm \Delta SRP + CCF - IRP - RG + ADJ$

with PC	=	Production of crude
CRP	=	Consumption of refined products (including losses during refining)
IC	=	Imports of crude from other regions
EC	=	Exports of crude to other regions
OBF	=	Oil balance factor
and ERP	=	Interregional exports of refined products
NEU	=	Non-energy uses
MAB	=	Maritime and aerial bunkers
$\pm \Delta$SCP	=	Variations of crude stocks
$\pm \Delta$SRP	=	Variations of refined product stocks
CCF	=	Direct Consumption of crude in final uses
− IRP	=	Interregional imports of refined products
ADJ	=	Adjustment term
− RG	=	Refining gain

This last factor is specific in the statistics for R_1 (North America) in terms of barrels. The variations of volume introduced by the transformation of crude into refined products were ignored in other regions (1 tonne of refined products in fact contains more barrels than 1 tonne of crude).
In our forecasts, ΔSCP and ΔSRP are nil;
CCF is almost always insignificant;
RG is specific to R_1.
One thus most often finds in our forecasts:

$$OBF = ERP + NEU + MAB - IRP + ADJ$$

In the following tables representing the 1978 position and the forecasts for 2000 and 2020 of the total balance factor, we have given estimates of non-energy consumption for coal, gas and new energies. For oil, we have provided the OBF total resulting from all available data and global projections. In some cases, this OBF was divided between non-energy uses, bunkers and export surplus figures for refined products (including also the adjustment term). We have added details of our hypotheses for two key regions: R5 and R10.
It will be noted that the balance factor is set at 98% oil in 1978, and that it remains at 93% in (I) and at 85–90% in (II). It will represent an important and growing percentage of global oil production (especially in II): 14% in 1978 but 14–16% in 2000 and 15–20% in 2020.
This is not surprising, in so far as this residual term essentially covers specific oil markets (non-energy uses and bunkers) towards which demand will primarily focus in the long term.

BALANCE FACTOR

- 1978 -

| (Mtoe) | SMF | PP | | | | NG | NS | BF |
	NEU	NEU	BUN	PPE	TOTAL	NEU	NEU	
R1					1			1
R2					67			67
R3					29			29
R4					139			139
R5		4	25	72	101	2		103
R6	1				11			12
R7	1				5			6
R8					20			20
R9					3	2		5
R10		7	10	44	61	2	2	65
WORLD	2				437	6	2	447

BALANCE FACTOR

- 2000 (I) -

| (Mtoe) | SMF | PP | | | | NG | NS | BF |
	NEU	NEU	BUN	PPE	TOTAL	NEU	NEU	
R1					178			178
R2					54			54
R3					28			28
R4					80			80
R5		16	41	8	75	11		86
R6	5				20			25
R7	2				12	5		19
R8					30			30
R9					9	8		17
R10		20	21	28	69	6	5	80
WORLD	7				555	30	5	597

BALANCE FACTOR

- 2000 (II) -

(Mtoe)	SMF NEU	PP NEU	PP BUN	PP PPE	PP TOTAL	NG NEU	NS NEU	BF
R1					177			177
R2					49			49
R3					25			25
R4					45			45
R5		9	34	52	95	11		106
R6	4				13			17
R7	2				10	5		17
R8					20			20
R9					7	4		11
R10		15	15	12	42	4	3	49
WORLD	6				483	24	3	516

BALANCE FACTOR

- 2020 (I) -

(Mtoe)	SMF NEU	PP NEU	PP BUN	PP PPE	PP TOTAL	NG NEU	NS NEU	BF
R1					299			299
R2					48			48
R3					27			27
R4					50			50
R5		28	51	-29	50	25		75
R6	10				35			45
R7	4				18	10		32
R8					35			35
R9					15	40		55
R10		30	32	28	90	15	12	117
WORLD	14				667	90	12	783

BALANCE FACTOR

- 2020 (II) -

(Mtoe)	SMF NEU	PP NEU	PP BUN	PP PPE	PP TOTAL	NG NEU	NS NEU	BF
R1					268			268
R2					40			40
R3					21			21
R4					30			30
R5		15	41	99	155	21		176
R6	8				15			23
R7	4				12	10		26
R8					15			15
R9					9	10		19
R10		20	22	0	42	10	8	60
WORLD	12				607	51	8	678

Region: North Africa/Middle East (R5*)

BALANCE FACTOR

1978 Total   103.2 MTOE
of which $\begin{cases} NG \\ PP \end{cases}$ 2.1
                                        101.1
  • NEU   3.8
  • Bunkers   25.3
  • Exports PP   72.3
  • Stocks   −0.3

2000 (I) Total   86 MTOE   **2020 (I) Total**   75 MTOE
of which $\begin{cases} NG \\ PP \end{cases}$ 11   of which $\begin{cases} NG \\ PP \end{cases}$ 25
                                  75   50
  • NEU   16   • NEU   28
  • Bunkers   41   • Bunkers   51
  • Exports PP   8   • Exports PP   −29

2000 (II) Total   106 MTOE   **2020 (II) Total**   176 MTOE
of which $\begin{cases} NG \\ PP \end{cases}$ 11   of which $\begin{cases} NG \\ PP \end{cases}$ 21
                                  95   155
  • NEU   9   • NEU   15
  • Bunkers   34   • Bunkers   41
  • Exports PP   52   • Exports PP   99

*06/1982 (2)

Region R10: Latin America

BALANCE FACTOR

1978 Total   64.6 MTOE
SMF   0.3
PP   61.3   =   exports refined
                         products   42.0
                         non-energy uses
                              7.2
                         adjustment on
                         crude   2.4
                         bunkers   9.7
NG   1.7
NE   0.3
VAW   1.0

2000 (I) **Total**   80 MTOE
PP   28   (exports refined products)
PP   20   (non-energy uses)
PP   21   (bunkers)
NG   6
NE   5

2020 (I) **Total**   117 MTOE
PP   28   (exports refined products)
PP   30   (non-energy uses)
PP   32   (bunkers)
NG   15
NE   12

2000 (II) **Total**   49 MTOE
PP   12   (exports refined products)
PP   15   (non-energy uses)
PP   15   (bunkers)
NG   4
Ne   3

2020 (II) **Total**   80 MTOE
PP   0   (exports refined products)
PP   20   (non-energy uses)
PP   22   (bunkers)
NG   10
NE   8

REGIONAL STRUCTURES: ENERGY DEPENDENCES -ELASTICITIES-INTENSITIES

Table   : 1
Appendix : 15

COMPARISON OF REGIONAL STRUCTURES

POPULATION - GNP

(%)	1960		1978		2000 I		2000 II		2020 I		2020 II	
	POP	GNP	POP	GNP	POP	GNP	POP	GNP	POP	GNP	POP	GNP
R1	6,7	31,3	5,7	26,4	4,9	20,7	4,9	22,8	4,5	17,4	4,5	20,8
R2	11,8	34,9	9,8	29,2	7,7	25,8	7,7	25,6	6,3	21,6	6,3	23,3
R3	3,6	6,1	3,1	11,4	2,5	12,0	2,5	12,3	2,0	11,1	2,0	11,0
R4	10,4	12,7	8,7	15,4	7,0	16,5	7,0	16,5	6,0	16,8	6,0	16,4
R5	3,7	1,6	4,3	3,3	5,3	5,3	5,3	4,9	6,0	8,1	6,0	6,6
R6	7,3	2,1	8,2	1,9	11,2	2,2	11,2	2,5	14,9	2,5	14,9	3,0
R7	18,6	2,1	19,9	1,7	21,3	1,9	21,3	1,9	21,3	2,1	21,3	2,1
R8	7,3	1,7	7,9	2,5	8,3	4,2	8,3	3,7	8,2	5,3	8,2	4,4
R9	23,7	2,6	24,3	2,8	22,5	2,8	22,5	2,1	20,4	3,5	20,4	2,2
R10	6,9	4,9	8,1	5,4	9,3	8,6	9,3	7,7	10,4	11,6	10,4	10,2
WIC	22,6	72,8	19,3	67,0	15,8	59,0	15,8	61,4	13,9	50,7	13,9	55,8
IC	33,0	85,5	28,0	82,4	22,8	75,5	22,8	77,9	19,9	67,9	19,9	72,2
TW	67,0	14,5	72,0	17,6	77,2	24,5	77,2	22,1	80,1	32,5	80,1	27,8
TW - R9	43,3	11,9	47,7	14,8	54,7	21,7	54,7	20,0	59,7	29,0	59,7	25,6
WORLD	100	100	100	100	100	100	100	100	100	100	100	100

COMPARISON OF REGIONAL STRUCTURES

CONSUMPTION - TOTAL ENERGY PRODUCTION

(%)	1960	1978		2000 I		2000 II		2020 I		2020 II	
	EC	EC	PR	EC	PR	EC	PR	EC	PR	EC	PR
R1	35	30	24	22	23	24	25	20	22	22	24
R2	20	18	9	16	9	17	10	14	8	15	10
R3	4	6	2	6	3	6	3	5	4	6	4
R4	17	21	22	21	21	21	21	19	19	20	20
R5	1	2	18	5	13	4	12	7	11	5	10
R6	3	3	4	3	5	4	5	5	6	5	6
R7	3	3	3	4	3	4	4	4	4	4	4
R8	3	3	3	5	4	4	4	6	4	5	4
R9	9	9	9	9	10	8	8	8	10	7	7
R10	5	5	6	9	9	8	8	12	12	11	11
WIC	60	55	36	45	36	48	39	41	36	45	40
IC	77	76	58	66	57	69	60	60	55	65	60
TW	23	24	42	34	43	31	40	40	45	35	40
TW - R9	14	15	33	25	33	23	32	32	35	28	33
WORLD	100	100	100	100	100	100	100	100	100	100	100

REGIONAL ENERGY DEPENDENCE

RATIO PRODUCTION / CONSUMPTION BY REGION

(total energy)

	1978	2000		2020	
		I	II	I	II
R1	84	106	108	112	113
R2	51	57	62	63	68
R3	38	60	57	74	71
R4	115	106	103	105	102
R5	945	299	348	155	211
R6	154	143	143	124	125
R7	94	96	98	98	100
R8	108	85	93	74	97
R9	103	113	105	123	104
R10	120	109	108	107	106
WIC	69	84	86	91	93
IC	81	91	91	96	96
TW	179	134	136	114	121
TW - R9	225	142	148	112	125
WORLD	105	109	105	103	105

/ ELASTICITIES /

INCREASE OF COMMERCIAL ENERGY CONSUMPTION / INCREASE OF TOTAL GNP

(% per year)

	1960-1978	1978-2000 I	1978-2000 II	2000-2020 I	2000-2020 II
R1	0,80	0,44	0,35	0,75	0,80
R2	0,95	0,63	0,70	0,70	0,73
R3	0,81	0,64	0,67	0,60	0,67
R4	0,89	0,65	0,67	0,57	0,65
R5	1,07	1,20	1,13	0,88	0,89
R6	1,02	1,18	0,98	1,49	1,50
R7	1,34	1,24	1,16	1,00	0,96
R8	1,22	1,05	1,00	0,81	0,80
R9	1,17	0,87	1,04	0,58	0,85
R10	1,18	0,97	1,00	0,89	0,94
WIC	0,82	0,53	0,54	0,74	0,77
IC	0,85	0,59	0,58	0,68	0,79
TW	1,14	0,96	0,87	0,82	0,88
TW - R9	1,17	1,07	1,05	0,91	0,94
WORLD	0,90	0,74	0,68	0,79	0,85

ENERGY INTENSITIES

(Koe/$)	PEC / GNP		2000		2020		CEC / GNP		2000		2020	
	1960	1978	I	II	I	II	1960	1978	I	II	I	II
R1	1,01	0,87	0,65	0,67	0,59	0,62	0,97	0,85	0,63	0,65	0,57	0,60
R2	0,50	0,47	0,37	0,41	0,32	0,38	0,47	0,46	0,36	0,40	0,32	0,37
R3	0,52	0,39	0,29	0,32	0,24	0,29	0,51	0,39	0,29	0,31	0,24	0,28
R4	1,23	1,03	0,75	0,82	0,58	0,71	1,13	1,00	0,74	0,81	0,58	0,70
R5	0,54	0,46	0,53	0,48	0,46	0,44	0,36	0,40	0,51	0,45	0,46	0,42
R6	1,28	1,07	0,92	0,91	1,00	0,96	0,46	0,47	0,56	0,46	0,78	0,62
R7	1,55	1,43	1,26	1,32	1,08	1,13	0,51	0,64	0,79	0,72	0,79	0,68
R8	1,41	1,00	0,71	0,75	0,56	0,61	0,41	0,54	0,57	0,54	0,50	0,48
R9	2,99	2,58	1,91	2,52	1,18	1,91	1,46	1,70	1,54	1,70	1,12	1,60
R10	0,88	0,74	0,61	0,67	0,53	0,61	0,51	0,60	0,57	0,60	0,52	0,57
WIC	0,73	0,62	0,46	0,50	0,42	0,47	0,70	0,61	0,45	0,48	0,41	0,45
IC	0,80	0,70	0,53	0,57	0,46	0,52	0,76	0,68	0,51	0,55	0,45	0,51
TW	1,42	1,11	0,82	0,89	0,64	0,73	0,62	0,70	0,67	0,64	0,58	0,59
TW - R9	1,07	0,83	0,68	0,71	0,57	0,62	0,43	0,51	0,56	0,53	0,51	0,51
WORLD	0,89	0,77	0,60	0,64	0,51	0,58	0,74	0,68	0,55	0,57	0,49	0,53

Table : 1
Appendix : 19

COMPARISON OF GLOBAL STRUCTURES : LARGE ZONES

1960 - 1978

(Total amounts)	1960				1978			
	Population (Minhb)	GNP (G $ 78)	Ener.consump. (Mtoe)	Ener.prod. (Mtoe)	Population (Minhb)	GNP (G $ 78)	Ener.consump. (Mtoe)	Ener.prod. (Mtoe)
(1) INDUSTRIALIZED COUNTRIES	996 (33 %)	3169 (86 %)	2545 (77 %)		1195 (28 %)	7371 (83 %)	5134 75 %	4175 (58 %)
(2) THIRD WORLD IN RAPID GROWTH (R5+R8+R10)	542 (18 %)	304 (8 %)	281 (8 %)		866 (20 %)	995 (11 %)	709 (11 %)	1930 (27 %)
(3) THIRD WORLD IN TRANSITION (R6+R7+R9)	1477 (49 %)	234 (6 %)	482 (15 %)		2201 (52 %)	519 (6 %)	975 (14 %)	1076 (15 %)
WORLD	3015 (100 %)	3707 (100 %)	3308 (100 %)		4262 (100 %)	8885 (100 %)	6818 (100 %)	7181 (100 %)

Table : 2
Appendix : 19

COMPARISON OF GLOBAL STRUCTURES : LARGE ZONES

- Scenario I -

(Total amounts)	2000 I				2020 I			
	Population (Minhb)	GNP (G $ 78)	Ener.consump. (Mtoe)	Ener.prod. (Mtoe)	Population (Minhb)	GNP (G $ 78)	Ener.consump. (Mtoe)	Ener.prod. (Mtoe)
(1) INDUSTRIALIZED COUNTRIES	1370 (23 %)	14790 (75 %)	7786 (66 %)	7070 (57 %)	1533 (20 %)	23541 (67 %)	10748 (60 %)	10281 (55 %)
(2) THIRD WORLD IN RAPID GROWTH (R5+R8+R10)	1387 (23 %)	3557 (18 %)	2175 (19 %)	3290 (27 %)	1898 (24 %)	8754 (25 %)	4523 (25 %)	5123 (28 %)
(3) THIRD WORLD IN TRANSITION (R6+R7+R9)	3273 (54 %)	1245 (7 %)	1783 (15 %)	2024 (16 %)	4291 (16 %)	2580 (8 %)	2686 (15 %)	3130 (17 %)
WORLD	6039 (100 %)	19592 (100 %)	11744 (100 %)	12384 (100 %)	7722 (100 %)	34875 (100 %)	17957 (100 %)	18534 (100 %)

COMPARISON OF GLOBAL STRUCTURES : LARGE ZONES

- Scenario II -

(Total amounts)	2000 II				2020 II			
	Population (Minhb)	GNP (G $ 78)	Ener.consump. (Mtoe)	Ener.prod. (Mtoe	Population (Minhb)	GNP (G $ 78)	Ener.consump. (Mtoe)	Ener.prod. (Mtoe)
(1) INDUSTRIALIZED COUNTRIES	1379 (23 %)	12362 (78 %)	7004 (69 %)	6390 (60 %)	1533 (20 %)	17147 (72 %)	8962 (65 %)	8591 (60 %)
(2) THIRD WORLD IN RAPID GROWTH (R5+R8+R10)	1387 (23 %)	2581 (16 %)	1629 (16 %)	2596 (25 %)	1898 (24 %)	5060 (21 %)	2814 (21 %)	3656 (25 %)
(3) THIRD WORLD IN TRANSITION (R6+R7+R9)	3273 (54 %)	916 (6 %)	1471 (15 %)	1625 (15 %)	4291 (56 %)	1558 (7 %)	1990 (14 %)	2146 (15 %)
WORLD	6039 (100 %)	15859 (100 %)	10104 (100 %)	10611 (100 %)	7722 (100 %)	23765 (100 %)	13766 (100 %)	14393 (100 %)

COMPARISON OF GLOBAL STRUCTURES : LARGE ZONES

1960 - 1978

(Total amounts)	1960 - 1978 INCREASE			
	Population (Minhb)	GNP (G $ 78)	Ener.consump. (Mtoe)	Ener. prod. (Mtoe)
(1) INDUSTRIALIZED COUNTRIES	+199 (16 %)	+4202 (81 %)	+2589 (74 %)	
(2) THIRD WORLD IN RAPID GROWTH (R5 + R8 + R10)	+324 (26 %)	+691 (13 %)	+428 (12 %)	
(3) THIRD WORLD IN TRANSITION (R6 + R7 + R9)	+724 (58 %)	+285 (6 %)	+493 (14 %)	
WORLD	+1247 (100 %)	+5178 (100 %)	+3510 (100 %)	

COMPARISON OF GLOBAL STRUCTURES : LARGE ZONES

- Scenario I -

(Total amounts)	1978 - 2000 I INCREASE				2000 - 2020 I INCREASE			
	Population (Minhb)	GNP (G $ 78)	Ener.consump. (Mtoe)	Ener.prod. (Mtoe)	Population (Minhb)	GNP (G $ 78)	Ener.consump. (Mtoe)	Ener.prod. (Mtoe)
(1) INDUSTRIALIZED COUNTRIES	+184 (10 %)	+7419 (69 %)	+2652 (54 %)	+2895 (56 %)	+154 (9 %)	+8751 (57 %)	+2962 (48 %)	+3211 (52 %)
(2) THIRD WORLD IN RAPID GROWTH (R5 + R8 + R10)	+521 (29 %)	+2562 (24 %)	+1466 (30 %)	+1360 (26 %)	+511 (30 %)	+5197 (34 %)	+2348 (38 %)	+1833 (30 %)
(3) THIRD WORLD IN TRANSITION (R6 + R7 + R9)	+1072 (61 %)	+726 (7 %)	+808 (16 %)	+948 (18 %)	+1018 (61 %)	+1335 (9 %)	+903 (14 %)	+1106 (18 %)
WORLD	+1777 (100 %)	+10707 (100 %)	+4926 (100 %)	+5203 (100 %)	+1683 (100 %)	+15283 (100 %)	+6213 (100 %)	+6150 (100 %)

COMPARISON OF GLOBAL STRUCTURES : LARGE ZONES

- Scenario II -

(Total amounts)	1978 - 2000 II INCREASE				2000 - 2020 II INCREASE			
	Population (Minhb)	GNP (G $ 78)	Ener.consump. (Mtoe)	Ener.prod. (Mtoe)	Population (Minhb)	GNP (G $ 78)	Ener.consump. (Mtoe)	Ener.Prod. (Mtoe)
(1) INDUSTRIALIZED COUNTRIES	+184 (10 %)	+4991 (71 %)	+1870 (57 %)	+2215 (65 %)	+154 (9 %)	+4785 (61 %)	+1958 (54 %)	+2201 (58 %)
(2) THIRD WORLD IN RAPID GROWTH (R5 + R8 + R10)	+521 (29 %)	+1586 (23 %)	+920 (28 %)	+666 (19 %)	+511 (30 %)	+2479 (31 %)	+1185 (32 %)	+1060 (28 %)
(3) THIRD WORLD IN TRANSITION (R6 + R7 + R9)	+1072 (61 %)	+397 (6 %)	+496 (15 %)	+549 (16 %)	+1018 (61 %)	+642 (8 %)	+519 (14 %)	+521 (14 %)
WORLD	+1777 (100 %)	+6974 (100 %)	+3286 (100 %)	+3430 (100 %)	+1683 (100 %)	+7906 (100 %)	+3662 (100 %)	+3782 (100 %)

Table   :   7
Appendix : 19

COMPARISON OF GLOBAL STRUCTURES : LARGE ZONES

TOTAL INCREASE 1960 - 2020

	POPULATION (Minhb)	GNP (G $ 78)		ENERGY CONSUMPTION (Mtoe)		ENERGY PRODUCTION (*) (Mtoe)	
		Sc. I	Sc. II	Sc. I	Sc. II	Sc. I	Sc. II
(1) INDUSTRIALIZED COUNTRIES	+537 (11 %)	+20372 (65 %)	+13978 (70 %)	+8203 (56 %)	+6417 (62 %)	+6106 (54 %)	+4416 (61 %)
(2) THIRD WORLD IN RAPID GROWTH	+1356 (29 %)	+8450 (27 %)	+4756 (24 %)	+4242 (29 %)	+2533 (24 %)	+3193 (28 %)	+1726 (15 %)
(3) THIRD WORLD IN TRANSITION	+2814 (60 %)	+2346 (8 %)	+1324 (6 %)	+2204 (15 %)	+1508 (14 %)	+2054 (18 %)	+1070 (15 %)
WORLD	+4707 (100 %)	+31168 (100 %)	+20058 (100 %)	+14649 (100 %)	+10458 (100 %)	+11353 (100 %)	+7212 (100 %)

(*) 1978 - 2020 period

Table   :   8
Appendix : 19

COMPARISON OF PER CAPITA STRUCTURES : LARGE ZONES

1960 - 1978

(average amounts per capita)	1960			1978		
	GNP ($ 78)	Ener.consump. (toe)	Ener. prod. (toe)	GNP ($ 78)	Ener.consump. (toe)	Ener. prod. (toe)
(1) INDUSTRIALIZED COUNTRIES	3181 (259 %)	2,56 (233 %)		6170 (296 %)	4,30 (269 %)	3,49 (208 %)
(2) THIRD WORLD IN RAPID GROWTH (R5 + R8 + R10)	561 (46 %)	0,52 (47 %)		1149 (55 %)	0,82 (51 %)	2,23 (133 %)
(3) THIRD WORLD IN TRANSITION (R6 + R7 + R9)	158 (13 %)	0,33 (30 %)		236 (11 %)	0,44 (28 %)	0,49 (29 %)
WORLD	1230 (100 %)	1,10 (100 %)		2085 (100 %)	1,60 (100 %)	1,68 (100 %)

COMPARISON OF PER CAPITA STRUCTURES : LARGE ZONES

- Scenario I -

(average amounts per capita)	2000 I			2020 I		
	GNP ($ 78)	Ener.consump. (toe)	Ener. prod. (toe)	GNP ($ 78)	Ener.consump. (toe)	Ener. prod. (toe)
(1) INDUSTRIALIZED COUNTRIES	10725 (331 %)	5,65 (291 %)	5,13 (250 %)	15356 (340 %)	7,01 (301 %)	6,71 (280 %)
(2) THIRD WORLD IN RAPID GROWTH (R5 + R8 + R10)	2565 (79 %)	1,57 (81 %)	2,37 (116 %)	4612 (102 %)	2,38 (102 %)	2,70 (113 %)
(3) THIRD WORLD IN TRANSITION (R6 + R7 + R9)	380 (12 %)	0,54 (28 %)	0,62 (30 %)	601 (13 %)	0,63 (27 %)	0,73 (30 %)
WORLD	3244 (100 %)	1,94 (100 %)	2,05 (100 %)	4516 (100 %)	2,33 (100 %)	2,40 (100 %)

COMPARISON OF PER CAPITA STRUCTURES : LARGE ZONES

- Scenario II -

(average amounts per capita)	2000 II			2020 II		
	GNP ($ 78)	Ener.consump. (toe)	Ener. prod. (toe)	GNP ($ 78)	Ener.consump. (toe)	Ener. prod. (toe)
(1) INDUSTRIALIZED COUNTRIES	8964 (341 %)	5,08 (304 %)	4,63 (263 %)	11185 (363 %)	5,85 (329 %)	5,60 (301 %)
(2) THIRD WORLD IN RAPID GROWTH (R5 + R8 + R10)	1861 (71 %)	1,17 (70 %)	1,87 (106 %)	2666 (87 %)	1,48 (83 %)	1,93 (104 %)
(3) THIRD WORLD IN TRANSITION (R6 + R7 + R9)	280 (11 %)	0,45 (27 %)	0,50 (28 %)	363 (12 %)	0,46 (26 %)	0,50 (27 %)
WORLD	2626 (100 %)	1,67 (100 %)	1,76 (100 %)	3078 (100 %)	1,78 (100 %)	1,86 (100 %)

ANALYSIS OF DIFFERENCES BETWEEN THE TWO STAGES OF THE STUDY

One of the objectives of the study, and a test of its proper functioning, is that the work of the RWT should produce real changes and corrections to the initial forecasts of the Central Team (these forecasts being only intended to play a dual role: that of providing a basis for discussion for the Team and that of a safeguard in case the decentralised process should partially fail).

The following tables measure the effective influence of the decentralized phase by calculating the total differences between the final and the initial forecasts for total primary energy consumptions (Annex 11 less Annex 9 in MTOE), according to regions (Table 1) and according to energy sources (Table 2).

1 REGIONAL DIFFERENCES

(A) At the global level, the gross differentials are relatively important: an additional 200 MTOE on the 1978 reference base, but nearly 1000 million less in 2000, and − 1.2/−1.7 GTOE in 2020. The corrections represent about +3% of the values submitted by the Central Team for 1978, −7/−9% for 2000 and −6/−11% for 2020.

A first remark should be made: the regional teams restructured the initial prospects by about −7% in (I) and −10% in (II). However, it must be noted that the R6 forecasting sheets in phase 1 excluded South Africa, which is nevertheless included in the final version of R6. The gross comparison thus contains an important differential which is about 55 MTOE in 1978, 150–120 MTOE in 2000 and 410–270 MTOE in 2020 (values which should be added to R6 in phase 1 to render the references homogeneous).

As a result the true differential between the two phases is more exactly about:

+ 135 MTOE in 1978;

but −1090/−1100 MTOE in 2000; −1650/−1980 MTOE in 2020, (or + 2% in 1978, −8% in I (2000-20) and −10% in 2000 (II) and −12% in 2020 (II)).

As an aside, we should like to note that the 2020 differential represents, at the rate of growth of energy consumption during that period, something of the order of 6–13 years of time difference on the global forecast: this is to say that, if phase (2) turns out to be correct, the levels for 2020 foreseen during phase (I) would in fact be reached in 2026 (I) and in 2033 (II).

Variations range widely in terms of regions:

(B) The forecasts were upgraded by the experts in all the western countries. However, it should be noted at the outset that the 1978 base was upgraded by more than 200 MTOE (+6%), while those for 2020 were only upgraded by +4%. One of the extraneous factors bearing on forecasting itself was the geographical transfer during the study (e.g. between phases (1) and (2)) of Turkey and Cyprus from R5 to R2 (Western Europe). This boundary change by itself explains an increase of +34 MTOE in 1978 for R2; of 90–80 MTOE in 2000, and of 190–140 MTOE in 2020 (or 15% of the increase to be found in 1978 for all the Western Countries together, 55% of it in 2000 and 67% in 2020).

Apart from changes in the boundaries of regions, the forecasting adjustment for the WIC was only: +180 MTOE in 1978; +70 MTOE in 2000 (I and II); +90 MTOE/+70 MTOE in 2020.

One then notices that if the level of the 1978 base is taken into account, the final forecasts are clearly lower than the initial forecasts in 2000 and 2020.

(C) For the Industrialised Nations of the East, the situation develops in the contrary direction. The 1978 base was confirmed, but the forecasts were markedly lowered: −16%/−14% in 2000, and −19%/−15% in 2020. It has to be pointed out that we had started off from the answers to the first questionnaire sent to the Member States of the Economic Commission for Europe of the United Nations (1980), those answers being linked to economic forecasts which manifestly did not cohere with the growing difficulties of the international environment.

(D) For the Third World excluding R9 (the Centrally Planned Asian Countries), the outlook was different. At first, it was feared that certain regions would outbid each other in their optimism. In fact, the modifications which were made remained within very reasonable limits which did not cast doubt on the general harmony of the study. The effect of the gross modifications in total only represented −2% in 1978, +9%/−6% in 2000, and +26%/−4% in 2020 on the initial total for the LDCs (excluding R9). We should immediately point out the effect of two external factors which must be borne in mind:

- the "removal" of Turkey and Cyprus from this zone between the two phases of the study,
- in phase 1, the omission of South Africa from R6.

As a result, a more exact comparison, excluding the two former countries from the initial reference and instead adding South Africa for the values indicated above, would produce final differentials of about: −40 MTOE in 1978; of +180 MTOE/−180 MTOE in 2000 (I) and (II); and of +990 MTOE/−280 MTOE in 2020 (that is to say, corrections of −4% in 1978; +6%/−7% in 2000; and +15%/−7% in 2020). Differentials are thus more marked in (I), in the sense of a significant upward re-evaluation. They are lower in II. Phase 2 thus contributed to the greater differentiation of the forecasts for each scenario.

(E) There remains the zone of the Centrally Planned Asian Countries (R9) where the adjustments were considerable, in this case. In absolute terms, these adjustments were of about $-870/-640$ MTOE in 2000 and of $-1930/-1260$ MTOE in 2020, of the order of the total global reduction recorded between the two phases. R9 is the region which experienced far and away the most dramatic developments, which were all negative. This means that, in this case, the RWT was much more conservative in its prognostics than our initial forecasts which were derived directly from the "Energy Horizons of the Third World" study of Munich.

The relative reduction compared to the propositions of phase (I) is thus: 0% in 1978 (and therefore not one which is attributable to a differential in the reference base), $-45\%/-43\%$ in 2000, and $57\%/-56\%$ in 2020.

Consumption perspectives have been reduced by half for 2000 and 2020. Demographic forecasts were confirmed by the experts: they do not, therefore, explain the decrease. By contrast, one notices that the final economic forecasts are very much reduced. Measured in terms of total GNP (R9), the decrease is of $-19\%/-41\%$ in 2000, and $-8\%/-40\%$ in 2020. One can thus automatically attribute 42% and 14% of the reduction in energy prospects for 2000 and 2020, respectively, in Scenario I to the economic slowdown (and 93% and 71% for 2000 and 2020 in Scenario II). The decrease of economic projections can therefore explain the larger part of the forecasting reduction in (II) and 50% of it in (I). By contrast, it only marginally operates in 2020 (I), where the reduction in forecasts is then explained overall by a Draconian reduction of income elasticity.

(F) When the differential in global forecasting excluding the readjustment relating to R9 is examined, it appears that, after a correction is made for South Africa, the differential between phases (1) and (2) is the following, (excluding R9): +135 MTOE in 1978; $-220/-460$ MTOE in 2000; $+280/-720$ MTOE in 2020 (or in terms of percentage of world forecast in phase (1) excluding R9: +2% in 1978; ±2% in (I) and -5% in (II), in 2000 as in 2020).

One can try to take a "mixed" view of the problem, by measuring forecasting differentials not by geographical zones but in terms of the various supply sources.

2 SUPPLY SOURCE DIFFERENTIALS

Table 2 provides a precise indication of the origin of differentials (give or take a few adjustments on the total of supplies).

Firstly, one must take care to rectify the bias introduced by the absence of South Africa in the values of phase (1). This reorganisation produces a total of +135 MTOE in 1978; almost -1100 MTOE in 2000, and of $-1700/-2000$ MTOE in 2020.

(A) The forecasts for certain sources were re-evaluated. This was done for natural gas, hydropower and wood, especially within the framework of Scenario I. In the case of hydropower and wood, this up-grading was partly on the 1978 base.

If one takes account of this evolution of the 1978 values, the forecasted growth is very attenuated, especially in the case of wood. One can thus ascertain that the strongest expansion affects gas and hydropower, and that mainly in (I) (about +250 MTOE in 2020).

(B) All the other sources experience a significant decrease in the forecasts instead.

Despite a net growth of the 1978 base (+180 MTOE), the oil forecasts in particular are very much diminished (a decrease of the order of −450/−320 MTOE in 2020). Values of a similar order are to be found in nuclear power (−450/−340), new energies (−280/−470) and waste (−260/−350). However, the most remarkable decrease is experienced by coal, whose forecast decreases by −780/−580 MTOE in 2020 (including South Africa); in 2020, coal will therefore account for 45% and 30% of the total decrease in world forecasts in (I) and (II).

It might be possible to envisage the further elaboration of this analysis, this time by mixing regions with energy sources to ascertain the differentials on regional supplies more accurately.

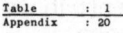

Table   : 1
Appendix   : 20

GROSS DEVIATIONS ON TOTAL PRIMARY ENERGY CONSUMPTIONS :

DECENTRALIZED PHASE (2) LESS CENTRALIZED PHASE (1)

PEC (2) - PEC (1) (Mtoe)	1978	2000		2020	
		I	II	I	II
R1	+62	+23	-31	+100	-98
R2	+102	+135	+116	+112	+183
R3	+48	+3	+64	+70	+126
R4	-1	-462	-354	-792	-507
R5	-20	+63	-82	+309	-189
R6	+38	+125	+87	+395	+224
R7	-25	-14	-66	+85	-124
R8	0	+23	-55	+95	-134
R9	0	-874	-641	-1934	-1257
R10	-14	+36	-21	+324	+70
WIC	+212	+161	+149	+282	+211
IC	+211	-301	-205	-510	-296
TW	-21	-641	-778	-726	-1410
TW - R9	-21	+233	-137	+1208	-153
WORLD	+190	-942	-983	-1236	-1706

GROSS DEVIATIONS ON WORLD SUPPLIES BY SOURCES :

DECENTRALIZED PHASE (2) LESS CENTRALIZED PHASE (1)

WORLD (Mtoe)	1978	2000		2020	
		I	II	I	II
SMF	-46	-588	-421	-413	-331
OIL	+193	-312	-117	-447	-316
NATURAL GAS	-18	+123	-79	+241	+82
HYDROPOWER	+40	+90	+22	+263	+36
NUCLEAR	+2	-155	-171	-436	-332
NEW SOURCES	+8	-25	-140	-259	-461
WOOD	+85	+92	+128	+16	-35
WASTES	-71	-167	-205	-261	-349
CEC	+179	-867	-906	-1051	-1322
NCEC	+14	-75	-77	-245	-384
PEC	+193	-942	-983	-1296	-1706

COMPARISON OF THE "MUNICH" AND "NEW DELHI" STUDIES

It is interesting to study the evolution of the forecasts of the Conservation Commission since Munich. What are the main inflexions? Which parts of the world and which supply sources do they affect? A comparison of the results of the present study with those published in ref. 42 as a result of work in Munich, will enable us to answer these questions by means of the two following tables. In these tables, one can find:

(a)   the comparative regional division of primary energy consumptions (PEC) in the related scenarios: I and B; II and C.
(b)   the analysis of this PEC according to energy source for Scenarios I and B since this had only been attempted in Munich in the case of B.

1   COMPARISON OF RESULTS

1.1. Total comparison of world *Primary Energy Consumptions*

One firstly observes a clear reduction of world energy consumption prospects in all four configurations studied for both 2000 and 2020. The decline is of $-1.2/-1.7$ GTOE in 2000, and of $-2.1/-4.1$ GTOE in 2020 (or $-9\%/-14\%$ in 2000, and $-11\%/-23\%$ in 2020) compared to the Munich hypotheses. The decline is therefore more marked in II than in I. In fact, it is noteworthy that the values of Scenario I are in line with those of C for 2000 and for 2020. Everything happens as if there had been a downward slide between Munich and New Delhi: the scenario thought to be "the most probable" in 1980 has disappeared; the "probable" scenario of 1983 corresponds to the "pessimistic" scenario of 1980. As for the "pessimistic" scenario of 1983, it is well below the levels of the "pessimistic" scenario of 1980. Undoubtedly, the continuing current economic recession must have rendered the RWTs much more cautious in their forecasts.

1.2. Comparison of regional structures

(A) If one then examines the source of differentials region by region, in the same way as in ref. 42, one observes that:

(i)   the total prospects for the Industrialised Countries (ICs) are confirmed (0 to 5%), and are even quite clearly upgraded in 2020 (II) (+1 GTOE, or +8% compared to Munich).

(ii) By contrast, the decline is spectacular for all the Third World: −25/−35% in 2000; −31/−47% in 2020.

(iii) Within the ICs, there is a substantial upturn for the "Western" nations, especially in 2020. Compared to Munich, one indeed notices a forecasted growth of PEC of +3%/+1% in 2000 and of +17%/+11% in 2020. By contrast, the Eastern Countries remain very stable, with variations of only +2%/−2% in 2000, and 0/−12% in 2020.

(iv) The omission of Region R9 (Centrally Planned Asian Countries or "CPAC") from the rest of the Third World reveals the main cause of the world decrease: the considerable downward inflexion of that region's prospects (that is to say, that of China in particular) is the dominant factor in the decrease in world forecasts. Indeed, this region R9 experiences a decline of −50%/−55% in 2000, and of −63%/−71% in 2020 on its previous forecasts. By itself, the decline of R9 forecasts accounts for 87%/62% in 2000 and 115%/61% in 2020 of the world deficit.

(B) The reorganisation between ICs and LDCs is therefore important. In Munich,the ICs only represented 59% of world demand in 2000 and 48-49% in 2020. They now resist the growth of the Third World much better, in that they still account for 68-69% of demand in 2000, and 60-65% in 2020, when they will easily retain their dominance.

Region R9 is the main victim of this decline. In Munich, 16% of the responsibility for world consumption was allocated to R9 in 2000 and 20% of that of 2020. In our study, R9 will be hard put to retain its share of 9-8% in 2000, and 8-7% in 2020.

However, an extraneous element operates to slant this regional comparison in a specific direction: the rectification of boundaries between the Third World and the Industrialised Countries because of the transfer of Turkey and Cyprus, from the LDCs to the ICs. In fact, it is necessary to correct the gross differentials caused by this transfer between groups, which are of the order of about 90-80 MTOE in 2000, and 190-140 MTOE in 2020.

If these amounts are added to the New Delhi forecasts for the LDCs and also deducted from the forecasts for Western ICs, the differential between these two studies is reduced to:

	2000		2020	
GTOE	I–B	II–C	I–B	II–C
Western ICs	+0.06	−0.03	+0.86	+0.42
ICs	+0.11	−0.08	+0.86	+0.02
LDCs	−0.26	−0.57	−0.51	−1.62
Third World	−1.26	−1.62	−3.01	−4.12

While this correction will not fundamentally modify the impression which one might have formed from the above, it does, nevertheless, introduce some nuances:

(i)   the forecast for ICs is more stable from study to the other, in that the change compared to Munich is only +1.4%/−1.1% in 2000, and +9%/+5% in 2020; and, as for the western nations, it diminishes sharply to +1.1%/−0.6% in 2000, and to +14%/+8% in 2020.

(ii) likewise, the collapse of the Third World is in fact a little less dramatic than appeared above: $-24\%/-34\%$ in 2000 and $-29\%/-45\%$ in 2020, or an increase of 1% in 2000 and of 2% in 2020.

1.3. Comparison of supply structures

If one next compares the respective supply structures, the results are also very clear (although, unfortunately, the comparison can only be made in the case of I and B). Even keeping a very approximate unit of measurement, one observes the following:

(i) There is a general decrease in commercial sources in 2000 (except in the case of new energies). This decrease is variable, however. In terms of quantities, it mainly affects gas (-0.6 GTOE) and (-0.5 GTOE), and represents a decrease of -21% and -14% for both of these sources. All the other sources vary by ±0.1 GTOE around their 1980 forecast level.

(ii) In 2020, the contrast sharpens. This time, nuclear power declines considerably: -2.2 GTOE (accounting for all the world decrease by itself, in that its prospects are reduced by 50% compared to Munich). All the non-commercial sources will lose 25% of their previous level. The most noteworthy variation affecting the other sources is that affecting natural gas: $+10\%$. Apart from nuclear power, the PEC remains stable.

(A) An important element should be noticed, however: the substantial reduction of consumption forecasts hardly affects the forecasts for oil. In fact, world oil consumption remains stable at a level of about 3.4–3.5 GTOE (starting from 2.7 GTOE in 1978), and is as such slightly lower than the forecasts made in 1980 (3.5–3.7 GTOE).

It would therefore appear that the decrease in consumption has no effect on the most worrying point (oil), but instead operates to the detriment of oil's most direct competitors: coal and gas in 2000, nuclear power in the longer term. Such a state of affairs is worrying: it means that, contrary to the state of the economic climate observed since the second oil shock of 1979, oil will remain very much in demand on the world market. The downward adjustments of energy consumption forecasts will only slightly reduce the pressures on oil demand (which thus appear to have a rather low elasticity to potential variations of energy prospects).

(B) If one calculates the increases in volume of each of the sources mobilised from 2000 to 2020 in Scenario I and B, and, if one then subtracts the previous increase in B from the actual increase in I, one observes that the most striking positive change affects natural gas ($+0.9$ GTOE), ahead of coal ($+0.6$ GTOE), while the most substantial decrease affects nuclear power (-2.1 GTOE) and non-commercial sources (-0.3 GTOE).

The 2000 gas and coal forecasts are clearly lower than previous forecasts. In 2020, they catch them up in the case of coal, and easily surpass them in the case of gas. From then on, coal contributes 40% of the increase in the volume of the world's PEC between 2000 and 2020 in I, gas 15% and nuclear power 20%. In Munich, coal only contributed 25%, gas 0% and nuclear 50%,

2 ANALYSIS OF DIFFERENTIALS

As in the case of §I, one thing is to ascertain the forecasting differentials between the successive studies. One other thing is to analyse the cause of these differentials.

The total procedure employed in New Delhi and partially in Munich, enables one to attribute the responsibility for the differentials to the three factors which account for energy consumption: demography, economic growth, and income elasticities. We will now attempt to delineate the respective influence of these three factors on the variation of consumption forecasts between Munich and New Delhi.

2.1. Demography

(A) If one first consults total demographic forecasts, one can measure the differentials on the world total. These are of more than -57 M inhabitants (inhs) in 2000, and of -578 M inhs in 2020 in the present study compared with the 1980 study.

Minh	2000			2020		
	New Delhi	Munich	Δ	New Delhi	Munich	Δ
Third World	4660	4770	– 110	6189	6835	– 646
ICs	1379	1326	+ 53	1533	1465	+ 68
WORLD	6039	6096	– 57	7722	8300	– 578

Affected by the average world per capita consumption in the Munich study, the differentials would thus justify an overall decrease of -121 MTOE/-111 MTOE in 2000 (I and II), and of -1400 MTOE/-1248 MTOE in 2020 (I and II), or 11%/7% of the differential observed on the world's PEC in 2000, and 65%/30% of that for 2020. While it remains a marginal effect in 2000, this factor will become decisive in 2020.

(B) However, this analysis must be refined a little. Indeed, by itself the average world per-capita PEC is hardly of significance. A more exact version of this phenomenon would be given by a division between ICs and LDCs.

If one then compares the demographic differentials with the average PEC of the ICs and the LDCs in the Munich study, one obtains the following picture:

MTOE	2000		2020	
	I–B	II–C	I–B	II–C
LDCs	– 123	– 110	– 982	– 859
ICs	+ 304	+ 280	+ 450	+ 408
WORLD	+ 181	+ 170	– 532	– 451

The "World" here is the algebraic addition of the separate effects of the Third World and the Industrialised Countries.

The importance of the demographic variable in the explanation of the total differential is now revealed to be very different from that indicated by the global measure: in 2000, it even has a positive impact. In 2020, it only accounts for 25%/11% of the PEC differential.

(C) So far, we have omitted one element. This is the effect of the geographical transfer of Turkey and Cyprus from the LDCs to the ICs.

The estimated population of Turkey and Cyprus put together is 71 million inh in 2000 and 95 minh in 2020. If these two nations are added to the Third World as defined in New Delhi and if one deducts them from the ICs, the demographic forecasts would then become on comparable bases:

	2000			2020		
Minh	New Delhi	Munich Δ		New Delhi	Munich Δ	
LDCs	4731	4770	−39	6284	6835	−551
ICs	1308	1326	−18	1438	1465	−27
WORLD	6039	6096	−57	7722	8300	−578

If these results are then compared to the respective per capita consumption levels of Munich, one then finds:

	2000		2020	
MTOE	I–B	II–C	I–B	II–C
LDCs	− 44	− 39	− 838	−733
ICs	− 103	− 95	− 179	− 162
WORLD	− 147	− 134	− 1017	−895

As a result, the variation of demographic forecasts "explains" semi-globally in reality: 13%/8% of the PEC differential in 2000, and 47%/22% in 2020.

A more refined analysis made on a region by region basis would enable one to define even more precisely the exact importance of the demographic variable, by comparing the population differentials of each of the 10 regions according to their per-capita primary consumptions.

2.2. Economic growth

(A) We have just measured very generally the effect of the reduction of demographic prospects on the variation of energy consumption (everything else being equal elsewhere). There now remains for us to estimate the impact of changes in economic perspectives.

We know that economic growth has been measured by the (classical) medium of GNP growth rates (or, more exactly, that of GRP: gross regional product). The demographic impact can be separated if one makes a comparison between the two studies on the basis of variations of per capita GNP.

We shall now attempt, as in 2.1, a semi-global analysis between the Industrialised Countries on the one hand and the Third World on the other.
(B) The per capita rates of economic growth appear to be the following:

GNP/NH	1978–2000						2000–2020					
	I	B	I/B	II	C	II/C	I	B	I/B	II	C	II/C
TW												
% year	3.4	3.9[1]		1.9	3.2[1]		2.9	2.7		1.8	2.4	
Index	209	233	0.90	152	201	0.76	371	403	0.92	216	325	0.66
IC												
% year	2.55	3.4[1]		1.7	2.5[1]		1.8	1.75		1.1	1.3	
Index	174	212	0.82	145	172	0.84	257	304	0.85	187	226	0.83
WORLD												
% year	2.0	2.75[1]		1.05	1.9[1]		1.7	1.4		0.8	1.05	
Index	156	185	0.84	126	153	0.82	223	249	0.90	152	190	0.80

(1) 1976–2000
(2) 100 base for 1978 in I and II (as in B and C where the landmark year 1978 is used with reference to the 1976–2000 progression). The indices are calculated from the basis of exact rates of growth (and not on approximate ones), and are then applied to 2000 and 2020. We have omitted here the regrouping of Turkey and Cyprus from the LDCs to the ICs.

From a reading of the indices, one can easily tell that all the economic forecasts of the New Delhi study show a decrease compared to those of Munich, and that they lead to globally lower per-capita GNP levels in 2020 of 10% (I) and 20% (II); and of 17% in 2000.

Poorer prospects, therefore, automatically account for a lowering of per-capita PEC in relation to constant income elasticity of the order of:

$-16\%/-18\%$ in 2000, and $-10\%/-20\%$ in 2020.

2.3. Global impact

One finds the following relations in the global models we have used:

(1) PEC = DEM × PEC/INH
    where PEC = total primary energy consumption
        DEM = population
    PEC/INH = total per-capita primary energy consumption

(2) $\dfrac{\Delta PEC/INH}{PEC/INH} = e \dfrac{\Delta GNP/INH}{GNP/INH}$

where the PEC/INH variation between two years is linked with the per-capita GNP variation between the same two years, by means of a coefficient e expressing the elasticity of energy consumption to income per capita.

As a result, by assuming, firstly, a constant elasticity between the two studies, one can synthesise the impact on total PEC of demographic and economic forecasts. This is the purpose of Table 3, where we provide the ratios of the separate indices of demographic and per capita GNP for each scenario, as well as their product.

The ratio of their difference at 1 of total PEC will express the share of the PEC variation which is explained by the combined effect of the variations of demography and economic growth. The complement is attributable to other factors which, in a global model, can be summarized by an elasticity coefficient.

For instance, for "Third World 2000" one finds the product "DEM × GNP/INH" = 0.89 in the case of I/B, whereas the ratio of the total PEC indices is 0.76. This can be interpreted as follows. In the overall decline of the PEC forecast of -24% between New Delhi and Munich, -11% (1 -0.89) is caused by the factor DEM × GNP/INH. That is to say that $-11\%/-24\% = 46\%$ of the total reduction of PEC can be explained by the combination of the reduction in population and of the hypotheses of economic growth. The remaining 54% are all to be globally attributed to the changes in income elasticities.

One therefore arrives at the following table by means of an approach based on scenarios and years. This table expresses the combined responsibility of the demographic and economic growth factors in the variations of total PEC forecasts:

		2000			2020		
		LDCs	ICs	W	LDCs	ICs	W
Share of DEM × GNP/INH in ΔPEC	I/B	46%	– 1900%	189%	52%	– 188%	145%
New Delhi–Munich	II/C	76%	1700%	136%	85%	– 1900%	113%

The calculation is most significant in the case of the Third World. The variation in demographic and economic forecasts accounts for about 50% of the decline of PEC in Scenario I, and some 3/4 or even 6/7 of that in Scenario II. For the Industrialised Countries, the measure is more fluid in that in three cases out of four, the PEC slightly increased between Munich and New Delhi, whereas the two factors studied declined rather sharply: that is to say that, conversely, the elasticities have been implicitly notably improved. Likewise, regarding the world total, the combined effect of the two factors would normally have led to a much sharper reduction of the PEC which is, in fact, partially compensated by the recovery of the other factors.

Let us note, however, that in Munich the most thorough forecasting work was made with regard to the Third World, and that the forecasts for the Industrialised Countries (and thus the entire world) were introduced without thorough research, with the sole aim of completing the overall panorama. The implicit values for income elasticities which resulted were not therefore delineated with any great care at the time (a fact which may explain the size of the alteration made to the hypotheses, in particular with regard to the ICs).

Table   : 1
Appendix : 21

COMPARISON BETWEEN WEC "NEW DELHI" AND "MUNICH" STUDIES

PRIMARY ENERGY CONSUMPTIONS : REGIONAL STRUCTURE

- Scenarios I/II and B/C -

(Gtoe)	Sc. I	Sc. B	I - B	Sc. II	Sc. C	II - C
2000						
WIC	5,35	5,20	+0,15	4,85	4,80	+0,05
EIC	2,45	2,40	+0,05	2,15	2,20	-0,05
IC	7,80	7,60	+0,20	7,00	7,00	0
CPAC	1,05	2,05	-1,00	0,85	1,90	-1,05
LDC	2,90	3,25	-0,35	2,25	2,90	-0,65
TW	3,95	5,30	-1,35	3,10	4,80	-1,70
WORLD	11,75	12,90	-1,15	10,10	11,80	-1,70
2020						
WIC	7,35	6,30	+1,05	6,20	5,60	+0,60
EIC	3,40	3,40	0	2,80	3,20	-0,40
IC	10,75	9,70	+1,05	9,00	8,80	+0,20
CPAC	1,45	3,95	-2,50	1,00	3,50	-2,50
LDC	5,75	6,45	-0,70	3,80	5,60	-1,80
TW	7,20	10,40	-3,20	4,80	9,10	-4,30
WORLD	17,95	20,10	-2,15	13,80	17,90	-4,10

COMPARISON BETWEEN WEC "NEW DELHI" AND "MUNICH" STUDIES

PRIMARY ENERGY CONSUMPTIONS : SUPPLY STRUCTURE

- Scenarios I and B -

WORLD (Gtoe)	SMF	PP	NG	HY	NU	NS	FW	VAW	CEC	NCEC	PEC
2000											
Sc. I	3,2	3,4	2,2	0,7	1,0	0,3	0,6	0,3	10,8	0,9	11,7
Sc. B	3,7	3,5	2,8	0,8	1,1	0,1	0,5	0,4	12,0	0,9	12,9
I - B	-0,5	-0,1	-0,6	-0,1	-0,1	+0,2	+0,1	-0,1	-1,2	0	-1,2
2020											
Sc. I	5,7	3,5	3,1	1,4	2,3	1,0	0,5	0,4	17,0	0,9	17,9
Sc. B	5,6	3,7	2,8	1,3	4,5	1,0	0,6	0,6	18,9	1,2	20,1
I - B	+0,1	-0,2	+0,3	+0,1	-2,2	0	-0,1	-0,2	-1,9	-0,3	-2,2
Δ 2000-2020											
Sc. I	+2,5	+0,1	+0,9	+0,7	+1,3	+0,7	-0,1	+0,1	+6,2	0	+6,2
Sc. B	+1,9	+0,2	0	+0,5	+3,4	+0,9	+0,1	+0,2	+6,9	+0,3	+7,2
I - B	+0,6	-0,1	+0,9	+0,2	-2,1	-0,2	-0,2	-0,1	-0,7	-0,3	-1,0

EXPLANATORY FACTORS OF THE NEW DELHI / MUNICH FORECAST DEVIATIONS

(index numbers ratio)

	2000			2020		
	TW	IC	WORLD	TW	IC	WORLD
DEMOGRAPHY						
I / B	0,99	0,99	0,99	0,92	0,98	0,93
II / C	0,99	0,99	0,99	0,92	0,98	0,93
GNP / p.c.						
I / B	0,90	0,82	0,84	0,92	0,85	0,90
II / C	0,76	0,84	0,82	0,66	0,83	0,80
DEM x GNP / p.c.						
I / B	0,89	0,81	0,83	0,85	0,83	0,84
II / C	0,75	0,83	0,81	0,61	0,81	0,74
PEC (1)						
I / B	0,76	1,01	0,91	0,71	1,09	0,89
II / C	0,67	0,99	0,86	0,54	1,01	0,77

NB : Calculation made after correction by the transfer of Turkey and Cyprus
from Third World to Western Europe.

(1) exactly calculated, not on the rounded up figures,with Turkey and
Cyprus transfer included.

ANNUAL WORLD AND CUMULATIVE 1978–2020–2050 CONSUMPTIONS

The purpose of this annex is to measure even very approximately some of the possible very-long-term evolutions of energy consumptions.

At most, we have parametered the growth rates of world consumption and supplies by assuming that the evolutions beyond 2020 would follow the pattern observed for 2000–2020, that is to say, in the line of a regular decline of growth rates.

This was done to create two pictures of 2050 which would be coherent with those for 2020, only as a point of reference enabling one to calculate the range of quantities required over the periods 1978–2020 and 2020–2050.

We then usefully compare the consumption volumes with the levels of proven and additional reserves (cf. Annex 23), so as to measure the following pressure on reserves which such consumption would create.

We then separate fuels and nuclear power (non-renewable sources) from the group of renewable energy sources (hydropower, new energies and non-commercial sources), since the concept of reserves does not have the same meaning for the latter group (annual fluxes).

Table   : 1
Appendix : 22

1978 – 2020 – 2050 YEARLY WORLD CONSUMPTION

| | Scenario I | | | | | Scenario II | | | | |
| | Rate of growth % | | World consumptions (Gtoe) | | | Rate of growth % | | World consumptions (Gtoe) | | |
	2000/2020	2020/2050	1978	2020	2050	2000/2020	2020/2050	1978	2020	2050
SMF	2,8	1,5	1,7	5,7	8,8	2,3	1,5	1,7	4,4	6,9
Oil	0,2	-1,5	2,65	3,55	2,3	-0,7	-3	2,65	2,4	1,0
Gas	1,9	1	1,15	3,2	4,3	1,5	0	1,15	2,4	2,4
Nuclear	4,4	3	0,15	2,3	5,6	3,7	2	0,15	1,7	3,0
Hydropower	3,3	2,5	0,4	1,35	3,0	2,4	1,5	0,4	1,0	1,6
New energies	5,6	4	ε	1,0	3,2	5,6	3	ε	0,8	2,0
Fuelwood	-0,5	-1	0,5	0,55	0,4	-0,2	-0,5	0,5	0,7	0,6
Wastes	0,4	0,5	0,25	0,35	0,4	0,9	1	0,25	0,4	0,5
PEC	2,1	1,5	6,8	18	28	1,6	0,8	6,8	13,8	18

Table   :   2
Appendix   : 22

CUMULATED WORLD CONSUMPTIONS

(Gtoe)	Scenario I		Scenario II	
	1978-2020	2020-2050	1978-2020	2020-2050
SMF	140	220	120	170
Oil	130	90	110	50
Gas	90	115	70	70
Nuclear	45	120	35	70
Hydropower	30	65	25	65
New energies	15	65	10	40
Fuelwood	20	15	25	20
Wastes	10	10	15	15
PEC	480	700	410	500

RESERVES OF NON-RENEWABLE ENERGY SOURCES
(From the WEC Survey of Energy Resources 1980)

Table   : 1
Appendix   : 23

RESERVES OF NON-RENEWABLE ENERGY SOURCES

(From the WEC Survey of Energy Resources 1980)

(Gtoe)	RESERVES		
	Proved	Additional	Total
SMF	460	670	1130
Oil	85	50	135
Gas	60	120	180
Ur (*)	45	60	105
Sub-total	650	900	1550
Tar sands	40	70	110
Oil shales	40	280	320
Sub-total	80	350	430

(*) 1 ton of uranium = 15000 toe (reactors presently commercialized)

BIBLIOGRAPHY

1 *World Energy: Looking ahead to 2020,* Report by the Conservation Commission of the World Energy Conference, WEC, 1978.
2 J.-R. FRISCH: *L'évolution des consommations et des sources d'énergie dans le monde: une rétrospective 1960–1976,* Electricité de France, 06/80.
3 R. J. EDEN and the Energy Research Group of the Cavendish Laboratory of Cambridge University: *World Energy Demand,* WEC, IPC, 1978.
4 *Facing the futures,* Report Interfutures, OECD, Paris, 1979.
5 *World Bank Atlas 1980,* 1978.
6 *Pétrole 1980,* Activite de l'Industrie Pétrolière, Comité Professionnel du Pétrole, Paris, 1980.
7 *FAO Yearbook of forest products,* 1978, 1971.
8 *FAO production yearbook,* 1978.
9 *Yearbook of world energy statistics,* UNO, 1979.
10 *World energy supplies 1950–1974,* ONU, 1978.
11 J.-R. FRISCH: *Third World energy horizons: 2000–2020,* Editions Techniques et Economiques, WEC, 1981.
12 J. K. PARIKH: *Energy and development,* World Bank PUN 43, 08/78.
13 D. TILLMAN: *Wood as an energy resource,* Academic Press, 1978.
14 J. GIRI, B. MEUNIER: *Evaluation des énergies nouvelles pour le développement des états africains,* SEMA, 1977.
15 B. A. STOUT: *Rapport sur l'énergie pour l'agriculture mondiale,* FAO, 1979.
16 *World population trends and prospects by country 1950–2025,* Population Division, ONU, 01/81.
17 *Questionnaire on selected energy issues,* UNECE, 10/80, 11/81.
18 T. L. SANKAR, G. SCHRAMM: *Regional Energy Survey,* Asian Development Bank, 03/81.
19 *Annual report to Congress,* 1980, 1981, EIA/DOE.
20 *1980 Energy Plan,* Ministry of Energy, Wellington, New Zealand.
21 *National Energy Data Report,* National Committee, Republic of China (T'ai-Wan), 1980.
22 *The future picture of a changing pattern of energy system,* Institute of Applied Energy, Japan, 1981.
23 *Perspectivas energeticas columbianas hasta el año 2000,* Bogota, 10/79.
24 P. ERBER, J. PORTO CARREIRO FILHO: *Evoluçao do balanco energetico 1980–2010,* Electrobras, 01/81.

25 K. SMITH, H. BROWN: *The energy problems of Asia*, East-West Center, Hawai, 1980.
26 *Energy balances for Latin America*, OLADE, 11/81.
27 *Energy Strategy 79*, Minister of Energy, Wellington, New Zealand, 12/79.
28 *Australian Energy Outlook*, ESSO — Australia, 11/80.
29 *The outlook for energy in South Africa*, Department of Planning and the Environment, 1977.
30 *Prospective expansion of black coal industry*, Joint coal board, Sydney, 10/80.
31 K. WOODARD: *The international energy relations of China*, Stanford, 1980.
32 J. K. PARIKH: *Energy systems and development*, Oxford University Press, 1980.
33 V. SMIL: *China's energy*, Praeger, 1976.
34 N. BAROUDI: *Arab energy: prospects to 2000*, McGraw Hill, ECWA, 1982.
35 R. LATTES, A. JEANBLANC: *Croissance économique, besoins d'énergie et économies d'énergie*, CEA, 11/81.
36 P. DAURES. J.-R. FRISCH: *Perspectives énergétiques pour le Tiers Monde 2000-2025*, Groupe Prospective de l'énergie, Maison des Sciences de l'Homme, Paris, 01/75.
37 P. PALMEDO, R. NATHANS: *Energy needs, uses and resources in developing countries*, Brookhaven National Laboratory, 03/78.
38 Reports presented at the 2nd Arab Energy Conference, Doha, Qatar, 03/82.
39 *World energy outlook*, OECD, 1982.
40 *Energy in a finite world*, IIASA, 1981.
41 WU ZONGHUA: *Solving the energy crisis from the viewpoint of energy science and technology*, Hongqi, no. 17, 09/80.
42 *World energy balance 2000-2020*, Conservation Commission,WEC, 1981.
43 R. K. PACHAURI: *Energy and economic development in India*, Praeger, 1977.
44 A. F. BEIJDORFF, P. STUERZINGER: *Improved energy efficiency: the invisible resource*, WEC, 1980.
45 T. R. GERHOLM; *Long range energy demand*, WEC, 1980.
46 J. S. FOSTER: *Prospective energy production*, WEC, 1980.

MINUTES OF ROUND TABLE NO 5 ENERGY 2000–2020: SUPPLY, DEMAND AND REGIONAL STRESSES

The report was presented and discussed during a Round Table of the 12th World Energy Conference, held in the afternoon of September 21, 1983. The specific results for the five Third-World regions were also presented at the LDC Round Table held the following day, that is September 22, 1983.

Composition of the Panel of the Round Table held on September 21, 1983:

Mr Marcel BOITEUX
Chairman of the Panel
Chairman of the Conservation
Commission
Chairman of the Board of Electricité
de France

Mr K. SAMBAMURTI
Secretary of the Panel
Member of the Central Electricity
Authority (India)

Prof T. R. GERHOLM
Professor of Physics
Stockholm University (Sweden)

Mr. J. W. HOPKINS
Deputy Executive Director
International Energy Agency

Ir ABDUL KADIR
Director of Power Industries
Ministry of Mines and Energy
(Indonesia)

Mr D. C. RAO
Deputy Director
Department of Energy
World Bank

Mr S. N. YATROV
Director
Research Institute of Complex
Energy and Power Problems of the
State Planning Committee (USSR)

The following regional experts assisted the Panel:

Mr K. BRENDOW
Head of General Energy Group
Energy Division
Economic Commission for Europe
(UN)

Dr I. IBRAHIM
Director of the Economic
Department
Organization of Arab Petroleum
Exporting Countries

MR K. MUTOMBO
Secretary General
Union of West African Electricity
Producers and Distributors

Dr M. A. ROZALI
Chief Planning Engineer
National Power Corporation of
Malaysia

Mr G. SANCHEZ-SIERRA
Planning Engineer
World Bank

Mr T. L. SHANKAR
Director
Public Enterprise Research Institute
(India)

Dr Y. UCHIYAMA
Research Engineer
Economic Research Center
Central Research Institute of
Electric Power Industry (Japan)

*   *   *

After the presentation of the main results of the report by the author, there
was a first discussion by the members of the panel.

First of all, Prof T. R. GERHOLM pointed out that the scenarios of the
economic type used in the study furnish generally higher results than the
scenarios of the technological type. Scenario II, situated well below the
trends generally admitted, represents an original approach, though it could
imply dramatic consequences.

Prof T. R. GERHOLM also pointed out the spectacular decrease of the
forecasts established by the Conservation Commission since it presented its
first world-wide study in Istambul in 1977. But he noted the importance of a
subject that appears here for the first time: the development of interregional
trade.

Mr S. N. YATROV said he was impressed by the scope of the research
work, as well as the detailed analysis it presents, He defended Scenario I,
because it alone corresponds to a political will for peace and cooperation.
Referring to the work of his Institute, he remarked that the forecast of
Scenario I is one GTOE lower than its present forecasts for 2000. He believes
it would be desirable to extend the approach to an analysis of the secondary
vectors, such as electricity.

Mr A. KADIR insisted upon the considerable changes that occurred in the
Third World during the last few years, and expressed misgivings about the
durability of the adopted regions in geopolitical terms. He also thinks that it
might be useful to increase the number of the scenarios, in order to cover more
completely the range of future situations.

Mr D. C. RAO compared the results with those of a recent World Bank
survey of the energy transition of the LDCs by 1995. In overall terms, the
results of Scenario I are consistent with those of this survey, but the
prospects of hydropower are significantly higher for the World Bank. On the
other hand, the Bank's oil forecasts are lower, because it assumes that oil
substitution could proceed faster.

Mr J. W. HOPKINS compared the results with those of the last study of
the International Energy Agency for 2000. He pointed out that the forecasts
are not divergent in overall terms but that there is a significant difference as
far as North America's oil consumption is concerned. In fact, the Agency
thinks that this consumption will remain much higher, because substitutions
will be slower, so that the region will remain a permanent net oil importer.

During the general discussion, the Author explained the differences between the methodologies used by the various regional teams, especially as far
as supply forecasts, price elasticities and estimates for non-commercial
energies are concerned.

Later on, the author opened the second part of the discussion by defining the main conclusions and orientations of the study.

Mr A. KADIR insisted upon the low levels of the Third World's future energy consumption per capita. He felt also difficult to assess to which extent the recession of the early '80s will affect the already mediocre prospects, especially as he doubted that situations might be more peaceful and easier to manage after 2000. Finally, he recalled that changes in life styles were bound to influence future developments.

Prof T. R. GERHOLM said that he is more optimistic for the long run. In fact, he believes that it will be possible to disconnect economic growth and energy growth. In his opinion, one of the major merits of the study is to bring out the vastness of the Third World's needs. Though the differences between the prospects of the two scenarios are very significant, he thinks it is essential that the report searched for realism in its projections.

Mr S. N. YATROV underlined the importance of the future role of coal, nuclear, and new energies. He asked for the reinforcement of international scientific cooperation in the research efforts made in the field of oil alternatives.

Mr D. C. RAO expressed doubts about the Third World's capacity to adapt rapidly its supply pattern to the increase of oil prices, as well as to the advance of new technologies. These considerations lead him to believe that the energy development of the whole of the Third World is possibly even more menaced than the report indicates.

Mr J. W. HOPKINS said that the potential of cooperation within each of the regions is very great, especially because of the differences of foreign trade situations. This is why it would be very useful to complete the broad picture by a study of intraregional flows.

The discussion then turned to the distinction between what is desirable and what is possible, between objectives and forecasts. One delegate proposed to analyse a purely normative scenario fixing the Third World's per-capita consumption levels on the basis of what is deemed acceptable and desirable, in order to bring out the constraints which have to be removed to arrive at these levels. In reply to a number of remarks, another delegate pointed out that the hypothesis of more pessimistic scenarios cannot be excluded if the industrial countries do not proceed to an adequate substitution of oil by coal and nuclear power, creating thus additional problems for the Third World. Finally, various delegates expressed the fear that the problem of financial constraints may have a serious impact on the development of countries with great debts.

In his concluding remarks, Mr Marcel BOITEUX noted that divergent opinions were expressed on the degree of optimism or pessimism implied in the scenarios. He said that not too much attention should be attached to the figures themselves, as a forecasting exercise's main purpose is to identify the problems and to define essential future constraints. He also recalled that the Conservation Commission has programmed a study of investment and financing problems.

Finally, in accordance with the wish expressed by numerous delegates, the Conservation Commission recommended to the International Executive Council to maintain in each of the regions a small permanent study group, in order to avoid the loss of the research investment resulting from the study and to ensure that the specific problems of each region will be further explored in greater depth.